STRUCTURAL STEELWORK DESIGN

STRUCTURAL STEELWORK DESIGN

L. J. MORRIS
Department of Engineering
University of Manchester

D. R. PLUM
Department of Civil Engineering
University of Newcastle upon Tyne

with a Forword by

P. A. Rutter
Partner, Scott Wilson Kirkpatrick and Partners
Member of BS 5950 Committee

Nichols Publishing
New York

Nichols Publishing
PO Box 96, New York, NY 10024

An imprint of GP Publishing, Inc
A General Physics Company

Copublished with Longman Scientific & Technical
Longman Group UK Limited, Longman House,
Burnt Mill, Harlow, Essex CM20 2JE, England

Library of Congress Cataloguing-in-Publication Data
Morris, L. J.
Structural steelwork design.

Includes bibliographies and index.
1. Building, Iron and steel. 2. Steel, Structural.
I. Plum, D. R. II. Title.
TA684.M7135 1988 624.1′821 88-15174
ISBN 0-89397-324-6

First published 1988

Set in Linotron 202 10/12pt Times

Printed and bound in Great Britain
at The Bath Press, Avon

Preface

Structural steelwork design is usually taught in degree and diploma courses after an initial grounding in theory of structures and strength of materials. The design teaching usually covers both simple structural elements and complete buildings. More complex elements and buildings are often covered in postgraduate courses, but the ideas and concepts outlined in this text still provide the basis for more complicated structures. This book has been prepared primarily for the student, but also for those engineers in practice who are not familiar with BS 5950: *Structural use of steelwork in buildings*.

This book falls naturally into two parts. Part I sets out in detail the design of elements (beams, columns, etc.) frequently found in a structural steel framework. Part II shows how these elements are combined to form a building frame, and should prove especially useful to the engineer in the context of practical design. Part II also develops other considerations such as overall stability of building structures. Those with some experience of element design may prefer to start with Part II using the cross-references to re-examine element design as necessary. A final chapter considers detailing practice, and the effects of a number of practical considerations such as fabrication and fire protection.

It is assumed that the reader has some knowledge of structural analysis and that a basic understanding of metallurgy has been gained elsewhere. The design examples concentrate on manual methods to ensure a proper understanding of steelwork behaviour, with suggestions where computing could be used. Detailed programs for specific microcomputers are increasingly being written, and a number of complete design packages are available commercially.

The principal documents required by the reader are:

BS 5950: Part 1 (1985): *Design in simple construction; hot rolled sections*. British Standards Institution.

Steelwork design. Vol 1: *Section properties; member capacities*. Steel Construction Institute.

The first of these documents is also available in abridged extract form from British Standards Institution as:

Extracts from British Standards for students of structural design.

The second document is available in extract form (dimensions and properties only) as;

A check list for designers Steel Construction Institute.

This extract is updated regularly and the latest edition should be used.

Throughout the book clause references and notation follow those given in BS 5950: Part 1, except that for composite construction the notation given in CIRIA Report 99 is used.

Whilst every effort has been made to check both calculation and interpretation of BS 5950 the authors cannot accept any responsibility for inadvertent errors.

LJM
DRP

Acknowledgements

The authors acknowledge the assistance of many structural engineers in industry and teaching, to whom details of both interpretation and current practice have been submitted, and whose helpful comments have been incorporated into the text. In particular, the collaboration with P. A. Rutter in the initial drafting of the text proved valuable.

Extracts from BS 5950 Part 1 1985 are reproduced by permission of the British Standards Institution. Complete copies can be obtained from BSI at Linford Wood, Milton Keynes, MK14 6LE.

CONTENTS

Foreword

In 1969, the British Standards Institution Committee B/20 responsible for BS 449, a permissible stress structural steelwork design code, instigated the preparation of a new draft code based on limit state principles and incorporating the latest research into the behaviour of structural components and complete structures. The draft was issued for public comment in 1977 and attracted considerable adverse comment from an industry long acquainted with the simpler design methods in BS 449.

It was realized by the newly constituted BSI Committee, CSB/27, that a redraft of the B/20 document would be necessary before it would be acceptable to the construction industry. The work of redrafting was undertaken by Constrado, partly funded by the European Coal and Steel Community, and the task was guided by a small steering group representing the interests of consulting engineers, steelwork fabricators and the Department of the Environment.

Prior to the completion of the redraft, calibration was carried out by the Building Research Establishment to derive suitable values for load and material factors, and design exercises to compare the design of whole structures to the draft code with designs to BS 449 were directed by Constrado. The object of these studies was to assess whether the recommendations to be contained within the new code would produce structural designs which would be no less safe than designs to BS 449 but would give an improvement in overall economy.

The resulting code of practice, BS 5950: Part 1, published in 1985, covering the design in simple and continuous construction of hot rolled steel sections, and Part 2, dealing with the specification of materials, fabrication and erection, achieved the greater simplicity sought by industry whilst allowing the design of building structures to be based on the more rational approach of limit state theory than the permissible stress method of BS 449. Part 3, which is in the course of preparation, will give recommendations for design in composite construction.

Whilst BS 5950: Part 1 is explicit in its design recommendations, the code is intended to be used by appropriately qualified persons who have experience in structural steelwork design and construction. There is a need, therefore, for a text for university and college students engaged on courses in civil and structural engineering which gives clear guidance on the application of the code to typical building structures by worked examples which set out the calculations undertaken in the design office. This is achieved in this book through explanatory text, full calculations and reference to the BS 5950: Part 1 clauses.

The first part of the book deals with the design of various types of structural members and the second part deals with complete designs of the most commonly encountered structures, namely, single-storey industrial buildings and multi-storey office blocks. Apart from the relevant codes and standards, references are given to well established publications commonly found in the designer's office. The book should also be useful to the design engineer requiring an understanding of the application of the limit state code.

P A Rutter
Partner, Scott Wilson Kirkpoatrick and Partners
Member of BS 5950 Committee

PART

I

THE DESIGN OF STRUCTURAL STEEL ELEMENTS

A simple basis for design is to consider a structural framework composed of a number of elements connected together. Loads are sustained by the element, and its reactions transferred to other elements via the connections. In this simple concept for design it is essential that overall action of the framework is considered. Therefore an introduction to the concept of overall stability of the structure is given in terms of bracing systems.

1 INTRODUCTION TO DESIGN IN STEELWORK

Structural steelwork can be either a single member or an assembly of a number of steel sections connected together in such a way that they perform a specified function. The function required by a client or owner will vary enormously but may include:

- **building frames** by which loads must be supported safely and without undue movement, and to which a weather-proof envelope must be attached;
- **chemical plant supports** by which loads must be supported but which commonly require no external envelope;
- **containers** which will retain liquids, granular materials or gases, and which may also be elevated as a further structural function;
- **masts** which must safely support mechanical or electrical equipment at specified heights, and in which the deflections, vibrations and fatigue must be controlled;
- **chimneys** which will support flues carrying waste gases to safe heights;
- **bridges** which must support traffic and other loads over greatly varying spans, and for which degrees of movement may be permitted;
- **temporary supports** used during the construction of some part of a structure, which may be of steelwork, concrete, brickwork, etc., in which safety for short periods and speed of assembly are important.

It will be noticed that both safety and movement of the structures described are important for proper function and, together with economy, these will be the main considerations when discussing the design method later. It should be noted that the design of only some of the above structures is covered by BS 5950 and hence discussed in the later chapters.

Steel sections are rolled or formed into a variety of cross-sections, a selection of which is shown in Fig. 1.1, together with their common descriptions. The majority of these cross-sections are obtained by the hot rolling of steel billets in a rolling mill, while a minority, sometimes involving complex shapes, are cold formed from steel sheet. Hollow sections are obtained by extrusion or by bending plates to the required cross-section, and seaming (welding) them to form tubes. The sections are usually produced in a variety of grades of steel having different strengths and other properties. The commonest grade is known as 43A, referred to sometimes as mild steel, having a yield strength in the range 245–275 N/mm^2. In some types of structure other grades (43B, 43C etc.), having the same yield stress, are more suitable due to their higher resistance to brittle fracture.

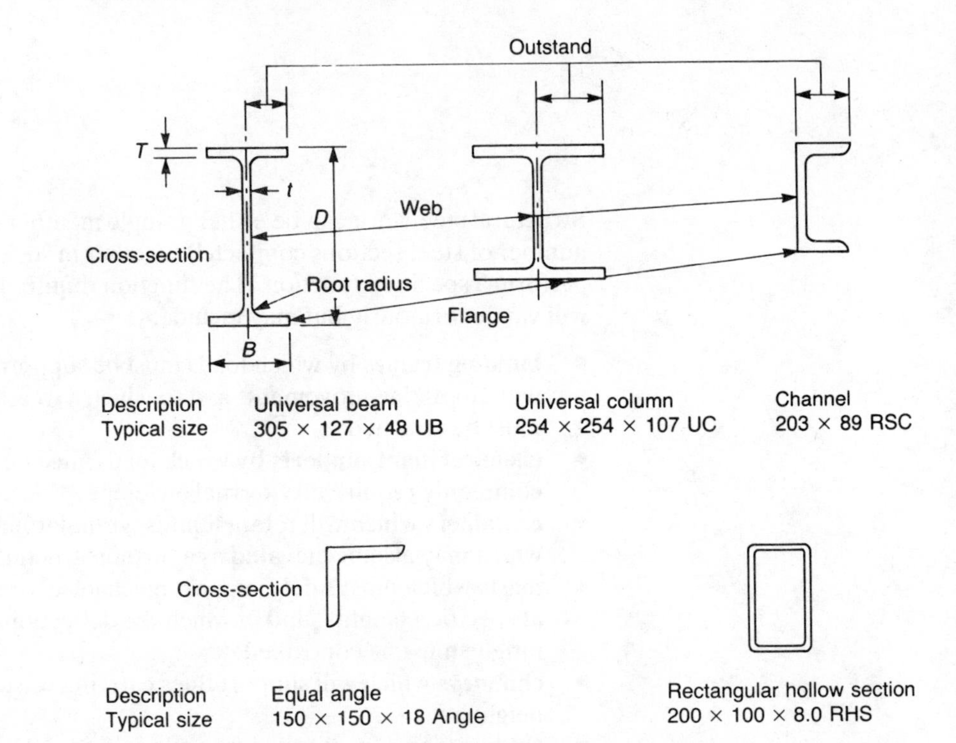

Fig 1.1 Steel sections

In addition to cross-section, the shape of a steel section will include reference to its length, curvature if required, cutting and drilling for connections etc., all of which are needed to ensure that each part fits accurately into the finished structure. These further shaping processes are known as fabrication and are carried out in a fabricating works or 'shop'. It is for this stage that drawings giving precise dimensions of the steelwork will be required, showing what the designer intends. In many cases these drawings are produced by the fabricators rather than by the designer of the structure.

The final stage of producing a structure in steelwork is the erecting or putting together of the various elements on site to form the required framework. At this final stage the safety of the partly finished structure must be checked, and prior thought given as to how the framework is to be erected in order to define the location of each part with precision. The steelwork usually forms only a skeleton to which other building elements (floors, walls, etc.) are fixed, but in frameworks supporting chemical or mechanical plant the steelwork may be the sole structural medium.

1.1 DESIGN REQUIREMENTS

The design of any structure must be judged by whether it fulfils the required function safely, can be built with economy and can maintain an acceptable appearance for its specified lifetime. It follows that the design of structural steelwork also will be assessed by these criteria of safety, economy and appearance.

Safety is assessed by considering the strength of the structure relative to the loads which it is expected to carry. In practice this assessment is applied to each structural element in turn, but these individual element checks are not sufficient without considering the overall safety of the framework. The strength of the structural element must always exceed the effects of the loads by a margin which is known as a factor of safety. The method of providing the factor of safety is discussed in Section 1.4. In the general sense assessment of the structure includes all the criteria by which its performance should be judged, e.g. strength, deflection, vibration, etc.

Whilst in practice economy of the design is of great importance to the owner of the finished structure, students are rarely required to make a full economic assessment. However, two basic matters should be taken into account. Firstly, the finished design should match, without excessively exceeding, as many of the design criteria as possible. Clearly the provision of excess strength in a structural element without reason will not be judged economic. Secondly, in structural steelwork construction only part of the cost is contained in the rolled steel sections, and a large part of the cost results from the fabrication and erection processes. Consequently economic design does not result from finding the smallest structural size and weight without considering the difficulties of fabrication. In many cases repetition of a member size and standardization of components can lead to substantial overall savings.

The appearance of the finished structure is generally of great importance due to the very size and impact of frames in structural steel. The achievement of an elegant design is desirable not only in complete structures but in small design details. It is here that the student should try to achieve stylish, neat and balanced solutions to problems set. In many cases these will prove to be the strongest and most economic solutions also.

1.2 SCOPE OF BS 5950 *STRUCTURAL USE OF STEELWORK IN BUILDINGS*

BS 5950 is subdivided into nine parts, each being published separately. Parts 3 and 5 to 9 inclusive are awaiting publication.

Part 1:	*Design in simple and continuous construction: hot rolled sections* (1985)
Part 2:	*Specification for materials, fabrication and erection: hot rolled sections* (1985)
Part 3:	*Design in composite construction*
Part 4:	*Design of floors with profiled steel sheeting* (1982)
Part 5:	*Design in cold formed sections*
Part 6:	*Design in light gauge sheeting, decking and cladding*
Part 7:	*Specification for materials and workmanship: cold formed sections*
Part 8:	*Design of fire protection for structural steelwork*
Part 9:	*Stressed skin design*

The purpose of BS 5950 is to define common criteria for the design of structural steelwork in building and allied structures, and to give guidance to designers on methods of assessing compliance with those criteria. Part 1 of this British Standard deals with design in simple and continuous construction for hot rolled sections. Part 2 covers the specification for materials, fabrication and erection. The following chapters give examples of the design of buildings principally covered by Parts 1 and 2 of BS 5950.

Use is also made of other Parts for particular design requirements such as composite construction, and these are referred to as appropriate in the following chapters. BS 5400 is the appropriate code for the design of bridges, and may also be a more appropriate basis for the design of other types of plated structure, e.g. bunkers.

1.3 LIMIT STATE DESIGN

In common with most current UK codes of practice, BS 5950 adopts a limit state approach to design. In this approach the designer selects a number of criteria by which to assess the proper functioning of the structure and then checks whether they have been satisified. The criteria are divided into two main groups based on whether assessment is made of the collapse (ultimate) condition, or normal working (serviceability) condition.

Ultimate limit state includes:

- strength (safety)
- stability (overturning)
- fatigue fracture (not normally considered in buildings)
- brittle fracture
- structural integrity (including accidental damage)

Serviceability limit state includes:

- deflection
- durability
- vibration

Limiting values for each criterion are given in BS 5950: Part 1 and their use is demonstrated in the following chapters. The designer should, however, always be aware of the need for additional or varied criteria.

1.4 PARTIAL SAFETY FACTORS

Safety factors are used in all design to allow for variabilities of load, material, workmanship and so on which cannot be assessed with absolute certainty. They must be sufficient to cover:

1. load variations;
2. load combinations;
3. design and detailing procedures;
4. fabrication and erection procedures;
5. material variations.

The safety factor can be applied at one point in the design (global or overall safety factor), or at several points (partial safety factors). In steelwork design a partial safety factor γ_f (the load factor) is applied to the loads (variations 1 to 3 above) and another factor γ_m to the material strengths (variations 4 and 5). BS 5950: Part 1 includes a factor γ_l for structural performance within the value of γ_f, and assumes a value of 1.0 for γ_m. The use of $\gamma_m = 1.0$ does not imply that no margin of safety for material has been included, but rather that a suitable allowance has been made in the design strengths given in, e.g. Table 1.2 of this chapter. Typical values of γ_f are given in Table 1.1 with further values given in BS 5950: Part 1, Table 2. Application of the factors to different loads in combination is given in Chapter 2 and throughout the design examples. The value of each load factor reflects the accuracy with which a load can be estimated, and the likelihood of the simultaneous occurrence of a given combination of loads.

Table 1.1 Partial safety factor for loads

Loading	Load factor γ_f
Dead load W_d	1.4
Dead load restraining uplift	1.0
Imposed load W_i	1.6
Wind load W_w	1.4
Combined loads ($W_d + W_i + W_w$)	1.2

1.5 LOADING

In most cases design begins with as accurate as possible an assessment of the loads to be carried. These may be given, or obtained from a British Standard[1] or other appropriate source. They will be used in idealized forms as either distributed loads or point (concentrated) loads.

Chapter 2 sets out typical loadings and gives examples of how they are combined in design.

These external loads, sometimes called actions, form only part of the total forces on a structure, or on a structural element. The reactions to the loads on each element must be obtained as design proceeds. These reactions must be carried through to supporting elements, so that all external loads, including self weight of members, are transferred through the structure by the shortest load path, until the foundation is reached. This process is essential to safe design. A simple example of this process is shown in Fig. 1.2, in which the load path for the external load (snow) on a section of roof cladding is traced to the foundation. The cladding (sheeting) carries snow load as well as self weight, and this combined load produces a reaction from a typical purlin. This constitutes a load W on the purlin, producing reactions P on the rafter. Similar purlin reactions together with rafter self weight constitute the rafter load, producing reactions R from the roof truss (at each node). Loading the roof truss produces reaction T from the column, and also a reaction S if the loading is non-vertical. These in turn, acting on the column, produce foundation reactions H, V and M.

1.6 INTERNAL FORCES AND MOMENTS

Loads with their reactions may be used to find internal forces and moments within any structural element. The usual method is to draw shear force and bending moment diagrams[2,3]. These diagrams are graphs showing how the internal forces vary along a structural element as a result of a set of stationary (static) loads. Influence lines and moment/force envelopes may be needed in cases of moving (dynamic) loads (see Ch. 2). It is also necessary to find the axial force present in a structural element, particularly vertical members, and in some cases the torsion as well. Diagrams may be used to advantage for these forces also.

In many cases design concentrates on specific values of maximum bending moment or shear force at a known position, e.g. mid-span. In these cases, formulae or coefficients may be useful and can be obtained from standard tables or charts[4]

1.7 STRESSES AND DEFORMATIONS

In the design of steelwork to BS 5950: Part 1 stresses are used to obtain capacities of structural sections in bending, shear, axial force, etc., and any combination of these forces. Stresses used are generally based on the yield stresses appropriate to the steel quality and maximum thickness required by the designer, and are detailed in table 6, BS 5950: Part 1, which is based on values stipulated in BS 4360[5]. The design strength p_y is used, for example, to calculate the moment capacity of a steel section.

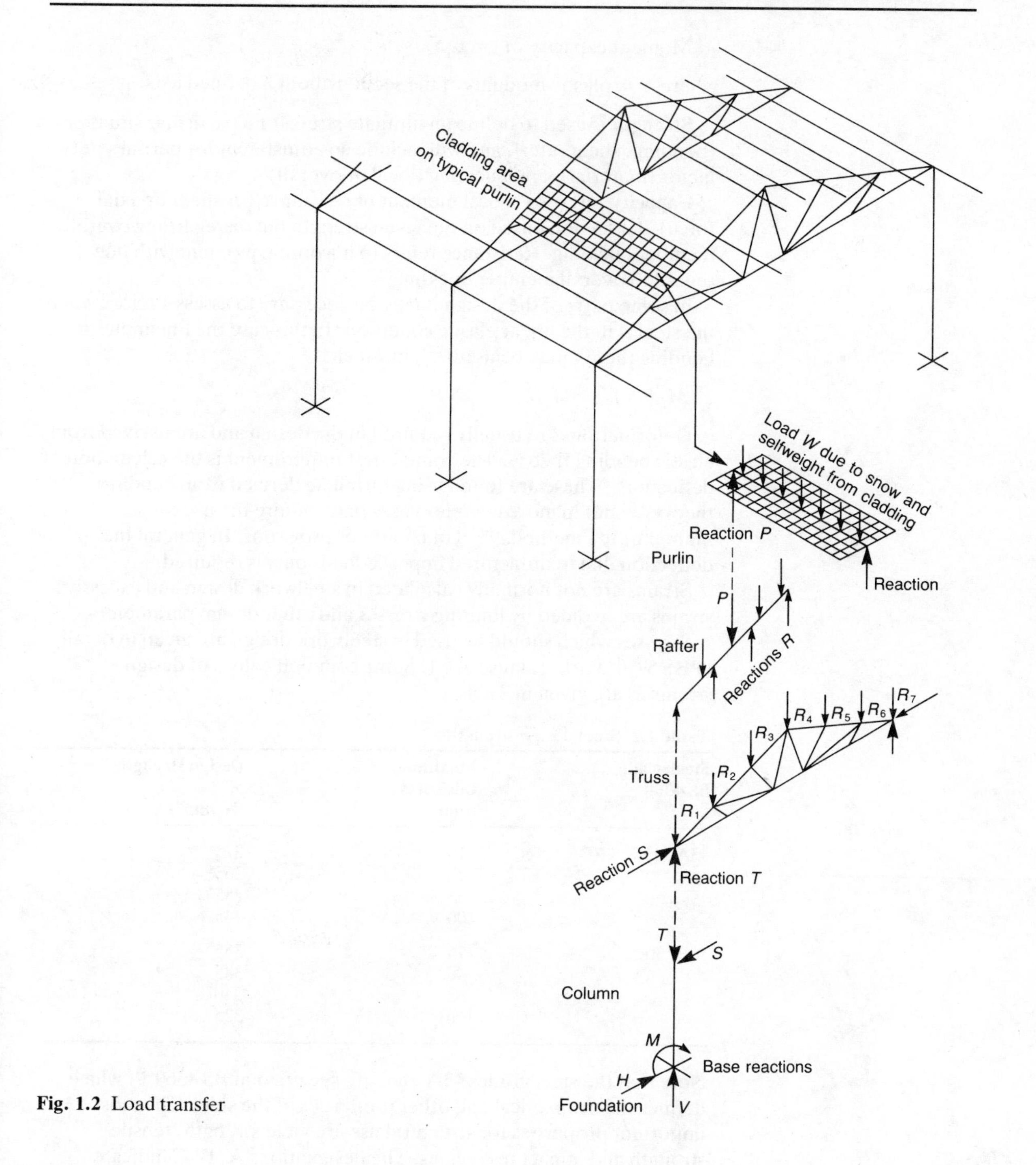

Fig. 1.2 Load transfer

Moment capacity $M_c = p_y S$

where S = plastic modulus of the section about a defined axis.

Strength is used to define an ultimate stress for a particular situation (bending, shear, etc.), and will include an adjustment for partial safety factor (material), and buckling (local or overall).

Capacity refers to a local moment of resistance (or shear or axial force) at a section based on the given strength but disregarding overall member buckling. **Resistance** refers to maximum moment with due regard to overall member buckling.

In some parts of the design it may be necessary to assess stresses when the steel is in the linear elastic condition. In this case the linear elastic bending theory may be used[6,7], in which

$$M/I = f/y = E/R$$

Deformations are usually required in the design and are derived from elastic bending theory. The commonest requirement is the calculation of deflections. These are found using formulae derived from bending theory[8,9] but in more complex cases may require the use of moment-area methods[10,11] or computer programs. In general the deflection due to unfactored imposed loads only is required.

Strains are not normally calculated in steelwork design and excessive strains are avoided by limiting stresses and other design parameters.

Stresses which should be used in steelwork design are given in detail in BS 5950: Part 1, clause 3.1.1. Some common values of design strengths are given in Table 1.2.

Table 1.2 Steel design strengths

Steel grade BS 4360	Maximum thickness (mm)	Design strength p_y (N/mm²)
43A, 43B, 43C	16	275
	40	265
	63	255
	100	245
50B, 50C	16	355
	40	345
	63	340
	100	325

Note that the steel grades 43A etc., are specified in BS 4360[5], which defines the mechanical and other properties of the steel. The most important properties for structural use are yield strength, tensile strength and impact test values. The designations A, B, C indicate increasing resistance to impact and brittle fracture, with no signficant change in other mechanical properties.

The cross-section of a structural member needs to be classified according to BS 5950: Part 1, clause 3.5 in order to assess the resistance to local buckling of the section. Cross-sections are classified as plastic, compact, semi-compact and slender by reference to the breadth/thickness ratios of flange outstands and webs (Fig. 1.1), and also to the design strength. Details are given in clause 3.5 and table 7 of BS 5950: Part 1. The classifications of most hot rolled sections in grades 43 and 50 steel are given with their section properties in reference (12).

Recommended values of maximum deflections are given in BS 5950: Part 1, table 5. Some common values are reproduced in Table 1.3. In cases where the steelwork structure is to support machinery, cranes and other moving loads, more stringent limitations on deflection may be necessary. Values of maximum deflections should be checked with manufacturers of any machinery to be used.

Table 1.3 Deflection limits

Structural element	Deflection limit due to unfactored imposed load
Cantilever	Length/180
Beam (brittle finishes)	Span/360
Other beam	Span/200
Purlin or sheeting rail	To suit cladding (but span/200 may be used)
Crane girder (vertical)	Span/600
Crane girder (horizontal)	Span/500

1.8 LAYOUT OF CALCULATIONS

Before design calculations are started, the designer must first interpret the client's drawings so that a structural arrangement can be decided to carry the loads down to the foundations. This structural arrangement must avoid intrusion into space required by the client's processes or operations. It is broken down into simple structural elements which are each given an individual code number by the steel designer (see Ch. 14). Calculations for an individual element can thus be identified, as in the design examples.

Clarity is essential in setting out calculations, and the designer should make sure that they can be checked without constant reference to himself. Designers develop their own individual styles for setting out their calculations. Design offices of consulting engineers, local authorities and contractors often use one particular format as a house style. The student should start using a basic framework such as that given here, but adapting it to suit the particular structure.

1.8.1 Subdivision

Subdivide the calculations into appropriate sections using subheadings such as 'dimensions', 'loading', 'moment capacity'. This makes checking of the basic assumptions and the results much easier, and helps the designer achieve a neat presentation.

1.8.2 Sketches

Engineers think pictorially, and should develop a spatial awareness. A sketch will clearly indicate what the designer intended, while in a string of numbers a serious omission can be overlooked. Sketches in the following chapters are placed in the left-hand margin where convenient.

1.8.3 References

Sources of information must be quoted as:

- loading — British Standard; manufacturer's catalogue; client's brief
- dimensions — drawing number
- stresses — BS 5950 clause number; research paper

This ensures that future queries about the calculations can be answered quickly and that subsequent alterations can be easily detected. It will also assist the designer when carrying out a similar design at a later date and this can be of great value to a student who will one day design in earnest.

1.8.4 Results

In design, results (or output), such as member size, load, moment, from one stage of the calculations may be used as input at a further stage. It is important therefore that such information is easily obtained from the calculations. Such results may be highlighted by placing in an output margin (on the right-hand side), or by placing in a 'box', or merely by underlining or using coloured marker pens; in this book, bold type has been used.

The student should attempt to maintain realism in calculation, and avoid quoting the eight or more digits produced by calculators and micro-computers. Loading is commonly no more than two-figure accurate, and section properties are given to only three figures. No amount of calculation will give results of higher accuracy.

1.8.5 Relationship with drawings

In most cases the final results of design calculations are member sizes, bolt numbers, connection details and so on, all of which information must be conveyed to the fabricator/contractor on drawings. It is, however, common in steelwork design for some of the drawing to be carried out by persons other than the design engineer, such as detailing draughtsmen, technician engineers, or even the fabricator. It is therefore essential that the final output should be clearly marked in the calculations. Specific requirements such as connection details must be clearly sketched.

1.9 STRUCTURAL THEORY

It is assumed in the following chapters that the reader will have available a copy of BS 5950: Part 1 or extracts from it. Tables and charts for design will not be reproduced in full in the text but extracts will be given where appropriate. In addition properties of steelwork sections will be required. These are available from the Steel Construction Institute[12]. The meaning of the properties given in these publications must be understood and may be studied in, for example, references (6,7).

It may be useful at some point for the student to examine the background to the steelwork design method and BS 5950. Reference is therefore made to the Steel Construction Institute publication[13] which is intended as explanatory to BS 5950: Part 1.

The designer of steelwork elements and structures must have a clear understanding of theory of structures and strength of materials. The student is referred to the relevant sections of textbooks such as those given in references (6) to (11) and further explanation is avoided.

1.10 FORMAT OF CHAPTERS

The following chapters provide design examples of structural steelwork elements (Part I) and structural steelwork frameworks (Part II).

The chapter order is intended to guide a student with a basic knowledge of theory of structures and strength of materials into steelwork design. It therefore starts (in Part I) with loading, including combination effects, and proceeds to simple elements with which the student is probably already familiar. Later more complex elements and those requiring special treatment are introduced. In Part II the simple elements are combined to form complete structures. While this chapter arrangement is preferred for teaching, in the actual design of structural steel elements the calculations are usually arranged in load order, i.e. as indicated in Section 1.5 and Fig. 1.2. This is sometimes known as reverse construction order.

Each chapter (in Part I) starts with basic definitions of structural members and how they act. General notes on the design of the element/frame follow and then the design calculations are set out for one or more examples demonstrating the main variations.

The calculations follow the layout suggested in Section 1.8. References to BS 5950: Part 1 are given merely by quoting the appropriate clause, e.g. 'clause 2.4.1', or table, e.g. 'BS table 13'. References for the structural theory required by the student, or for background to BS 5950, are given as study topics at the end of each chapter with a numerical reference in the text, e.g. (3).

STUDY REFERENCES

Topic	*References*
1. Loading	BS 6399 *Loading for buildings* Part 1: *Dead and imposed loads* (1984) Part 2: *Wind loads* (to be published; presently CP3 Ch. V Part 2) Part 3: *Snow loading* (to be published; presently BS 5502: Part 1 and *BRE Digest* 290)
2. BM and SF diagrams	**Marshall W.T. & Nelson H.M.** (1977) Bending moment and shearing force diagrams, *Structures,* pp. 24–30. Longman
3. BM and SF diagrams	**Coates R.C., Coutie M.G. & Kong F.K.** (1988) Shear forces and bending moments, *Structural Analysis,* pp. 58–71. Van Nostrand Reinhold
4. BM and SF coefficients	(1972) *Steel Designers' Manual,* Chs 2, 3 and 4. Crosby Lockwood Staples
5. Steel quality	BS 4360 (1979) *Specification for weldable structural steels*
6. Theory of bending	**Marshall W.T. & Nelson H.M.** (1977) Pure bending of straight uniform beams, *Structures,* pp. 67–71. Longman
7. Theory of bending	**Hearn E.J.** (1985) Bending, *Mechanics of Materials,* vol. 1 pp. 62–8. Pergamon
8. Deflections	**Marshall W.T. & Nelson H.M.** (1977) *Structures,* pp. 110–17. Longman
9. Deflections	**Hearn E.J.** (1985) Slope and deflection of beams, *Mechanics of Materials,* vol. 1 pp. 92–107. Pergamon
10. Moment-area method	**Marshall W.T. & Nelson H.M.** (1977) The moment-area method, *Structures,* pp 122–5. Longman
11. Moment-area method	**Coates R.C., Coutie M.G. & Kong F.K.** (1988) Moment-area methods, *Structural Analysis,* pp. 176–81. Van Nostrand Reinhold
12. Section properties	(1985) *Steelwork Design* vol. 1, Section properties, member capacities. Steel Construction Institute
13. Background to BS 5950	*Steelwork Design* vol. 3, Commentary on BS 5950: Part 1. Steel Construction Institute (to be published)

2

LOADING AND LOAD COMBINATIONS

The loading for most structures is obtained from the appropriate British Standard[1], manufacturer's data, and similar sources. The loads obtained must, however, be combined to simulate what is perceived by the designer to occur in practice, and be multiplied by appropriate load factors. The process of combining loads and including the load factors is carried out for simple structural elements when deriving the bending moments, shear forces, etc., which will occur. For more complex structures it is advantageous to include the load factors after deriving the bending moments etc., so that specific combinations can be examined more readily.

2.1 DEAD LOADS

These will include the following:

- own weight of steel member (kg/m of steel section)
- other permanent parts of building, etc., not normally moveable (e.g. concrete floor slabs, finishes, brick/block walls, cladding)

They are calculated either from density of material (kg/m^3) or specific weight (kN/m^3), or from manufacturer's data contained in catalogues or manuals. Table 2.1 shows some typical values; these are all permanent loads and are combined with the appropriate dead load partial safety factor (see Section 1.4).

Table 2.1 Typical values of common structural materials

Material	Density (kg/m^3)	Specific weight (kN/m^3)
Steel	7850	77
Reinforced concrete	2420	23.7
Brickwork	2000–2300	20–23
Timber	500–900	5–9

2.2 IMPOSED LOADS

These will include the following temporary loads:

- snow on roofs[(1)]
- people
- furniture
- equipment such as cranes and other machinery
- semi-permanent partitions which are moveable

Imposed loads vary with the function of the room or building[(1)], and some typical values are shown in Table 2.2. All imposed loads are based on experience within the construction industry and the statistical analyses of observed cases. These are all temporary loads and are combined with the imposed load partial safety factor (see Section 1.4).

Table 2.2 Typical values of imposed loads

Building usage	Imposed load (kN/m²)
Residential (self-contained dwellings)	1.5
Offices (depending on room usage)	2.5–5.0
Educational (classrooms)	3.0
Theatres (area with fixed seating)	4.0
Warehousing (general storage)	2.4 per m height
Industrial workshops	5.0

2.3 WIND LOADS

Wind loading on buildings is derived by use of BS 6399: Part 2[(1)], presently CP 3 Ch. V Part 2. Basic wind speed, appropriate to the location of the building, is selected and reduced to a design wind speed using factors which take into consideration topography, surrounding buildings, height above ground level, component size and period of exposure. The design wind speed is equated to a dynamic pressure q(kN/m^2). Due to building and roof shape, openings in walls, etc., pressures and suctions both external and internal will arise. Pressure coefficients external (C_{pe}) and internal (C_{pi}) may be used as shown in Section 2.7.

$$\text{Force on any element} = (C_{pe} - C_{pi})q \times \text{area of element}$$

Wind data with suitable factors and coefficients are given in reference (1). The method of obtaining the quasi-static wind load used in design is given in greater detail in Sections 2.7 and 11.4.3.

2.4 LOAD COMBINATIONS

Loads on any structure must be arranged in design so that the maximum force or moment is achieved at the point in the structure being considered. Hence all realistic load combinations must be considered to ensure that all peak values have been calculated at every point. Only in simple cases will one arrangement of maximum loads be sufficient to produce maximum moments or forces for design purposes.

2.5 EXAMPLE 1. LOADING OF A SIMPLY SUPPORTED GANTRY GIRDER

(a) Dimensions

Simply supported span	6.0 m
Crane wheels centres	3.6 m

(See further description in Section 5.1.)

(b) Loads specified

Self weight of girder (uniformly distributed)	1.5 kN/m
Maximum crane wheel load (static)	220 kN
Weight of crab	60 kN
Hook load to be lifted	200 kN

Dynamic effects to be included in accordance with BS 6399: Part 1[(1)] at 25%
(For further details of dynamic effects and the derivation of maximum wheel loads see Section 5.1.)

(c) Moments and forces (due to u.d.l.)

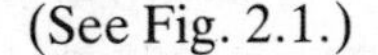

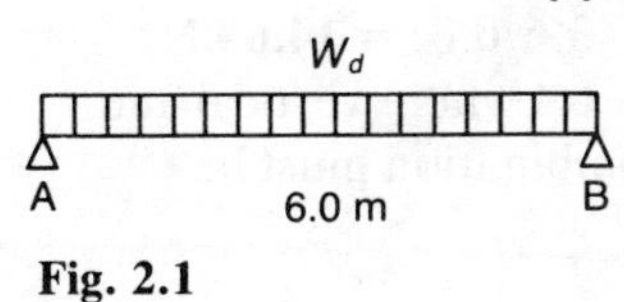

Fig. 2.1

(See Fig. 2.1.)

Self weight W_d (factored by γ_f) $= 1.4 \times 1.5 \times 6.0 = 12.6$ kN

Ultimate mid-span BM $= W_d L/8$

$= 12.6 \times 6.0/8 = 9$ kNm

Ultimate reactions $R_A = R_B = WL/2$

$= 12.6/2 = 6$ kN

(d) Moments and forces (due to vertical wheel load)

Wheel load W_c (including γ_f and 25% impact) $= 1.6 \times 1.25 \times 220 =$ 440 kN

The positions of moving loads to give maximum values of moment and shear force are given in Section 5.2 and references (2,3).

The maximum values of each case are now given (and see Fig. 2.2).

Ultimate BM under wheel (case 1):

$= 2W_c(L/2 - c/4)^2/L$

$= 2 \times 440\,(6.0/2–3.6/4)^2/6.0 = 647$ kNm

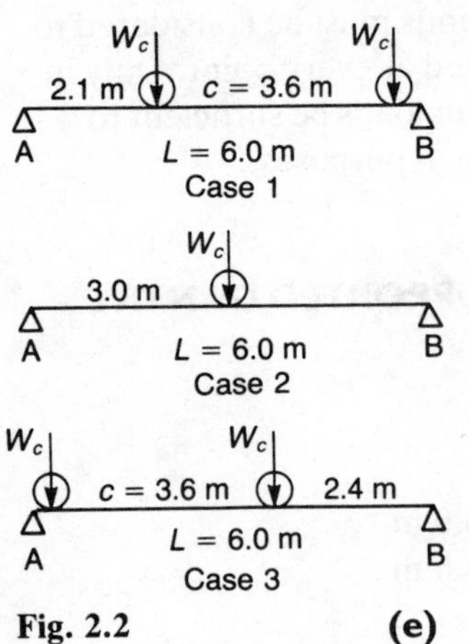

Fig. 2.2

Ultimate BM under wheel (case 2):

$= W_c L/4$

$= 440 \times 6.0/4 = 660$ kNm

Case 2 gives maximum ultimate BM of 660 kNm

Ultimate reaction R_A (case 3):

$= W_c(2 - c/L)$

$= 440(2 - 3.6/6.0) \quad = 616$ kN

Total ultimate BM $\quad = 9 + 660 =$ **669 kNm**

Total ultimate reaction $= 6 + 616 =$ **622 kN**

(e) Moments and forces (due to horizontal wheel load)

In addition to the vertical loads and forces calculated above, horizontal surge loading due to movement of the crab and hook load, gives rise to horizontal moments and forces (see Section 5.1), equal to 10% of these loads.

Horizontal surge load W_{hc} (incl. γ_f) $= 1.6 \times 0.10(200 + 60)$

$= 41.6$ kN

This is divided between 4 wheels

Horizontal wheel load $W_{hc} = 41.6/4 = 10.4$ kN

Using calculations similar to those for vertical moments and forces:

Ultimate horizontal BM (case 2) $= W_{hc}L/4$

$= 10.4 \times 6.0/4 =$ **15.6 kNm**

Ultimate horizontal reaction (case 3)

$= W_{hc}(2 - c/L)$

$= 10.4(2 - 3.6/6.0) =$ **14.6 kN**

Note that the γ_f value of 1.6 may be reduced to 1.4 where vertical and horizontal crane loads act together, and this combination must be checked in practice (see Section 5.3).

2.6 EXAMPLE 2. LOADING OF CONTINUOUS SPANS

Obtain the maximum values of bending moment, shear force and reaction for a continuous beam at the positions given in (c), (d), (e) and (f).

(a) Dimensions

Main beams, spaced at 4.5 m centres supporting a concrete slab (spanning one way only) for office accommodation. Steel beams to have four continuous spans of 8.0 m. Assume uniform section properties.

(b) Loading

Self weight of beam (assumed)	1.0 kN/m
Concrete slab and finish	5.4 kN/m^2
Imposed loading (offices)	5.0 kN/m^2

The dead loading (self weight, slab and finishes) is fixed, but the imposed loading is movable. The dead loading must be present ($\gamma_f = 1.4$) whilst the imposed loading may be present ($\gamma_f = 1.6$) or absent.

For one span:

Self weight = 1.0 × 8.0	= 8 kN
Slab + finishes = 5.4 × 4.5 × 8.0	= 194 kN
Dead load W_d (total)	= 202 kN
Imposed load W_i = 5.0 × 4.5 × 8.0	= 180 kN
Maximum span load = $1.4W_d + 1.6W_i$	= 571 kN
Minimum span load = $1.4W_d$	= 283 kN

Arrangements of loading to produce maximum moments and forces may be found by inspection of the appropriate influence lines. The use of influence lines to give the required arrangements (patterns) of loading is described in references (4,5). An influence line shows the effect, say bending moment, due to a unit moving load. Hence the maximum BM is obtained by placing the imposed loads where the influence is of one sign only; e.g. in Fig. 2.3 the maximum BM due to imposed load at the middle of span 1 is obtained by placing the imposed load on spans 1 and 3. Load on spans 2 and 4 would produce a BM of opposite sign at the point considered.

In the design office standard loading arrangements are used to speed up this process of selection, with influence lines used for more complex cases.

(c) Load pattern for mid-span moment M_1

Maximum value of M_1 is produced when spans 1 and 3 have the maximum loading and spans 2 and 4 have the minimum loading. Using the influence lines shown, the loading patterns producing maximum effect may be obtained, and are summarized below.

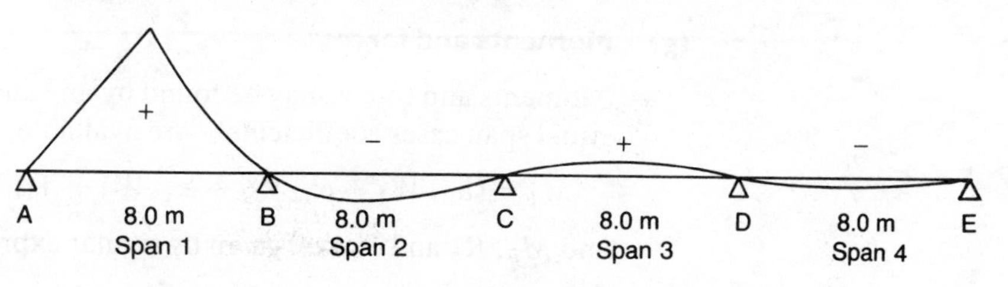

Fig. 2.3 Influence line for M_1

(d) Load pattern for support moment M_b

Maximum value of M_b is produced when spans 1, 2 and 4 have the maximum loading and span 3 has the minimum loading (Fig. 2.4).

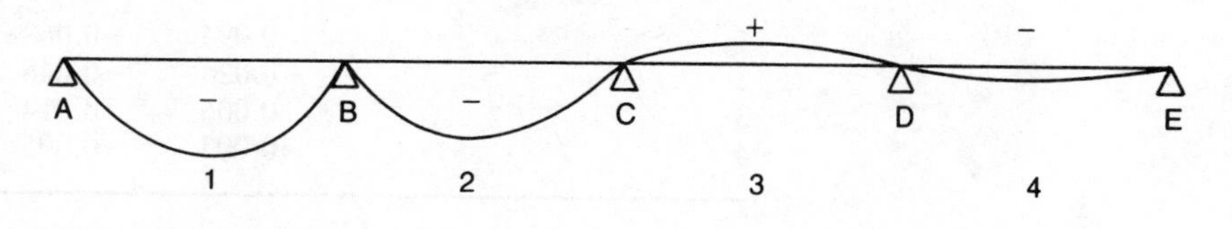

Fig. 2.4 Influence line for M_b

(e) Load pattern for reaction R_b

Maximum as in (d) (Fig. 2.5).

(f) Load pattern for shear force S_{ab}

Maximum as in (c) (Fig. 2.6).

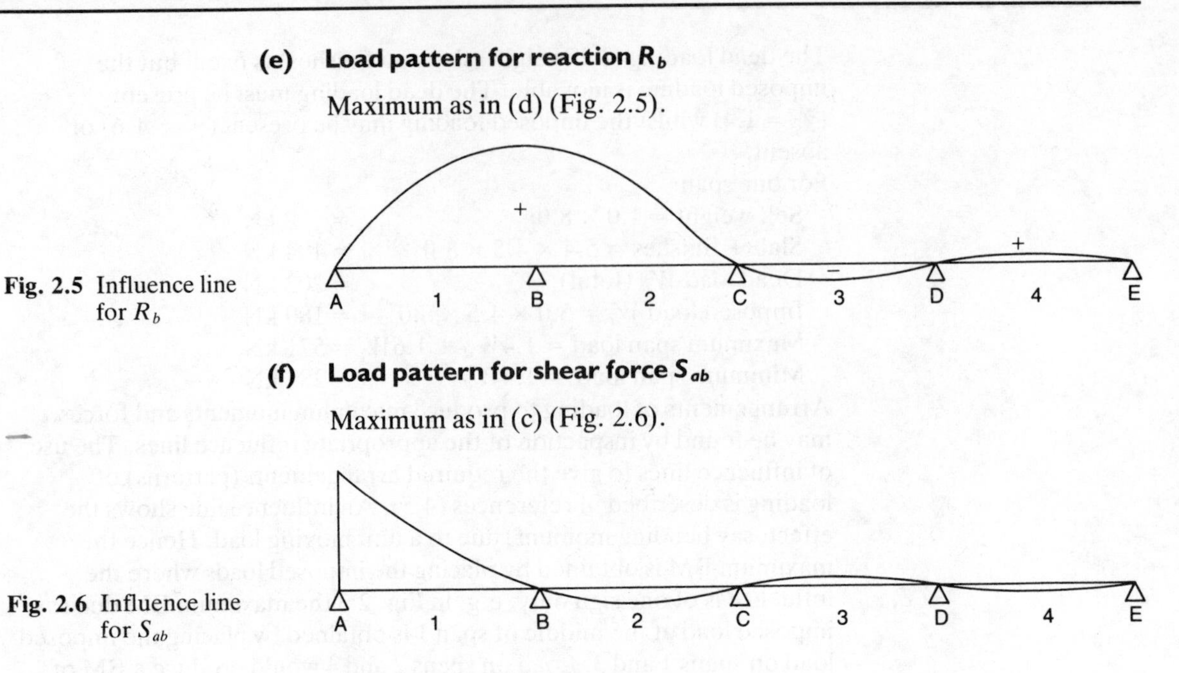

Fig. 2.5 Influence line for R_b

Fig. 2.6 Influence line for S_{ab}

Arrangements of loading for maximum moments and forces:

	Span loading (kN)			
	W_1	W_2	W_3	W_4
For moment M_1	571	283	571	283
For moment M_b	571	571	283	571
For reaction R_b	571	571	283	571
For shear force S_{ab}	571	283	571	571

(g) Moments and forces

Moments and forces may be found by any analytical method, but for equal span cases coefficients[6] are available, where, for example

$$M_1 = (\alpha_{11}\ W_1 + \alpha_{12}\ W_2 + \alpha_{13}\ W_3 + \alpha_{14}\ W_4)L$$

and M_b, R_b and S_{ab} are given by similar expressions.

The coefficients α may be summarized:

Span loaded	Moment or force coefficients α			
	M_1	M_b	R_b	S_{ab}
1	0.094	−0.068	0.652	0.433
2	−0.024	−0.048	0.545	−0.049
3	0.006	0.014	−0.080	0.013
4	−0.003	−0.005	0.027	−0.005

Hence the maximum value of M_1 occurs with the load arrangement shown and

$$M_1 = (0.094 \times 571 - 0.024 \times 283 + 0.006 \times 571 - 0.003 \times 283)\,8.0$$
$$= \mathbf{396\ kNm}$$

Similarly $M_b = \mathbf{-521\ kNm}$

$R_b = \mathbf{676\ kN}$

$S_{ab} = \mathbf{238\ kN.}$

2.7 EXAMPLE 3. LOADING OF A PORTAL FRAME

(a) Frame

See Fig. 2.7: pitched portal with pinned feet; span 38 m, spaced at 6 m centres.

Assume uniform section properties.

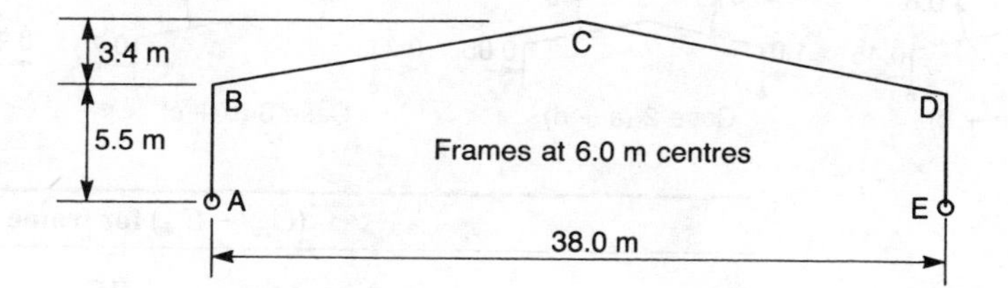

Fig. 2.7 Portal frame

(b) Loading

Self weight of frame	0.90 kN/m
Cladding (roof and walls)	0.09 kN/m²
Snow and services	0.75 kN/m²
Wind pressure q (walls)	1.20 kN/m²
Wind pressure q (roof)	1.20 kN/m²

Wind pressures are based on a basic wind speed of 50 m/s for a location in Scotland.

Factors[(1)] S_1 and S_2 are both taken as 1.0 and factor S_3 as 0.88 for a height of 10 m.

The design wind speed is hence 44 m/s, giving a dynamic pressure of 1.20 kN/m².

It is possible to use a lower wind pressure below a height of 5 m but this makes the analysis more complex for very little reduction in frame moments.

(c) Pressure coefficients

External pressure coefficients (C_{pe}) may be found and are summarized below. These values are obtained from reference (1).

	C_{pe} for frame member			
	AB	BC	CD	DE
Wind on side	0.7	−1.2	−0.4	−0.25
Wind on end	−0.5	−0.6	−0.6	−0.5

Internal pressure coefficients (C_{pi}) should be obtained[(1)], and are taken in this example as +0.2 (maximum) and −0.3 (minimum) which are combined algebraically with the values of C_{pe} above.

Figure 2.8 shows the individual pressure coefficients, and Fig. 2.9 shows the various combination cases.

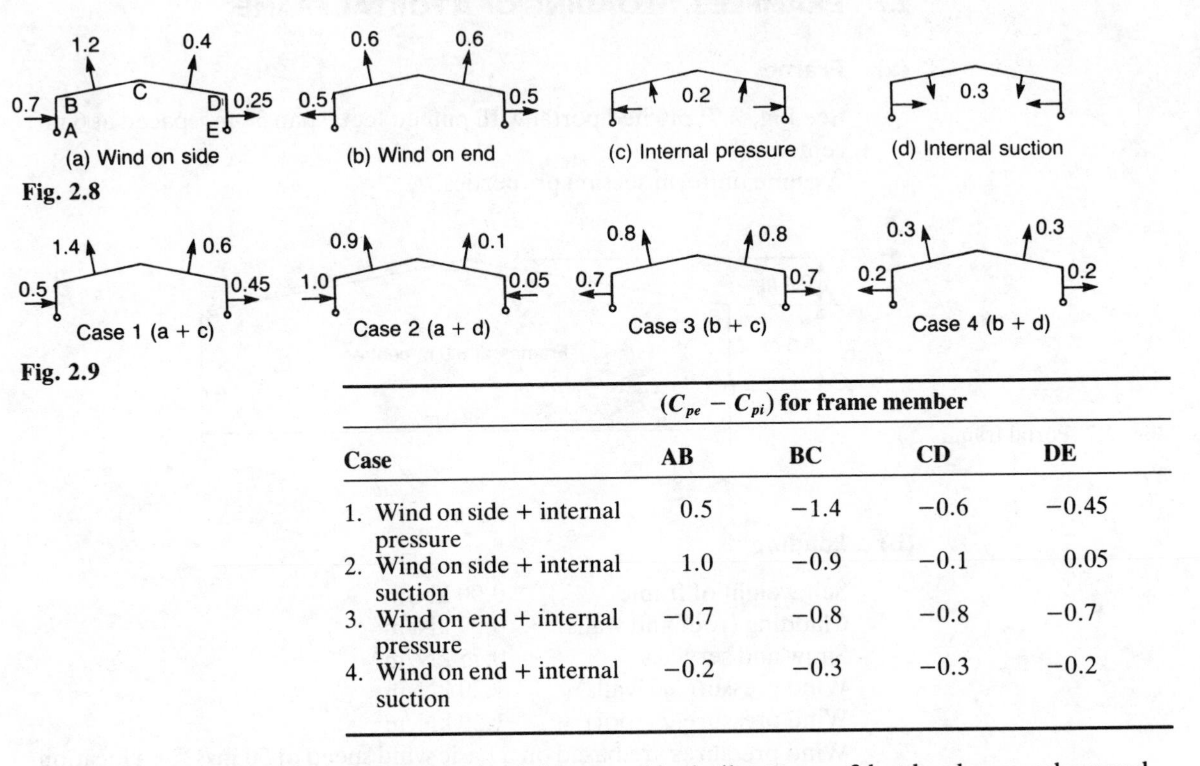

Fig. 2.8

Fig. 2.9

	$(C_{pe} - C_{pi})$ for frame member			
Case	AB	BC	CD	DE
1. Wind on side + internal pressure	0.5	−1.4	−0.6	−0.45
2. Wind on side + internal suction	1.0	−0.9	−0.1	0.05
3. Wind on end + internal pressure	−0.7	−0.8	−0.8	−0.7
4. Wind on end + internal suction	−0.2	−0.3	−0.3	−0.2

It may be noted that case 4 is similar to case 3 but has lower values and may be discarded.

(d) Member loads

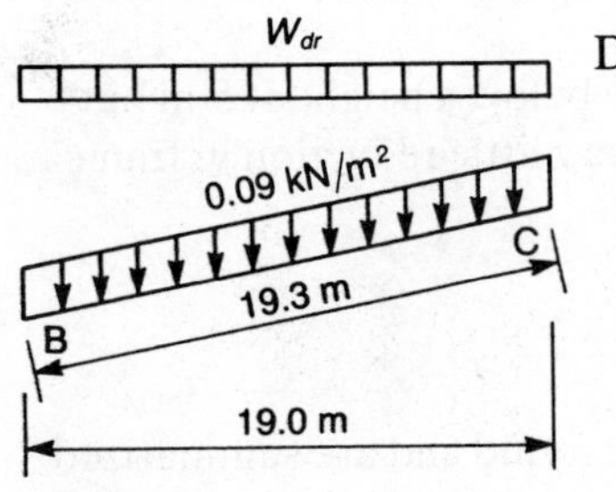

Fig. 2.10

Dead load on roof is calculated on the projected area (Fig. 2.10):

Self weight = 0.9 × 19.3 = 17.5 kN
Cladding = 0.09 × 19.3 × 6.0 = 10.5 kN
Total W_{dr} = 28.0 kN

Dead load on walls (Fig. 2.11):

Self weight = 0.9×5.5 = 5 kN

Cladding = $0.09 \times 5.5 \times 6.0$ = 3 kN

Total W_{dw} = 8 kN

Imposed load on roof given on plan area (and services) (Fig. 2.12):

Snow load $W_i = 0.75 \times 19.0 \times 6.0 = 86$ kN

Wind load on roof must be divided into vertical and horizontal components.

Case (1) wind on side + internal pressure (Fig. 2.13):

Vertical component $W_{wv} = -1.4 \times 1.20 \times 6.0 \times 19.0 = -192$ kN

Horizontal component $W_{wh} = -1.4 \times 1.20 \times 6.0 \times 3.4 = -34$ kN

Wind load on wall (Fig. 2.14), $W_{ww} = 0.5 \times 1.20 \times 6.0 \times 5.5 = 20$ kN

In the same manner wind load for each member and each case may be calculated. The loading due to dead, imposed, and wind loads may be summarized using the positive notation in Fig. 2.15.

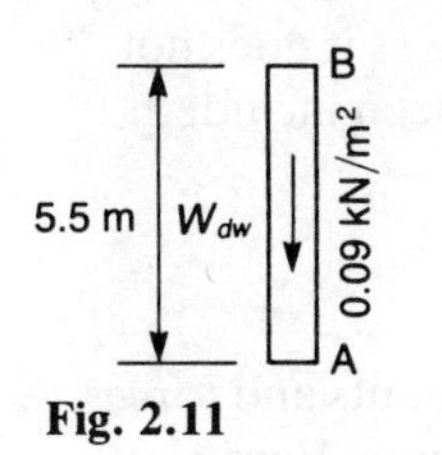

Fig. 2.11

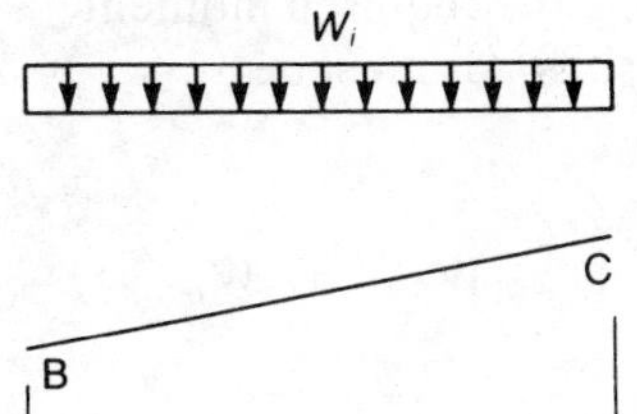

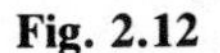

Fig. 2.12

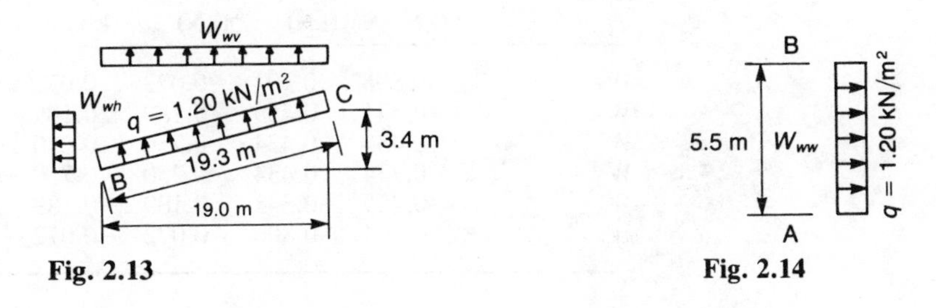

Fig. 2.13

Fig. 2.14

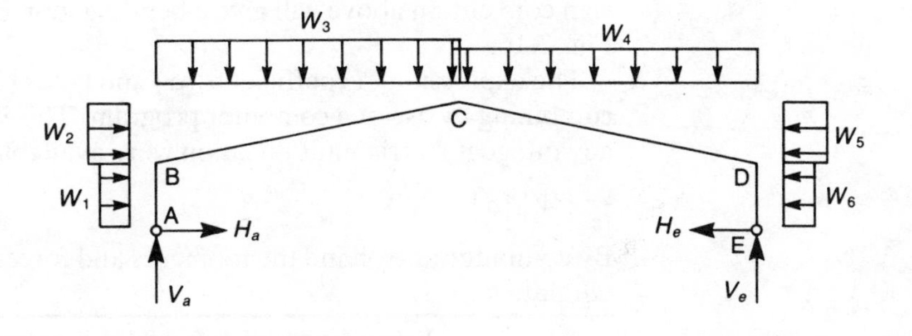

Fig. 2.15 Loading summary

	Unfactored loads (kN)					
	W_1	W_2	W_3	W_4	W_5	W_6
Dead load W_d	0	0	28	28	0	0
Imposed load W_i	0	0	86	86	0	0
Wind case (1) W_{w1}	20	−34	−192	−82	−15	−18
Wind case (2) W_{w2}	40	−22	−123	−14	−2	−2
Wind case (3) W_{w3}	−28	−20	−109	−109	−20	−28

Note that the wall dead load is not included at this stage as it does not produce a bending moment. Values of γ_f have not yet been included.

(e) Forces and moments

The loads given in Fig. 2.15 may be used to obtain moments and forces by any analytical method or by use of charts or coefficients. Each load W has an appropriate set of coefficients α giving the required moment or force[5]. For any load case the effects of all six loads must be summated.

For dead load:

$$H_a = \alpha_{11} W_1 + \alpha_{21} W_2 + \alpha_{31} W_3 + \alpha_{41} W_4 + \alpha_{51} W_5 + \alpha_{61} W_6$$

and similar for each force or moment.

The coefficients for elastic analysis may be calculated and tabulated:

Load	Coefficient α for moment or force						
	H_a (kN)	H_e (kN)	V_a (kN)	V_e (kN)	M_b (kNm)	M_c (kNm)	M_d (kNm)
W_1	−0.806	0.194	−0.072	0.072	1.683	−0.351	−1.067
W_2	−0.541	0.459	−0.189	0.189	2.975	−0.485	−2.525
W_3	0.434	0.434	0.750	0.250	−2.385	0.890	−2.385
W_4	0.434	0.434	0.250	0.750	−2.385	0.890	−2.385
W_5	0.459	−0.541	0.189	−0.189	−2.525	0.485	−2.975
W_6	0.194	−0.806	0.072	−0.072	−1.067	−0.351	1.683

Each force or moment may be calculated using equations similar to that given for H_a above. The positive notation for loads (Fig. 2.15) and the sign convention above will give a bending moment sign convention as Fig. 2.16.

The expression of coefficients (α) and loads (W) as arrays allows for combining by use of a computer program. This would be of particular advantage if matrix multiplication were available.

$$\{W\}\,[\alpha] = \{F\}$$

By computer or by hand the moments and forces are calculated and tabulated:

	Value of moment or force due to unfactored loads						
	H_a (kN)	H_e (kN)	V_a (kN)	V_e (kN)	M_b (kNm)	M_c (kNm)	M_d (kNm)
W_d	24.3	24.3	28.0	28.0	−133.6	49.8	−133.6
W_i	74.6	74.6	86.0	86.0	−410.2	153.1	−410.2
W_{w1}	−127.0	−108.0	−163.6	−110.4	643.1	−220.8	643.1
W_{w2}	−81.1	−59.1	−95.0	−42.0	335.8	−123.6	330.3
W_{w3}	−75.8	−75.8	−109.0	−109.0	493.7	−155.0	493.7

(f) Combinations and load factors

Combinations of load must now be considered and at this stage the values of γ_f need to be included. Possible combinations are:

Group 1 Dead + imposed	$1.4W_d + 1.6W_i$
Group 2 Dead + wind	$1.4W_d + 1.4W_{w1}$
	$1.4W_d + 1.4W_{w2}$
	$1.4W_d + 1.4W_{w3}$
Group 3 Dead + imposed + wind	$1.2W_d + 1.2W_i + 1.2W_{w1}$
	$1.2W_d + 1.2W_i + 1.2W_{w2}$
	$1.2W_d + 1.2W_i + 1.2W_{w3}$
Group 4 Dead (restraining uplift) + wind	$1.0W_d + 1.4W_{w1}$
	$1.0W_d + 1.4W_{w3}$

Group 4 is intended for use when considering restraint against uplift or overturning, i.e. maximum wind load plus minimum dead load. Some combinations may be eliminated by inspection, but care must be taken to retain combinations giving maximum values of opposite sign. Some of the combinations in groups 2 and 3 have been discarded in the following table.

For group 1 combination $1.4W_d + 1.6W_i$:

$$H_a = 1.4 \times 24.3 + 1.6 \times 74.6 = 153 \text{ kN}$$

Frame forces (kN) and moments (kNm) for factored loads

Group	H_a	H_e	V_a	V_e	M_b	M_c	M_d
1. $1.4\,W_d + 1.6\,W_i$	**153**	**153**	**177**	**177**	**−843**	**315**	**−843**
2. $1.4\,W_d + 1.4\,W_{w1}$	−134	−117	−190	−116	**713**	**−239**	**713**
3. $1.2\,W_d + 1.2\,W_i + 1.2\,W_{w2}$	21	48	23	86	−250	95	−256
4. $1.0\,W_d + 1.4\,W_{w1}$	−154	−127	**−201**	**−127**	764*	−259*	764*
$1.0\,W_d + 1.4\,W_{w3}$	−82	−82	−125	−125	557	−167	557

* While these values are maxima (opposite sign) the group 4 combinations (BS 5950) are to cover uplift and overturning only. It is necessary, however, for all the effects of a load combination to be considered (see also Ch. 11).

Maximum and minimum values may be selected from the table. Bending moment diagrams may be drawn as shown in Figs 2.16 and 2.17.

Finally the effect of wall dead load can be added if axial force in the columns is required (combination 1):

Max. axial force at A = 177 + 1.4 × 8 = **188 kN**

Min. axial force at A = −201 + 1.0 × 8 = **−193 kN**

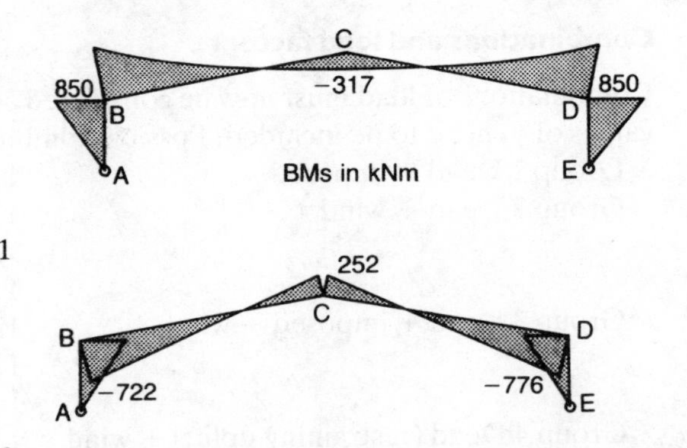

Fig. 2.16 Bending moments for combination 1

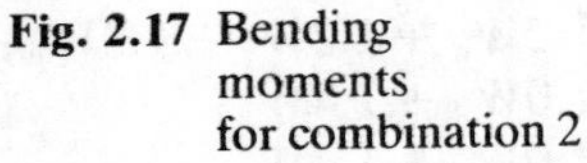

Fig. 2.17 Bending moments for combination 2

The alternative method of analysis is by application of plastic theory (see Ch. 12). In this elastic analysis of a portal frame the loading combinations can be added, but in Chapter 12 each load combination produces its own unique collapse mechanism, i.e. each load combination is analysed independently when applying plastic theory.

STUDY REFERENCES

Topic	*Reference*
1. Loading	BS 6399 *Loading for buildings* Part 1: *Dead and imposed loads* (1984) Part 2: *Wind loads* (to be published; presently CP3 Ch. V Part 2) Part 3: *Snow loading* (to be published; presently BS 5502: Part 1 and *BRE Digest* 332)
2. Moving load effects	**Marshall, W.T. & Nelson, H.M.** (1977) Maximum BM and SF due to a system of point loads, *Structures,* pp. 36–41. Longman
3. Moving load effects	**Wang C.K.** (1983) Influence lines for statically determinate beams, *Intermediate Structural Analysis,* pp. 459–67. McGraw-Hill
4. Influence lines	**Coates, R.C., Coutie, M.G. & Kong, F.K.** (1988) Mueller-Breslau's principle. Model analysis, *Structural Analysis,* pp. 127–31. Van Nostrand Reinhold
5. Influence lines	**Wang C. K.** (1983) Influence lines for statically indeterminate beams, *Intermediate Structural Analysis,* pp. 496–503. McGraw-Hill
6. BM and SF coefficients	(1972) Continuous beams, *Steel Designers' Manual,* pp. 51–9. Crosby Lockwood Staples
7. BM and SF coefficients	(1972) Formulae for rigid frames, *Steel Designers' Manual,* pp. 299–344. Crosby Lockwood Staples

3

BEAMS IN BUILDINGS

Most buildings are intended to provide load carrying floors, and to contain these within a weather-tight envelope. In some structural frameworks the weather-tight function is not needed, and the load carrying function only is required, e.g. supporting chemical plant. The loads to be supported are often placed on floor slabs of concrete, or on steel or timber grid floors, and these are in turn supported on the steelwork beams. In some cases, especially in industrial buildings, loads from equipment may be placed directly on to the beams without the use of a floor slab. Wind loads also must be carried to the beams by provision of cladding of adequate strength, and by secondary members such as purlins and side rails.

Beams which carry loads from floors or other beams to the columns are generally called main beams. Secondary beams will be provided to transfer load to the main beams, or in some cases just to give lateral stability to columns, while themselves only carrying their self weight. The manner in which loads are distributed from the floors on to the beams needs careful consideration so that each beam is designed for a realistic share of the total load. Examples of load distribution for one-way and two-way spanning slabs are shown in Fig. 3.1.

3.1 BEAMS WITH FULL LATERAL RESTRAINT

Many beams in a steel framework will be restrained laterally by the floors which transmit the loads to them. Concrete floor slabs, and wall or roof cladding, are generally able to give this lateral support or restraint. Timber floors and open steel floors are less certain in providing restraint. The degree of attachment of the flange to the floor may need to be assessed[1].

Alternatively lateral restraint may be provided by bracing members at specific points along the beams[1]. If adequate bracing or floor slab restraint is present then lateral torsional instability will be prevented. In

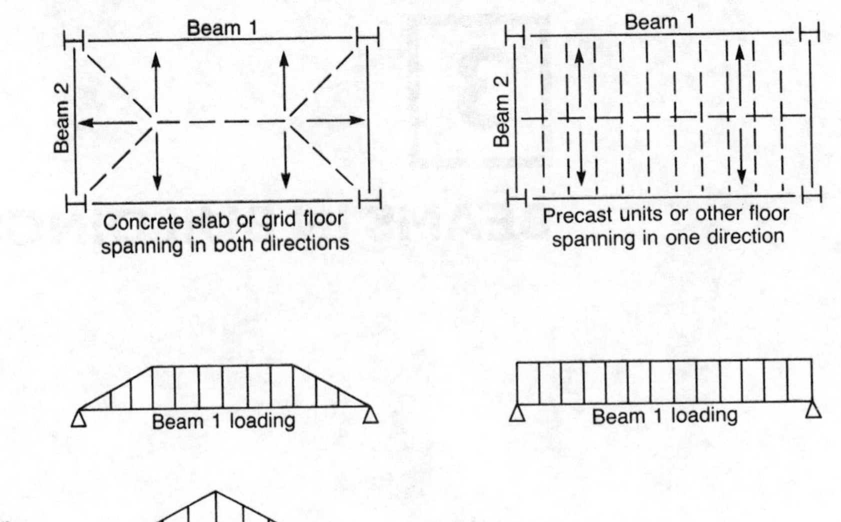

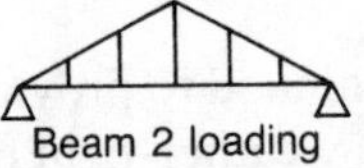

Fig. 3.1 Load distribution (Excluding selfweight)

addition this need not be considered for beams in which:

- the section is bent about a minor axis, or
- the section has a high torsional stiffness, e.g. a rectangular hollow section.

Beams in which lateral torsional instability will not occur are classed as restrained and are designed in the manner given in Section 3.7.

3.2 BEAMS WITHOUT FULL LATERAL RESTRAINT

An understanding of the behaviour of struts[2,3] will be useful in appreciating the behaviour of beams where full lateral restraint is not provided. The compression flange of such members will show a tendency to fail by buckling sideways (laterally) in the most flexible plane. Design factors which will influence the lateral stability can be summarized as:

- the length of the member between adequate lateral restraints;
- the shape of the cross-section;
- the variation of moment along the beam;
- the form of end restraint provided;
- the manner in which the load is applied, i.e. to tension or compression flange.

These factors and their effects are discussed in detail in study reference (1), and are set out in clause 4.3.7. The buckling resistance moment (M_b) of a beam may be found by use of a number of parameters and factors:

- **Effective length** (L_E), which allows for the effects of end restraint as well as type of beam, and the existence of destabilizing forces.
- **Minor axis slenderness** (λ), which includes lateral stiffness in the form of r_y, and is defined by $\lambda = L_E/r_y$.
- **Torsional index** (x) which is a measure of the torsional stiffness of a cross-section.
- **Slenderness factor** (v) which allows for torsional stiffness and includes the ratio of λ/x.
- **Slenderness correction factor** (n) which is dependent on the moment variation along the beam.
- **Buckling parameter** (u) which allows for section type and includes a factor for warping.
- **Equivalent slenderness** (λ_{LT}) which combines the above parameters and from which the bending strength (p_b) may be derived:

$$\lambda_{LT} = nuv\lambda$$

In addition an equivalent moment factor (m) is used which allows for the effect of moment variation along the beam.

3.3 SIMPLIFIED DESIGN PROCEDURES

The design of many simple beams will not require the calculation of all the above parameters. In particular, simply supported beams carrying distributed loads and not subjected to destabilizing loads will use:

$m = 1.0$

$n = 0.94$

$u = 0.9$ (for I, H and channel sections) or from published tables

$u = 1.0$ (for other sections)

A conservative approach is allowed in BS 5950 and may be used in the design of I and H sections only, basing the bending strength (p_b) on the design strength, the minor axis slenderness (λ) and the torsional index (x), which for this method may be approximated to D/T. This approach is useful in the preliminary sizing of members and is given in clause 4.3.7.7.

3.4 MOMENT CAPACITY OF MEMBERS (LOCAL CAPACITY CHECK)

The local moment capacity (M_c) at any critical point along a member must not be less than the applied bending moment at that point. The moment capacity will depend on:

- the design strength and the elastic or plastic modulus;

- the coexistent shear;
- the possibility of local buckling of the cross-section.

Provided the applied shear force is not more than 0.6 of the shear capacity no reduction in moment capacity is needed, and:

$$M_c = p_y S$$
$$\text{but } M_c \ngtr 1.2 p_y Z$$

where S = plastic modulus

Z = elastic modulus

Note that the limitation $1.2p_yZ$ is to prevent the onset of plasticity below working load. For UB, UC and joist sections the ratio S/Z is less than 1.2 and the plastic moment capacity governs design. For sections where $S/Z > 1.2$ then the constant 1.2 is replaced by the ratio (γ_o) of factored load/unfactored load. The limitation $1.2p_yZ$ is therefore purely notional and becomes in practice $\gamma_o p_y Z$.

If 0.6 of shear capacity is exceeded some reduction in M_c will occur as set out in clause 4.2.6.

Local buckling can be avoided by applying a limitation to the width/thickness ratios of elements of the cross-section. This leads to the classification of cross-sections discussed in Section 1.7.

Where members are subjected to bending about both axes a combination relationship must be satisfied:

(a) *For plastic and compact sections:*
For UB, UC and joist sections $(M_x/M_{cx})^2 + M_y/M_{cy} \ngtr 1$
For RHS, CHS and solid sections $(M_x/M_{cx})^{5/3} + (M_y/M_{cy})^{5/3} \ngtr 1$
For channel, angle and all other sections $M_x/M_{cx} + M_y/M_{cy} \ngtr 1$
where M_x, M_y are applied moments about x and y axes
M_{cx}, M_{cy} are moment capacities about x and y axes

(b) *For semi-compact and slender sections:*
(and as a simplified method for compact sections in (a) above)
$M_x/M_{cx} + M_y/M_{cy} \ngtr 1$

3.5 BUCKLING RESISTANCE (MEMBER BUCKLING CHECK)

Members not provided with full lateral restraint (Section 3.2) must be checked for lateral torsional buckling resistance (M_b) as well as moment capacity. The buckling resistance depends on the bending strength (Section 3.2) and the plastic modulus:

$$M_b = p_b S$$

Where members are subjected to bending about both axes (without axial load) a combination relationship must be satisfied:

$$m_x M_x/M_b + m_y M_y/M_{cy} \ngtr 1$$

where m_x, m_y are equivalent uniform moment factors
M_{cy} is the moment capacity about the y axis (as in Section 3.4)

but without the restriction of $1.2p_yZ$

This is described as a 'more exact' approach (clause 4.8.3.3.2) which gives an advantage over the 'simplified' approach (clause 4.8.3.3.1). The simplified approach is more conservative and, for bending about two axes (without axial load), does not reduce the calculations.

3.6 OTHER CONSIDERATIONS

In addition to the above requirements for moment capacity and buckling resistance, a member is usually required to meet some deflection criterion. These are outlined in Section 1.5 and reference (1).

The application of heavy loads or reactions to a member may produce high local stresses and it is necessary to check that web bearing and web buckling requirements are satisified. These requirements are generally only significant in beams carrying heavy point loads such as crane girders (Chapter 5), or beams supporting column members within the span.

Connections must be provided at junctions between members and must safely transmit the calculated loads from one member to the next. A variety of connection details exist for most common situations and are fully described in the BCSA handbook[4]. Design constants for connections are given in section 6 BS 5950, which includes some guidance on bolt spacing and edge distances. Bolt and weld sizes and capacities are given in reference (5).

3.7 EXAMPLE 4. BEAM SUPPORTING CONCRETE FLOOR SLAB (RESTRAINED BEAM)

(a) Dimensions

(See Fig. 3.2)

Beam centres	6.0 m
Span (simply supported)	7.4 m
Concrete slab (spanning in two directions)	250 mm thick
Finishing screed	40 mm thick

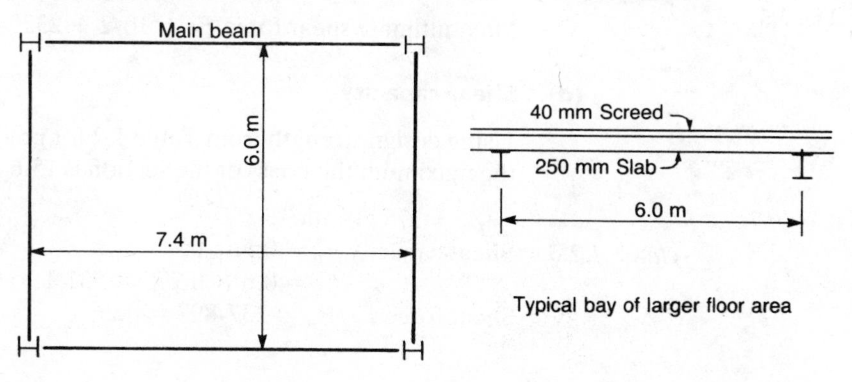

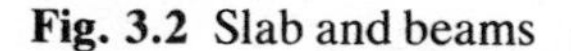
Fig. 3.2 Slab and beams

(b) Loading

Concrete slab	23.7 kN/m^3
Screed (40 mm)	0.9 kN/m^2
Imposed load	5.0 kN/m^2

Member size must be finally confirmed after all the design checks have been carried out.
For preliminary calculation an estimated self weight is included.
Assume beam to be **533 × 210 × 92 UB** (grade 43A). It is sufficiently accurate to take the beam weight of 92 kg/m = 0.92 kN/m.

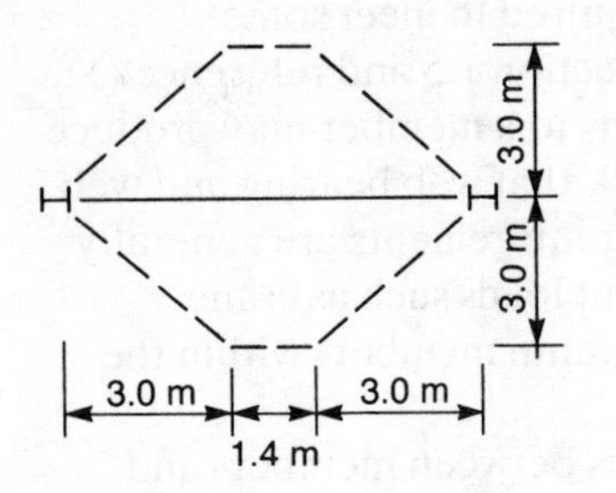

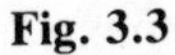
Fig. 3.3

Self weight 0.92×7.4 = 7 kN
Dead load slab 23.7×0.250 = 5.93 kN/m^2
Dead load screed = 0.90 kN/m^2
Total = 6.83 kN/m^2
Area of slab supported by beam (Fig. 3.3):
 rectangles $2 \times 1.4 \times 3.0$ = 8.4 m^2
 triangles $4 \times 3.0 \times 3.0/2$ = 18.0 m^2
Dead load W_d:
 on rectangles 6.83×8.4 = 57 kN
 on triangles 6.83×18.0 = 123 kN
Imposed load W_i:
 on rectangles 5.0×8.4 = 42 kN
 on triangles 5.0×18.0 = 90 kN
Ultimate load (factored) (Fig. 3.4):
 uniformly distributed 1.4×7 = 10 kN
 on rectangles $1.4 \times 57 + 1.6 \times 42$ = 147 kN
 on triangles $1.4 \times 123 + 1.6 \times 90$ = 316 kN

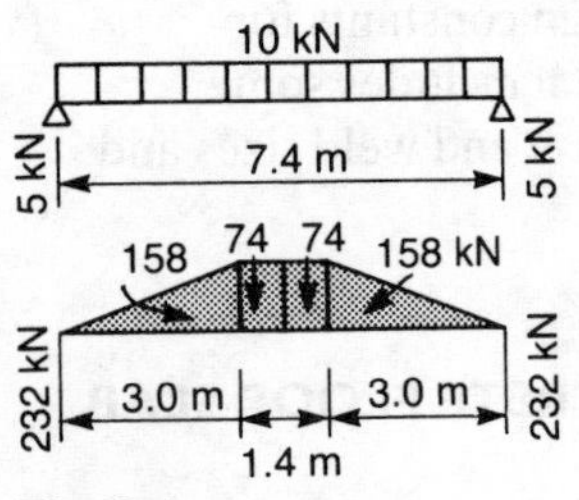

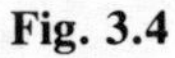
Fig. 3.4

(c) BM and SF

(See Fig. 3.5):
Max. ultimate moment M_x (mid-span)

$$= 10 \times 7.4/8 + 232 \times 3.7 - 158 \times 1.7 - 74 \times 0.35 = \mathbf{573\ kNm}$$

Max. ultimate shear force $F_x = 10/2 + 232 =$ **237 kN**

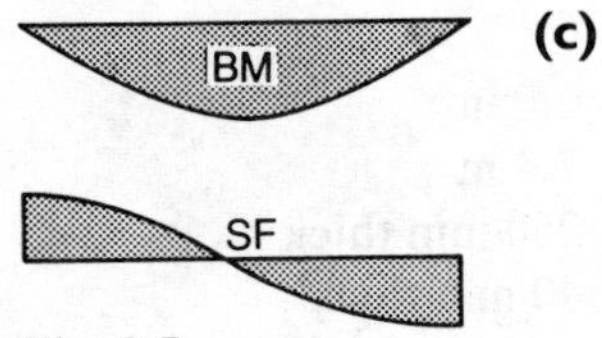

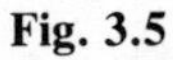
Fig. 3.5

(d) Shear capacity

Using design strength from Table 1.2 for grade 43A steel, noting that the maximum thickness of the section is 15.6 mm

$p_y = 275$ N/mm^2

clause 4.2.3 Shear capacity $p_v = 0.6 p_y A_v$

$$= 0.6 \times 0.275 \times 531.1 \times 10.2 = \mathbf{897\ kN}$$

Shear force $F_x/P_v = 237/897 = 0.26$

(e) Moment capacity

The concrete slab provides full restraint to the compression flange (Fig. 3.6), and lateral torsional buckling is not considered. The chosen UB is a plastic section ($b/T = 6.7$) and at mid-span shear force is zero.

clause 3.5.2
clause 4.2.5

$$M_{cx} = p_y S_x = 275 \times 2370 \times 10^{-3} = \mathbf{652\ kNm}$$

Note that 10^{-3} must be included for M_{cx} kNm (p_y N/mm², S_x cm³).
Alternatively p_y may be given as 0.275 kN/mm².
But $M_{cx} \not> 1.2 p_y Z_x = 1.2 \times 275 \times 2080 \times 10^{-3} = 686$ kNm
Note that for I and H sections bent about the x axis the expression $p_y S_x$ governs the design. For bending about the y axis, however, the expression $1.2 p_y Z_y$ governs the design. The factor 1.2 in this expression may be increased to the ratio factored load/unfactored load (clause 4.2.5).

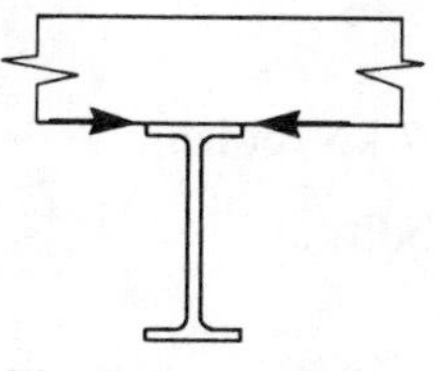

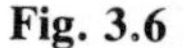
Fig. 3.6

$$M_x/M_{cx} = 573/652 = \mathbf{0.88}$$

Section is satisfactory

(f) Deflection

Deflection (which is a serviceability limit state) must be calculated on the basis of the unfactored imposed loads:

$$W_x = 90 + 42 = 132 \text{ kN}$$

Assume the load is approximately triangular and hence formulae are available for deflection calculations[6].

$$\delta_x = W_x L^3/60EI_x$$
$$= 132 \times 7400^3/(60 \times 205 \times 55400 \times 10^4)$$
$$= \mathbf{7.8\ mm}$$

BS table 5

Deflection limit = 7400/360 = 20.6 mm

(g) Connection

The design of connections which are both robust and practicable, yet economic, is developed by experience. Typical examples may be found in references[4,7].
The connection at each end of the beam must be able to transmit the ultimate shear force of 237 kN to the column or other support. The connection forms part of the beam, i.e. the point of support is the column to cleat interface. Design practice assumes that the column bolts support shear force only, while the beam bolts carry shear force together with a small bending moment.

$$\text{BM} = 237 \times 0.05 = 11.9 \text{ kNm}$$

Assume **9 bolts 22 mm diameter** (grade 4.6) as shown (Fig. 3.7).

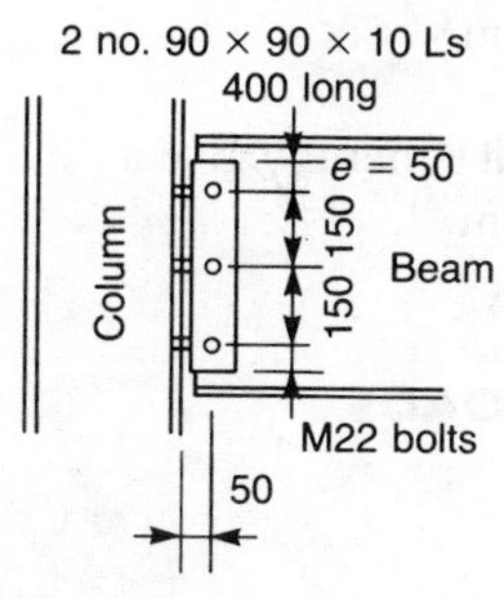

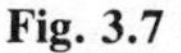
Fig. 3.7

(i) COLUMN BOLTS

Vertical shear/bolt = 237/6 = 39.5 kN

clause 6.3.2 Shear capacity $P_s = p_s A_s$

where A_s is the cross-sectional area at the root of the thread

$$P_s = 0.160 \times 303 = 48.5 \text{ kN/bolt}$$

clause 6.3.3.2 Bearing capacity of bolts $p_{bb} = dtp_{bb}$

$$= 22 \times 10 \times 0.435 = 95.7 \text{ kN/bolt}$$

clause 6.3.3.3 Bearing capacity of plates will be greater than that of bolts as $p_{bs} = 460 \text{ N/mm}^2$

But bearing capacity of plates $P_{bs} \ngtr etp_{bs}/2$

$$= 50 \times 10 \times 0.460/2 = 115 \text{ kN/bolt}$$

Note that the column flange will require checking also if less than 10 mm thick. Column bolt connection is satisfactory. Capacities of bolts and bearing values may alternatively be obtained from reference (5).

(ii) BEAM BOLTS

(See Fig. 3.8)

Shear capacity (double shear) $P_s = 160 \times 2 \times 0.303 = 97.0$ kN/bolt

Vertical shear/bolt = 237/3 = 79.0 kN

Bolt forces due to bending moment are discussed in Section 8.4.

Horizontal shear/bolt due to the bending moment

$$= Md_{max}/\Sigma d^2$$

$$= 11.9 \times 0.15/(2 \times 0.15^2) = 39.7 \text{ kN}$$

Resultant shear/bolt = $\sqrt{(79.0^2 + 39.7^2)}$ = **88.4 kN**

Bearing capacity of bolt $P_{bb} = dtp_{bb}$

$$= 22 \times 10.2 \times 0.435 = 97.6 \text{ kN}$$

clause 6.3.3.3 Bearing capacity of web plates $P_{bs} = dtp_{bs}$

$$= 22 \times 10.2 \times 0.460 = 103 \text{ kN}$$

But $P_{bs} \ngtr etp_{bs}/2$

$$= 89 \times 10.2 \times 0.460/2 = 209 \text{ kN}$$

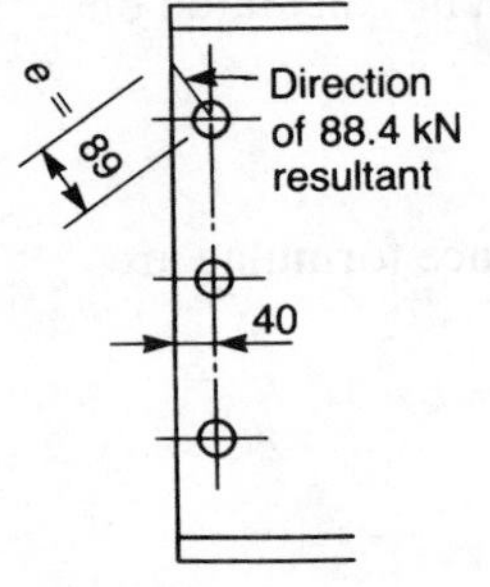

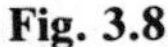

Fig. 3.8

(iii) ANGLE CLEAT

clause 4.2.3 Shear area of cleats (allowing for 24 mm holes)

$= 0.9(400 \times 10 \times 2 - 3 \times 2 \times 10 \times 24) = 5904 \text{ mm}^2$

Shear capacity $P_v = 0.6 \times 0.275 \times 5904 = 974$ kN

A check for bending may also be carried out but will generally give a high bending capacity relative to the applied moment(4).

3.8 EXAMPLE 5. BEAM SUPPORTING PLANT LOADS (UNRESTRAINED BEAM)

(a) Dimensions

Main beam span 9.0 m simply supported (Fig. 3.9).

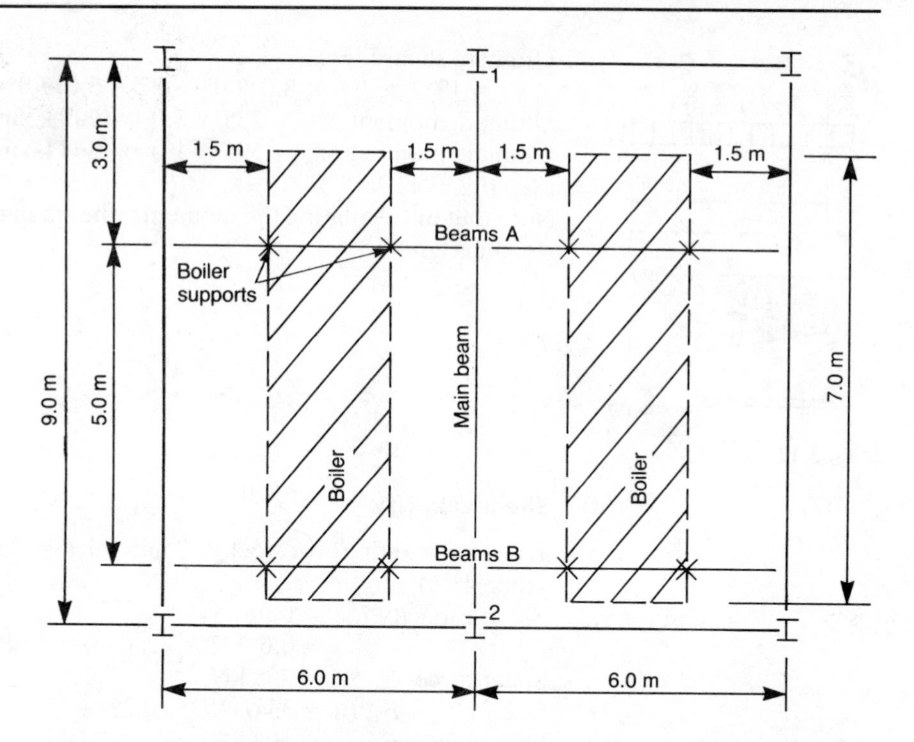

Fig. 3.9 Plant loads

Boilers supported symmetrically on secondary beams of span 6.0 m, which are at 5.0 m centres.

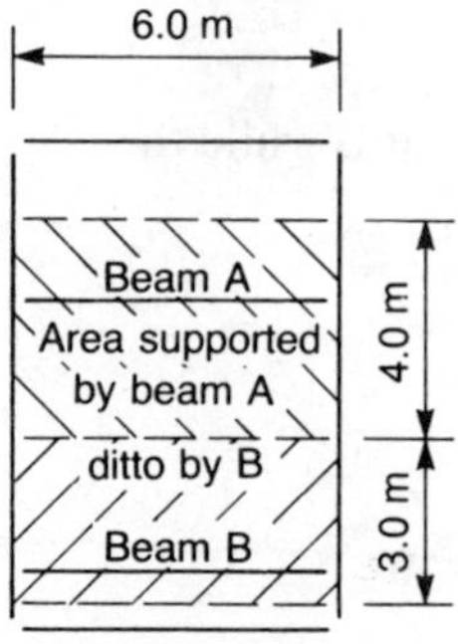

Fig. 3.10

(b) Loading

Boiler loading (each) 400 kN

Open steel flooring (carried on secondary beams A and B) $0.3\ kN/m^2$

Imposed load (outside boiler area) $4.5\ kN/m^2$

The boilers give reactions of 100 kN at the end of each secondary beam. Allow 4.0 kN for the self weight of each secondary beam (not designed here).

Assume **610 × 305 × 149 UB** (grade 43)

Self weight (ultimate) $= 1.4 \times 1.49 \times 9.0 = 18.8$ kN

Flooring on beam A $= 0.3 \times 4.0 \times 6.0 = 7.2$ kN

Imposed load on beam A $= 4.5 \times 3.0 \times 4.0 = 54.0$ kN

(See Fig. 3.10.)

Flooring on beam B = 5.4 kN

Imposed load on beam B = 40.5 kN

(See Fig. 3.11.)

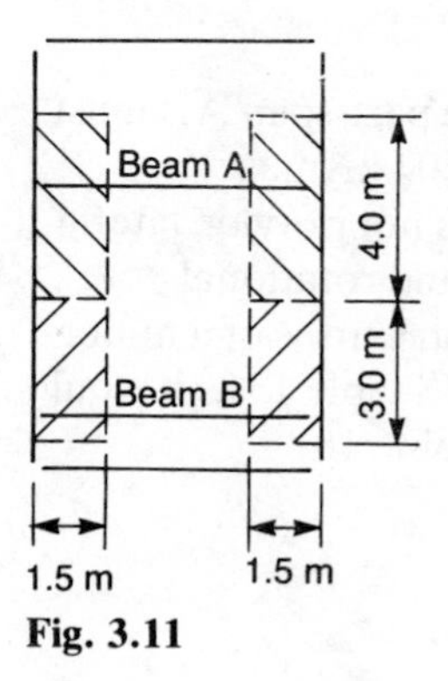

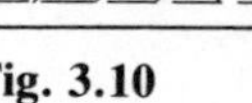

Fig. 3.11

Ultimate point load W_A = end reactions from two beams A

$$= 1.4(4.0 + 7.2) + 1.6(200 + 54) = 422 \text{ kN}$$

Ultimate point load W_B = end reactions from two beams B

$$= 1.4(4.0 + 5.4) + 1.6(200 + 40.5) = 398 \text{ kN}$$

(c) BM and SF

Ultimate shear force

$$F_1 = 19/2 + 422 \times 6.0/9.0 + 398 \times 1.0/9.0 = \mathbf{335\ kN}$$

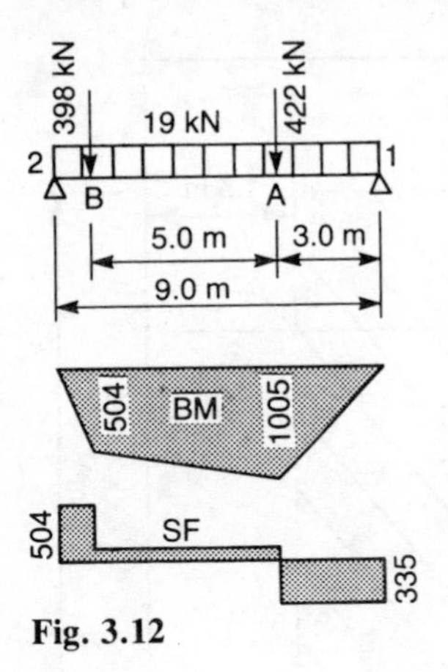

Fig. 3.12

Ultimate shear force
$F_2 = 19/2 + 398 \times 8.0/9.0 + 422 \times 3.0/9.0 =$ **504 kN**
Ultimate moment $M_A = 335 \times 3.0 =$ **1005 kNm**
Ultimate moment $M_B = 504 \times 1.0 =$ **504 kNm**
(See Fig. 3.12.)
Note that in calculating the moments, the small reduction due to the self weight is ignored.

(d) Shear capacity

Design strength p_y for steel 19.7 mm thick = 265 N/mm² (Table 1.2. grade 43A)

clause 4.2.3 Shear capacity $P_v = 0.6 p_y A_v$
$= 0.6 \times 265 \times 609.6 \times 11.9 \times 10^{-3} = 1153$ kN

Shear force $F_1 = 335$ kN
$F_1/P_v = 335/1153 = 0.29$
Shear force $F_2 = 504$ kN
$F_2/P_v = 504/1153 = 0.44$
i.e. $\leqslant 0.6P_v$
This maximum coexistent shear force is present at point B while the maximum moment occurs at point A.

(e) Moment capacity

clause 3.5.2 The chosen section is a plastic section ($b/T = 7.7$).
Moment capacity $M_c = p_y S_x$
$= 265 \times 4570 \times 10^{-3} =$ **1210 kNm**

Moment $M_A/M_c = 1005/1210 = 0.83$
Section is satisfactory.

(f) Buckling resistance

The buckling resistance moment of the beam in the part-span AB must be found, and in this part the moment varies from 504 kNm to 1005 kNm. It is assumed that the steel flooring does not provide lateral restraint, but the secondary beams give positional and rotational restraint at 5.0 m spacing. Loading between the restraints is of a minor nature (self weight only) and is ignored for use of BS table 13 as it would affect the moments by less than 10% (see also BS table 16).

clause 4.3.5 $L_E = 5.0$ m

Slenderness $\lambda = L_E/r_y$
$= 5000/69.9 = 72$

where r_y is given in mm, hence L_E is in mm.

Torsional index $x = 32.5$

$\lambda/x = 2.2$

BS table 14 $v = 0.94$ (for $N = 0.5$, i.e. equal flanges)

BS table 13 $n = 1.0$ (for a member not loaded between restraints)

clause 4.3.7.5, tables of properties(5) $u = 0.886$

$$\lambda_{LT} = nuv\lambda$$
$$= 1.0 \times 0.886 \times 0.94 \times 72 = 60$$

BS table 11 Bending strength $p_b = 207\ \text{N/mm}^2$

Buckling resistance $M_b = p_b S_x$

$$= 207 \times 4570 \times 10^{-3} = \mathbf{946\ kNm}$$

clause 4.3.7.2, BS tables 13, 18 Equivalent uniform moment factor m needs to be obtained for a member not loaded between restraints:

$$\beta = 504/1005 = 0.50$$

$$m = 0.76$$

Equivalent uniform moment $\bar{M} = mM_A$

$$= 0.76 \times 1005 = \mathbf{764\ kNm}$$

Hence $\bar{M}/M_b = 764/946 = \mathbf{0.81}$ and section is satisfactory.

Trying a smaller section (610 × 229 × 140 UB):

$$S_x = 4150\ \text{cm}^4$$
$$\lambda = 99.4$$
$$P_b = 161\ \text{N/mm}^2$$
$$M_b = 668\ \text{kNm}$$
$$\bar{M}/M_b = \mathbf{1.14}$$

which is not satisfactory.

This comparison indicates the sensitivity of the buckling resistance moment to small changes in section properties, particularly to a reduction in flange width.

(g) Deflection

Calculation of the deflection for the serviceability imposed loading cannot be carried out easily by the use of formulae, which become complex for non-standard cases. Serviceability point load W_A (unfactored) = 200 + 54 = 254 kN

$$W_B = 200 + 40.5 = 240\ \text{kN}$$

(See Fig. 3.13.)

Mid-span deflection may be found by the area-moment method(8,9):

$$\delta = \int (M(x)/EI)dx$$

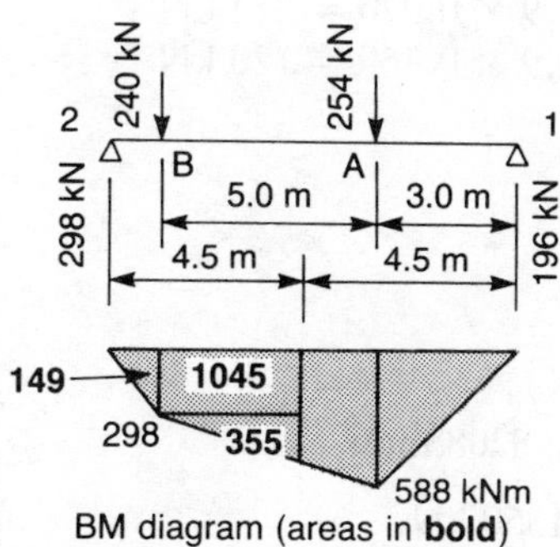

Fig. 3.13

$$\delta = (149 \times 0.667 + 1045 \times 2.75 + 355 \times 3.333)/(205 \times 124660 \times 10^{-5}) = \mathbf{16.3\ mm}$$

Calculation by Macaulay's method(10,11) gives the point of maximum deflection 4.65 m from support 2 with a value of 17.6 mm.
Deflection limit = 9000/200 = 45 mm
An approximate estimate of deflection is often obtained by treating the load as an equivalent u.d.l.

(h) Connection

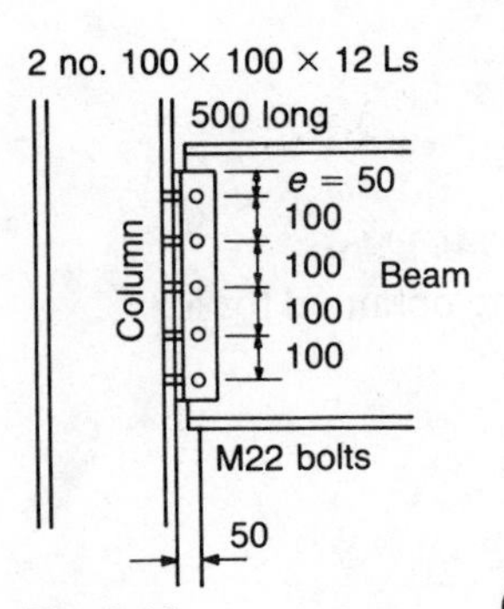

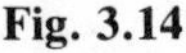
Fig. 3.14

The connection at support 2 of the main beam must transmit an ultimate shear force of 504 kN and follows the method given in Section 3.7(g)

$$M = 504 \times 0.05 = 25.3 \text{ kNm}$$

Assume 22 mm bolts (grade 8.8) as shown (Fig. 3.14).

(i) COLUMN BOLTS

Vertical shear/bolt = 504/10 = 50.4 kN
clause 6.3.2 Shear capacity P_s = 375 × 303 = 114 kN/bolt
clause 6.3.3.3 Bearing capacity of plates P_{bs} = 22 × 12 × 460 = 121 kN/bolt
But $P_{bs} \not> 50 \times 12 \times 460/2$ = 138 kN/bolt
Note that the column flange will require checking if less than 12 mm thick.
Column bolt connection is satisfactory.

(ii) BEAM BOLTS

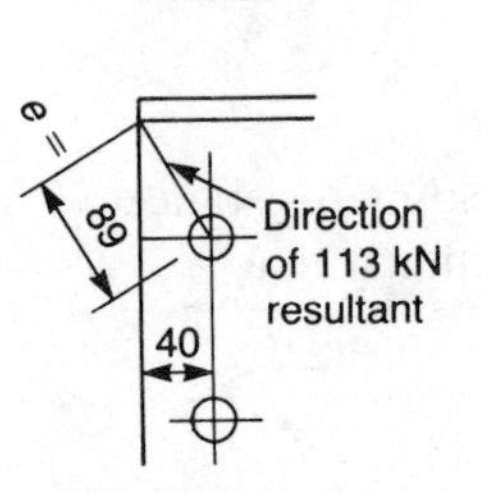

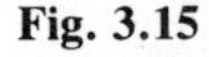
Fig. 3.15

(See Fig. 3.15.)
Vertical shear/bolt = 504/5 = 100.8 kN
Horizontal shear/bolt = $Md_{max}/\Sigma d^2$
= 25.3 × 0.20/2(0.10² + 0.20²) = 50.6 kN
Resultant shear/bolt = $\sqrt{(101.2^2 + 50.6^2)}$ = **113 kN**
Shear capacity (double shear) = 0.375 × 2 × 303 = 227 kN
Bearing capacity of bolt P_{bb} = 22 × 11.9 × 0.970 = 254 kN
clause 6.3.3.3 Bearing capacity of web plate P_{bs} = 22 × 11.9 × 0.460 = 120 kN
But $P_{bs} \not> 89 \times 11.9 \times 0.460/2$ = 243 kN
Beam bolt connection is satisfactory.

(iii) ANGLE CLEAT

Shear area of cleats (24 mm holes)

$$= 0.9(500 \times 12 \times 2 - 5 \times 2 \times 12 \times 24) = 8208 \text{ mm}^2$$

clause 4.2.3 Shear capacity $P_v = 0.6 \times 0.275 \times 8208 = 1350$ kN

$$F_v/P_v = 506/1350 = 0.38$$

Connection cleat is satisfactory.

STUDY REFERENCES

Topic	*Reference*
1. Lateral restraint	*Steelwork Design,* vol. 3, Commentary BS 5950: Part 1. Steel Construction Institute (to be published)
2. Strut behaviour	**Marshall W.T. & Nelson H.M.** (1977) Compression members and stability problems, *Structures,* pp. 329–64. Longman
3. Strut behaviour	**Coates R.C., Coutie M.G. & Kong F.K.** (1988) Instability of struts and frameworks, *Structural Analysis,* pp. 304–361. Van Nostrand Reinhold
4. Connections	**Pask J.W.** (1982) Angle cleat connections, *Manual on Connections,* pp. 18–27. BCSA
5. Bolt details	(1985) *Steelwork Design,* vol. 1, Section properties, member capacities. Steel Construction Institute
6. Deflection formulae	(1972) Simply supported beams, *Steel Designers' Manual,* pp. 17–32. Crosby Lockwood Staples
7. Connections	**Needham F.H.** (1980) Connections in structural steelwork for buildings, *Structural Engineer,* vol. 58A (no. 9) pp. 267–77.
8. Moment-area method	**Marshall W.T. & Nelson H.M.** (1977) The moment-area method, *Structures,* pp. 122–5. Longman
9. Moment-area method	**Coates R.C., Coutie M.G. & Kong F.K.** (1988) Moment-area methods, *Structural Analysis,* pp. 176–181. Van Nostrand Reinhold
10. Deflection	**Marshall W.T. & Nelson H.M.** (1977) Simply supported beams by Macaulay's method, *Structures,* pp. 114–17. Longman
11. Deflection	**Hearn E.J.** (1985) Slope and deflection of beams, *Mechanics of Materials,* vol 1 pp. 102–7. Pergamon

4

PURLINS AND SIDE RAILS

In the UK purlins and side rails used in the construction of industrial buildings are often manufactured from cold formed sections. These sections can be designed in accordance with Part 5 of BS 5950, but the load tables for these sections are frequently based on test data. They are marketed by companies specializing in this field who will normally give the appropriate spans and loadings allowed in their catalogues. Sections of this kind are commonly of zed or [form as illustrated in Fig. 4.1. Although their design is not covered in this chapter the selection of cold formed sections is discussed in Chapter 11.

Hot rolled sections may be used as an alternative, and in some situations may be preferred to cold formed sections. The design of angles and hollow sections may be carried out by empirical methods which are covered by clause 4.12.4. The full design procedure (i.e. non-empirical) is set out in this chapter.

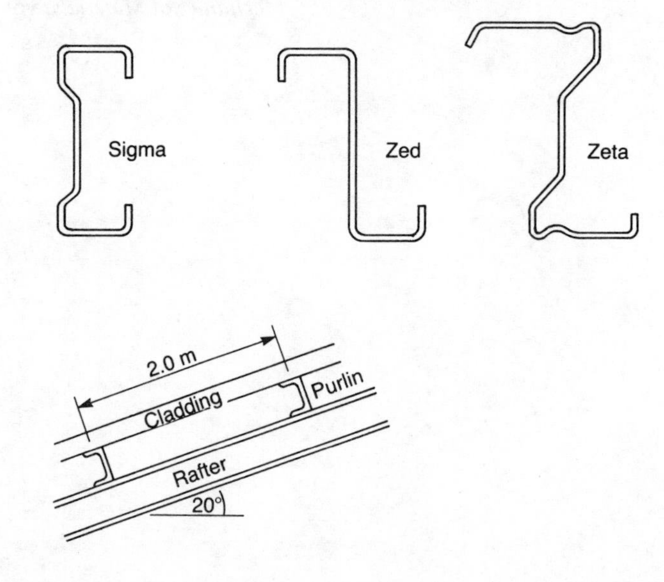

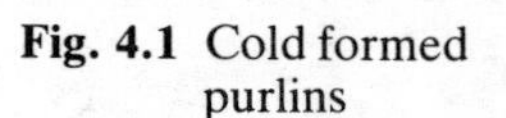
Fig. 4.1 Cold formed purlins

Fig. 4.2

4.1 DESIGN REQUIREMENTS FOR PURLINS AND SIDE RAILS

The design of steelwork in bending is dependent on the degree of lateral restraint given to the compression flange and the torsional resistance of the beam, and also on the degree of lateral/torsional restraint given at the beam supports. These are given in detail in clause 4.3 and have been discussed and illustrated in Chapter 3. Side rails and purlins may be considered to have lateral restraint of the compression flange due to the presence of the cladding, based on adequate fixings (clause 4.12.1). Loads will be transferred to the steel member via the cladding (see Fig. 4.2), and the dead, imposed and wind pressure loads will cause the flange restrained by the cladding to be in compression. Wind suction load can, however, reverse this arrangement, i.e. the unrestrained flange will be in compression. Torsional restraint to a beam involves both flanges being held in position and for purlins and side rails this will be true only at the supports.

Side rails are subjected to both vertical loading (cladding) and horizontal loading (wind pressure/suction), but in general the vertical loading is considered to be taken by the cladding acting as a deep girder. Consequently only moments in the horizontal plane (due to wind) are considered in design. In the design of new construction where the cladding is penetrated by holes for access, ductwork or conveyors, the design engineer should satisfy himself that the cladding and fixings are capable of acting in this manner.

Sag rods are sometimes used to reduce the effective length of purlins and side rails, and result in continuous beam design (see later). Where sag rods are used, provision must be made for the end reaction on eaves or apex beams. As is shown in the examples given in the following pages, there is no reason why purlins and side rails should not be designed as beams subject to biaxial bending in accordance with the normal design rules.

4.2 EXAMPLE 6. PURLIN ON SLOPING ROOF

(a) Dimensions

See Fig. 4.2: purlins at 2.0 m centres; span 6.0 m simply supported; rafter slope 20°.

(b) Loading

Dead load (cladding + insulation panels)	0.15 kN/m^2
Imposed load	0.75 kN/m^2 (on plan)
Wind load	0.40 kN/m^2 (suction)

Reference should be made to Chapter 2 for the derivation of loads, the direction in which each will act, and the area appropriate to each load. Maximum values of bending moment and shear force must be found at

the ultimate limit state making due allowance for the slope angle, and including the γ_f factors.
Assume purlin to be **152 × 76** channel section, grade 43A steel (See Fig. 4.3.)

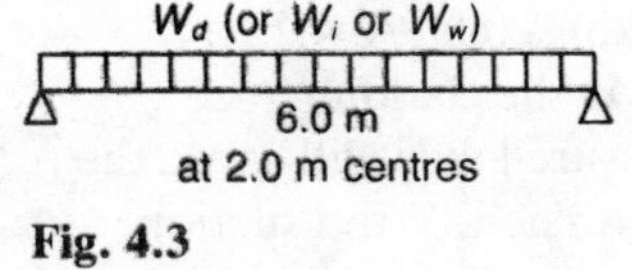

Fig. 4.3

Cladding 2.0 × 6.0 × 0.15 = 1.80 kN
Self weight 6.0 × 0.18 = 1.08 kN
Total dead load W_d = 2.88 kN

Imposed load = 2.0 cos 20° × 6.0 × 0.75
W_i = 8.46 kN

Wind load = 2.0 × 6.0 × (−0.40)
W_w = −4.80 kN

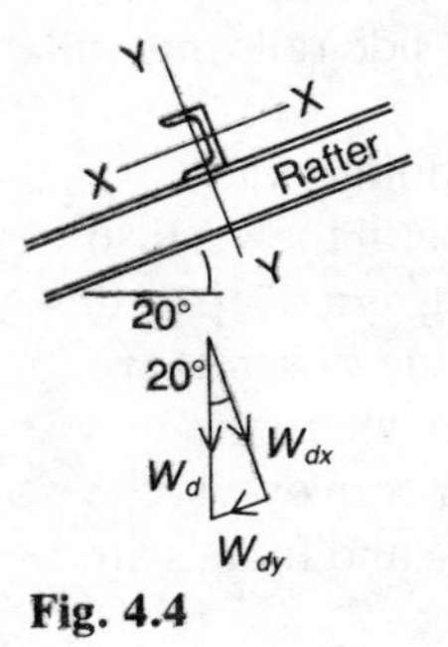

Fig. 4.4

The rafter slope of 20° results in purlins at the same angle. Components of load are used to calculate moments about the *x* and *y* axes, i.e. normal and tangential to the rafter (Fig. 4.4). As with side rails it would be possible to ignore bending in the plane of the cladding, but in practice, biaxial bending is usually considered in purlin design.

W_{dx} = 2.88 cos 20° = 2.71 kN
W_{dy} = 2.88 sin 20° = 0.99 kN
W_{ix} = 8.46 cos 20° = 7.95 kN
W_{iy} = 8.46 sin 20° = 2.89 kN
W_{wx} = −4.80 kN

Note that W_{wy} is zero as wind pressure is perpendicular to the surface on which it acts, i.e. normal to the rafter.

Ultimate load W_x = 1.4 × 2.71 + 1.6 × 7.95 = 16.5 kN
where 1.4 and 1.6 are the appropriate γ_f insert factors. (Section 1.7)

(c) BM and SF

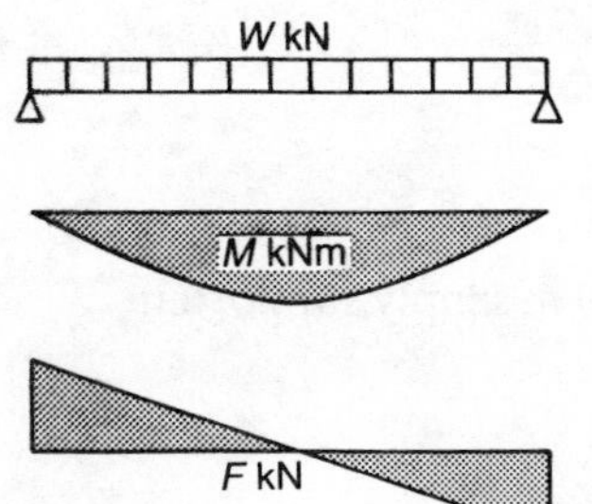

Fig. 4.5

(See Fig. 4.5.)
Max. ultimate moment M_x = 16.5 × 6.0/8 = **12.4 kNm**
Max. ultimate shear force F_x = 16.5/2 = **8.3 kN**
Similarly W_y = 6.0 kN
M_y = **4.5 kNm**
F_y = **3.0 kN**

(d) Shear capacity

Design strength p_y is given in Table 1.2 and for the selected purlin section is 275 N/mm^2
Shear area A_{vx} = 152.4 × 6.4 = 975 mm^2

clause 4.2.3 Shear capacity $P_{vx} = 0.6p_yA_{vx}$
$= 0.6 \times 275 \times 975 \times 10^{-3} = 161$ kN
Shear area $A_{vy} = 0.9A_o$
$= 0.9 \times 2 \times 76.2 \times 9.0 = 1234$ mm^2
Shear capacity $P_{vy} = 0.6 \times 275 \times 1234 \times 10^{-3} = 204$ kN
It may be noted that in purlin design shear capacity is usually high relative to shear force.

(e) Moment capacity

The section classification of a channel subject to biaxial bending depends on b/T and d/t which in this case are 8.47 and 16.5

BS table 7 respectively. The channel is therefore a plastic section.
Hence moment capacity:

clause 4.2.5 $M_{cx} = p_yS_x = 275 \times 130 \times 10^{-3} = 35.8$ kNm
But M_{cx} must not exceed $1.2p_yZ_x$
$= 1.2 \times 275 \times 112 \times 10^{-3} = 37.0$ kNm
Note that 10^{-3} must be included for M_{cx} kNm (p_y N/mm^2, S_x cm^3).
Alternatively p_y may be expressed as 0.275 kN/mm^2, but this requires care later when axial forces and stresses are used.
Moment capacity $M_{cy} = p_yS_y$
$= 0.275 \times 41.3 = 11.4$ kNm
The ratio S_y/Z_y is greater than 1.2 and hence the constant 1.2 is replaced by the ratio factored load/unfactored load
$(= 6.0/[0.99 + 2.89] = 1.55)$
M_{cy} must not exceed $1.55p_yZ_y$
$= 1.55 \times 275 \times 21.0 = 9.0$ kNm
The local capacity check may now be carried out (see Section 3.4):

$$M_x/M_{cx} + M_y/M_{cy} \not> 1 \text{ (for a channel section)}$$

$$12.4/35.8 + 4.5/9.0 = \mathbf{0.85}$$

The local capacity of the section is therefore adequate.

(f) Buckling resistance

The buckling resistance moment M_b of the section does not need to be found due to the fact that the beam is restrained by the cladding in the x plane (Fig. 4.6) and instability is not considered for a moment about the minor axis (Fig. 4.7) (Section 3.1).

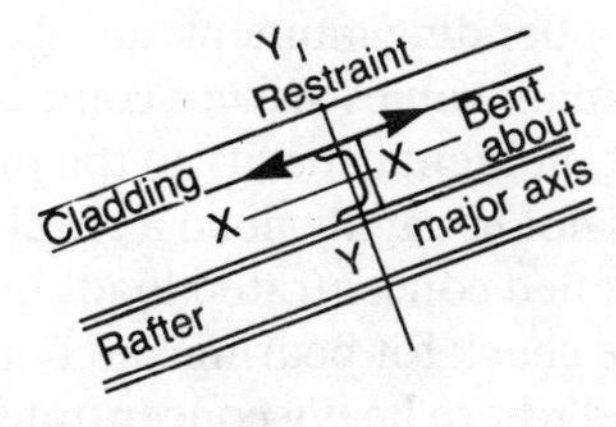

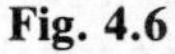
Fig. 4.6

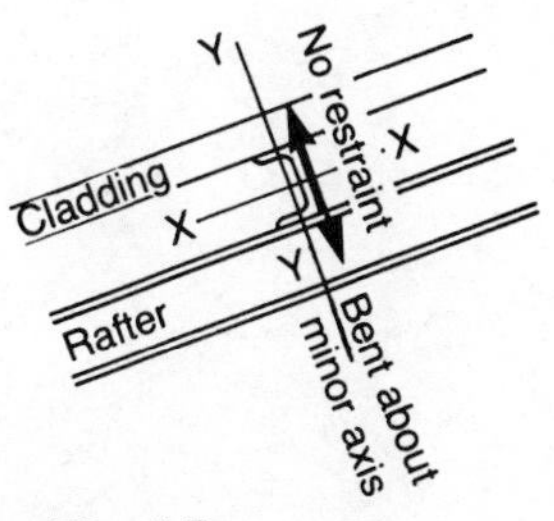

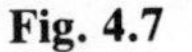
Fig. 4.7

(g) Wind suction

The effect of the wind suction load has so far not been considered, and in some situations could be critical. In combination with loads W_d and W_i, a lower total load W is clearly produced.

Ultimate load $W_x = 1.0 \times 2.71 - 1.4 \times 4.8 = -4.01$ kN
$M_x = -4.01 \times 6.0/8 = -3.01$ kNm
$W_y = 1.0 \times 0.99 = 0.99$ kN
$M_y = 0.99 \times 6.0/8 = 0.75$ kNm

The value of M_x is much lower than the value 12.4 kNm used above, but the negative sign indicates that the lower flange of the channel is in compression and this flange is not restrained. The buckling resistance M_b must therefore be found.

The effective length L_E of the purlin may be found from BS table 9.

clause 4.3.5 $L_E = 1.0 \times 6.0 = 6.0$ m

Slenderness $\lambda = L_E/r_y = 6000/22.4 = 268$ (which is less than 350 as required by clause 4.7.3.2) where L_E and r_y are both in mm.

Equivalent slenderness λ_{LT} allowing for lateral torsional buckling is given by:

$\lambda_{LT} = nuv\lambda$

Torsional index $x = 14.5$

$\lambda/x = 268/14.5 = 18$

BS table 14 $v = 0.49$

$u = 0.902$

BS table 16 $n = 0.94$ (for $\beta = 0$ and $\gamma = 0$)

$\lambda_{LT} = 0.94 \times 0.902 \times 0.49 \times 268 = 111$

Clause 4.3.7.4 Bending strength p_b may be obtained:

BS table 11 $p_b = 108\ \text{N/mm}^2$

Buckling resistance $M_b = p_b S_x$
$= 108 \times 130 \times 10^{-3} = 14.0$ kNm

BS table 13 The overall buckling check may now be carried out using an equivalent uniform moment factor (m) equal to 1.0 (member loaded between restraints):

$$mM_x/M_b + mM_y/M_{cy} \not> 1$$

$$3.01/14.0 + 0.75/(275 \times 41.3 \times 10^{-3}) = \mathbf{0.28}$$

The overall buckling resistance is therefore satisfactory.

The diagrams for bending moment and shear force shown in Fig 4.5. indicate that maximum values are not coincident and it is not therefore necessary to check moment capacity in the presence of shear load. Purlin design does not normally need a check on web bearing and buckling as the applied concentrated loads on low – note the low values of shear force. The check for bearing and buckling of the web is particularly needed where heavy concentrated loads occur, and reference may be made to Chapter 5 for the relevant calculations.

(h) Deflection

Deflection limits for purlins are not specified in BS table 5 but a limit of span/200 is commonly adopted.
Deflection $\delta_x = 5W_xL^3/384EI_x$
where W_x is the serviceability imposed load, i.e. 7.95 kN, and E is 205 kN/mm^2

$$\delta_x = 5 \times 7.95 \times 6000^3/(384 \times 205 \times 852 \times 10^4) = \mathbf{12.8\ mm}$$

$$\delta_y = \mathbf{28.5\ mm}$$

Deflection limit $= 6000/200 = 30$ mm

(i) Connections

The connection of the purlin to the rafter may be made by bolting to a cleat as shown in Fig. 4.8. The design of these connections is usually nominal due to the low reactions at the ends of the purlins. However, the transfer of forces between the purlin and rafter should be considered. For the channel section chosen W_x and W_y transfer to the rafter through a cleat. Bolts must be provided but will be nominal due to the low reactions involved (8.3 kN and 3.0 kN). Chapter 3 gives calculations for a bolted connection in more detail.

Multi-span (continuous) purlins may be used and minor changes in design are considered in Section 4.4.

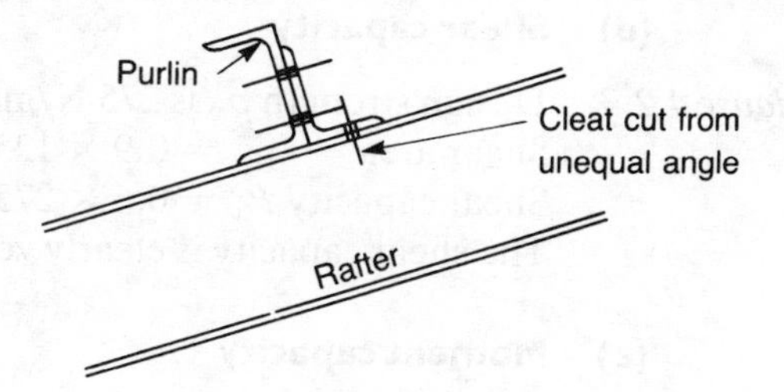

Fig. 4.8 Purlin connection

4.3 EXAMPLE 7. DESIGN OF SIDE RAIL

(a) Dimensions

See Fig. 4.9: side rails 2.0 m centres; span 5.0 m simply supported.

(b) Loading

Dead load (cladding/insulation panels) $= 0.18$ kN/m^2
Wind load (pressure) $= 0.80$ kN/m^2
Maximum values of bending moment and shear force must be found allowing for the wind loading (horizontal) only (Fig. 4.9) and including the safety factor γ_f.
Assume side rail to be **125 × 75 × 10** unequal angle, grade 43 steel. An

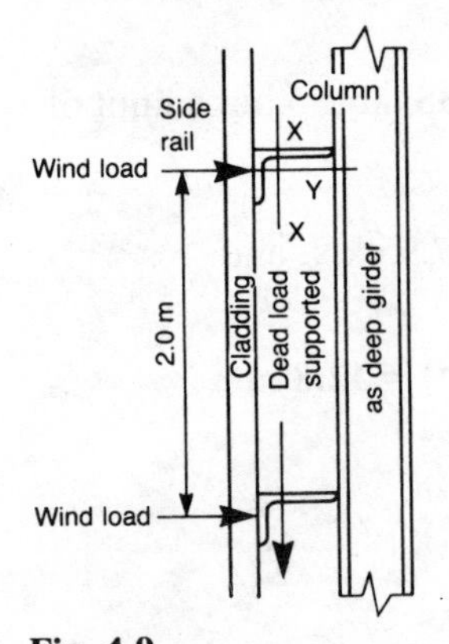

Fig. 4.9

angle such as that chosen provides greater resistance to bending (higher section properties) about the x axis than the y axis, and than an equal angle of the same area.

Cladding $2.0 \times 5.0 \times 0.18 = 1.80$ kN
Self weight $5.0 \times 0.15 = 0.75$ kN
Total dead load $W_d = 2.55$ kN
Wind load W_w $2.0 \times 5.0 \times 0.80 = 8.0$ kN

The loads W_w and W_d act in planes at right angles producing moments about x and y axes of the steel section, but only moments about x are used in design (as discussed in Section 4.1).

Ultimate load $W_x = 1.4W_w$
$= 1.4 \times 8.0 = 11.2$ kN

(c) BM and SF

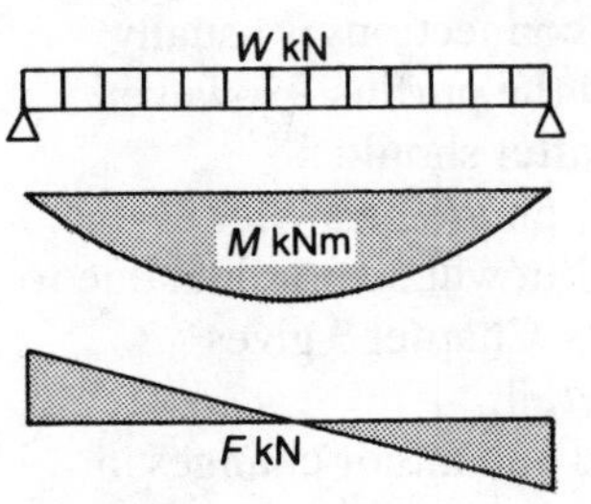

Fig. 4.10

Max. moment $M_x = 11.2 \times 5.0/8 =$ **7.0 kNm**
Max. shear force $F_x = 11.2/2 =$ **5.6 kN**
(See Fig. 4.10.)

(d) Shear capacity

clause 4.2.3

Design strength p_y is 275 N/mm² (Section 1.5)
Shear area $= 0.9 \times 125 \times 10 = 1125$ mm²
Shear capacity $P_v = 0.6 \times 275 \times 1125 \times 10^{-3} = 186$ kN
The shear capacity is clearly very large relative to the shear force.

(e) Moment capacity

For single angles lateral restraint is provided by the cladding, which also ensures bending about the x axis, rather than about a weaker axis (Fig. 4.11). The moment capacity only of the section is therefore checked. The section chosen is defined as semi-compact having $b/T = 7.5$ and $d/T = 12.5$ (both < 15) and $(b + d)/T = 20$ (< 23), hence:

BS table 7

$$M_{cx} = p_y Z_x$$
$$= 275 \times 36.5 \times 10^{-3} = 10.0 \text{ kNm}$$

$M_x/M_{cx} = 7.0/10.0 =$ **0.70**

Section is satisfactory.
The design of side rails does not normally include a check on web bearing and buckling, as discussed in Section 4.2(g).

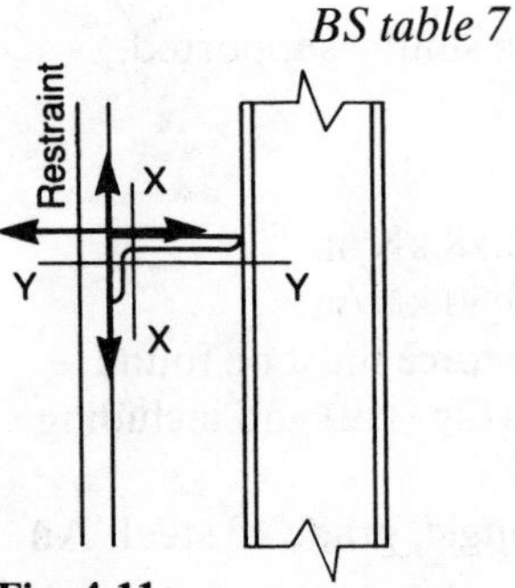

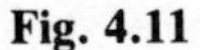

Fig. 4.11

(f) Deflection

Calculation of deflection is based on the serviceability condition, i.e. with unfactored imposed loads.

$W_y = 8.0$ kN

$\delta_y = 5W_yL^3/384EI_x$

$= 5 \times 8.0 \times 5000^3/(384 \times 205 \times 302 \times 10^4)$

$=$ **21.0 mm**

Although clause 4.12.2 avoids specifying any value use deflection limit say $L/200 = 25$ mm.

4.4 EXAMPLE 8. DESIGN OF MULTI-SPAN PURLIN

Continuity of a structural element over two or more spans may be useful in order to reduce the maximum moments to be resisted, and hence the section size, and to improve the buckling resistance of the member.

1. In general the bending moments in a continuous beam are less than those in simply supported beams of the same span. It should be noted, however, that a two-span beam has the same moment ($WL/8$) at the middle support as the mid-span moment of a simply supported beam.
2. The resistance of a member to lateral torsional buckling is improved by continuity and this is reflected in BS table 16.

Continuity may be achieved by fabricating members of length equal to two or more spans. Length will, however, be limited by requirements for delivery and flexibility during site erection. For the purlin designed in Section 4.2 a length of not more than two spans (12 m) would be acceptable (they can be delivered bundled together to reduce flexibility). Continuity can also be arranged by use of site connections capable of transmitting bending moments. Such connections are costly to fabricate and to assemble and are rarely used in small structural elements such as purlins and side rails. Using the same example as in Section 4.2 the design is repeated for a purlin continuous over two spans of 6.0 m.

(a) Dimensions

As Section 4.2(a).

(b) Loading

As Section 4.2(b); assume purlin to be **152 × 76** channel grade 43 steel

$W_d = 2.88$ kN

$W_{dx} = 2.71$ kN

$W_{dy} = 0.99$ kN

Ultimate load W_x = 16.5 kN
Ultimate load W_y = 6.0 kN

(c) BM and SF

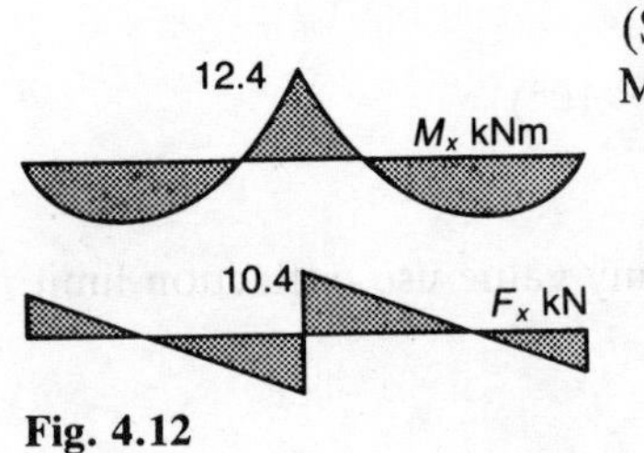

Fig. 4.12

(See Fig. 4.12.)
Max. ultimate moment (at central support)

$M_x = 16.5 \times 6.0/8 = 12.4$ kNm
$F_x = 0.625 \times 16.5 = 10.4$ kN
$M_y = 4.5$ kNm
$F_y = 3.8$ kN

(d) Shear capacity

Shear force is less than 0.6 shear capacity, as Section 4.2(d).

(e) Moment capacity

BS table 7 The 152 × 76 channel is a compact section ($b/T = 8.47$)
Hence $M_{cx} = 275 \times 130 \times 10^{-3} = 35.8$ kNm
$M_{cy} = 9.0$ kNm

$M_x/M_{cx} + M_y/M_{cy} \not> 1$

$12.4/35.8 + 4.5/9.0 = \mathbf{0.85}$

(f) Buckling resistance

$\lambda = L_E/r_y = 6000/22.4 = 268$

The factor n is obtained from BS table 16 for $\beta = 0$ and $\gamma = M/M_o = -1.00$ as the end moment M and simply supported moment M_o are equal (but opposite sign).

Hence $n = 0.66$
$m = 1.0$
$u = 0.902$

$\lambda/x = 268/14.5 = 18.5$

where x is the torsional index

BS table 14 $v = 0.49$
$\lambda_{LT} = 0.66 \times 0.902 \times 0.49 \times 268 = 78$

BS table 11 $p_b = 170$ N/mm^2
$M_b = 170 \times 130 \times 10^{-3} = 22.1$ kNm

$mM_x/M_b + mM_y/M_{cy} \not> 1$

$12.4/22.1 + 4.5/(275 \times 41.3 \times 10^{-3}) = \mathbf{0.96}$

(g) Deflection

W_x at serviceability limit state (imposed load only) = 7.95 kN

$\delta_x = 7.95 \times 6000^3/(185 \times 205 \times 852 \times 10^4) =$ **5.3 mm**

$W_y = 2.89$ kN

$\delta_y =$ **14.5 mm**

Deflection limit 6000/200 = 30 mm

5

CRANE GIRDERS

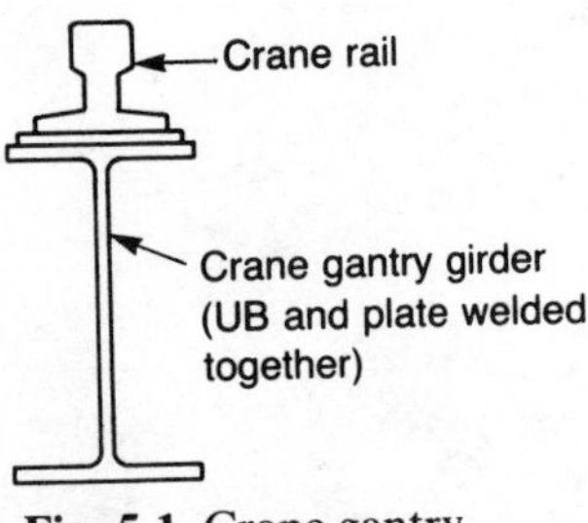

Fig. 5.1 Crane gantry girder

Industrial buildings commonly house manufacturing processes which involve heavy items being moved from one point to another during assembly, fabrication or plant maintenance. In some cases overhead cranes are the best way of providing a heavy lifting facility covering virtually the whole area of the building. These cranes are usually electrically operated, and are provided by specialist suppliers. The crane is usually supported on four wheels running on special crane rails. These rails are not considered to have significant bending strength, and each is supported on a crane beam or girder (Fig. 5.1). The design of this girder, but not the rail, is part of the steelwork designer's brief. However, the position and attachment of the rail on the crane girder must be considered, as a bad detail can led to fatigue problems, particularly for heavy duty cranes. The attachment of the rail should allow adjustment in the future to be carried out, as continuous movement of the crane can cause lateral movement of the rail.

5.1 CRANE WHEEL LOADS

Parts of a typical overhead crane are shown in Fig. 5.2. The weight or load associated with each part should be obtained from the crane supplier's data and then be combined to give the crane wheel loads. Alternatively wheel loads may be given directly by the crane manufacturer. Reference may be made to BS 6399: Part 1[1] for full details of loading effects. The following notes apply to single crane operation only.

The crab with the hook load may occupy any position on the crane frame up to the minimum approach shown in Fig. 5.2. Hence the vertical load on the near side pair of wheels can be calculated, adding an amount for the crane frame, which is usually divided equally between the wheels. Maximum wheel loads are often provided by the crane manufacturer.

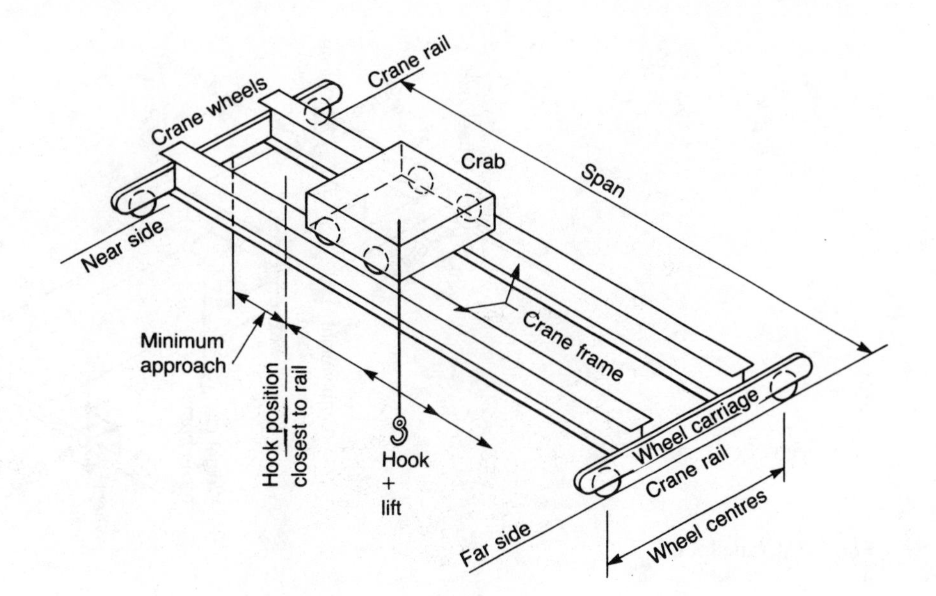

Fig. 5.2 Overhead crane

An allowance for impact of 25% is made for most light/medium duty cranes (class Q1 and Q2), and this is added to each vertical wheel load. For heavy duty cranes (class Q3 and Q4) reference should be made to BS 2573[(2)] and to supplier's data for appropriate impact values.

In addition to the vertical loads transferred from the wheels to the crane rail, horizontal loads can also develop. The first of these is called surge and acts at right angles (laterally) to the girder and at the level of the rail. This surge load covers the acceleration and braking of the crab when moving along the crane frame, together with the effects of non-vertical lifting. The value of this load is assessed in BS 6399[(1)] at 10% of the sum of the crab weight and hook load. It is usually divided equally between the four crane wheels and can act in either direction.

The second horizontal load (longitudinal) is the braking load of the whole crane, and in this case acts along the crane girder at the level of the top flange. The value of this load is assessed at 5% of each wheel load, and is therefore a maximum when the wheel load is a maximum. As before, the braking load covers acceleration and non-vertical lifting as well.

The loads are summarized in Fig. 5.3. In addition gantry girders intended to carry class Q3 and Q4 cranes (as defined in BS 2573: Part 1) should be designed for the crabbing forces given in clause 4.11.2.

The load factor γ_f for crane loads (ultimate limit state) is taken as 1.6, i.e. as for imposed loads generally (Section 1.7). Whenever the vertical load and the surge load are combined in the design of a member, the load factor should however be taken as 1.4 for both loads (BS table 2). Further detailed provisions for gantry girders are given in clauses 4.11 and 2.4.1.2.

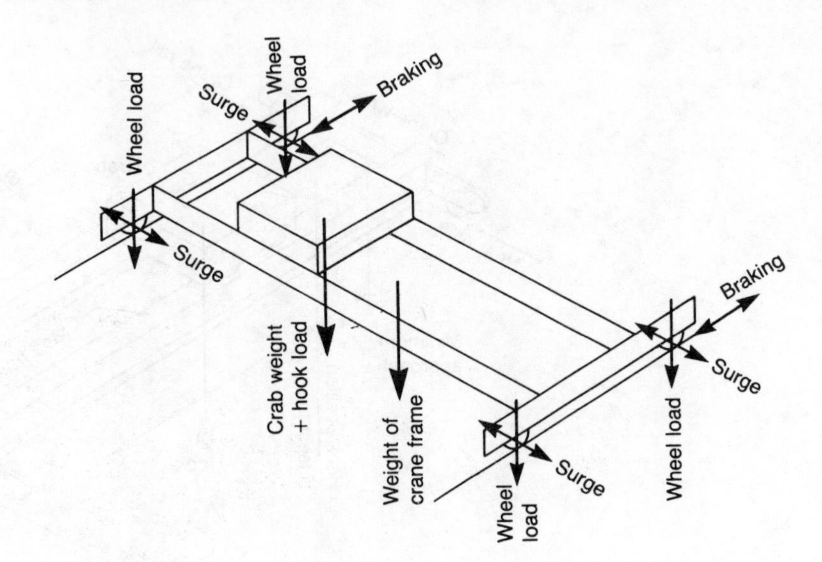

Fig. 5.3 Crane loads

5.2 MAXIMUM LOAD EFFECTS

Moving loads, such as crane wheels, will result in bending moments and shear forces which vary as the loads travel along the supporting girder. In simply supported beams the maximum shear force will occur immediately adjacent to a support, while the maximum bending moment will occur near, but not necessarily at, mid-span. In general, influence lines[3,4,5] should be used to find the load positions producing maximum values of shear force and bending moment.

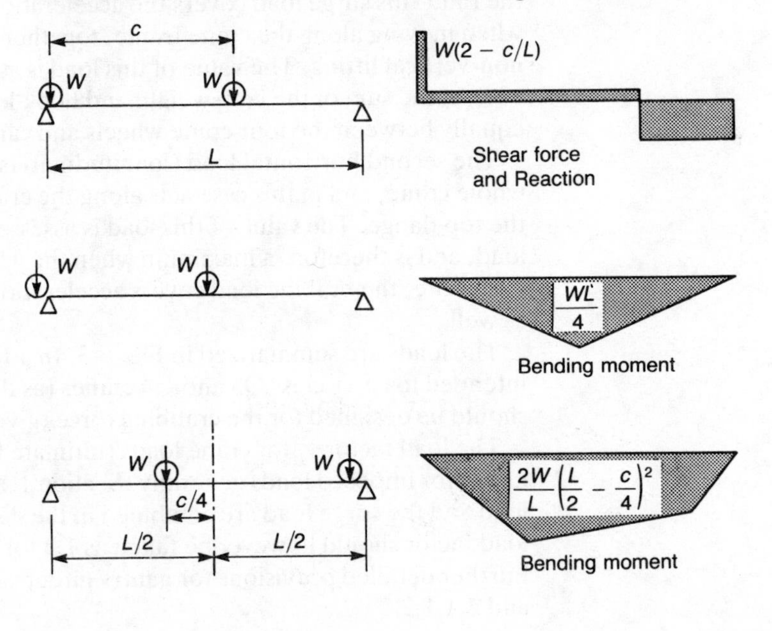

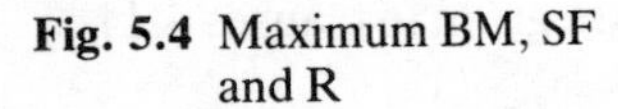
Fig. 5.4 Maximum BM, SF and R

The maximum effects of two moving loads may be found from formulae[3] as demonstrated in Section 2.5. For a simply supported beam the load positions shown in Fig. 5.4 give maximum values:

Shear force (max) $= W(2 - c/L)$

Bending moment (max) $= WL/4$

or $= 2W(L/2 - c/4)^2/L$

The greater of the bending moment values should be adopted.

The design of the bracket supporting a crane girder uses the value of maximum reaction from adjacent simply supported beams, as in Fig. 5.4. Where adjacent spans are equal, the reaction is equal to the shear force, i.e.

Reaction (max) $= W(2 - c/L)$

In all cases the effect of self weight (uniformly distributed) of the girder must be added.

5.3 EXAMPLE 9. CRANE GIRDER WITHOUT LATERAL RESTRAINT ALONG SPAN

(a) Dimensions

Span of crane	15.0 m
Wheel centres	3.5 m
Minimum hook approach	0.7 m
Span of crane girder	6.5 m (simply supported)

(b) Loading

Class Q2

Hook load	200 kN
Weight of crab	60 kN
Weight of crane (excluding crab)	270 kN

(c) Wheel loads

Vertical load/wheel from:

hook load, $200(15.0 - 0.7)/(15.0 \times 2)$	$= 95.3$ kN
crab load, $60(15.0 - 0.7)/(15.0 \times 2)$	$= 28.6$ kN
crane load, $270/4$	$= 67.5$ kN
Total vertical load	$= 191.4$ kN per wheel

Vertical load W_c including impact allowance and γ_f)

$= 1.25 \times 1.4 \times 191.4 = 335$ kN

Where vertical load is considered acting alone, then γ_f is 1.6 and W_c becomes 383 kN.

Lateral (horizontal) surge load is 10% of hook + crab load

$= 0.10(200 + 60) = 26$ kN

Total lateral load = 26/4 = 6.5 kN per wheel
Surge load W_{hc} (including γ_f) = 1.4 × 6.5 = 9.1 kN

Longitudinal (horizontal) braking load is 5% of wheel load and including γ_f is:

$$0.05 \times 1.6 \times 191.4 = 15.3 \text{ kN per wheel}$$

In the following design it will become clear that the critical considerations are lateral buckling and web bearing at the support. Hence first sizing of the girder would be based on these criteria. Assume a **610 × 305 × 179 UB** (grade 43) with extra plate (grade 43) to top flange. This plate (Fig. 5.5) is used to give additional strength to the top flange which is assumed alone to resist the lateral (surge) loading. A channel section may be preferred instead of the flat plate, and this type of section was used commonly in the past. Dead load due to self weight of girder (1.79 + 0.36 kN/m) and rail (0.25 kN/m) including γ_f is:

$$W_d = 1.4 \times 2.40 \times 6.5 = 21.8 \text{ kN}$$

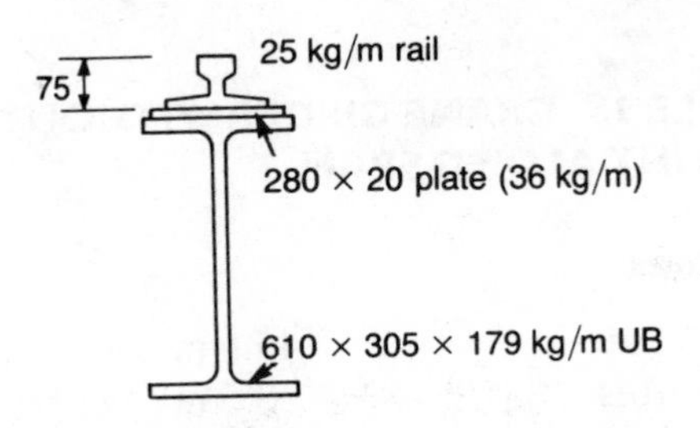

Fig. 5.5

(d) BM and SF

Moment due to vertical wheel loads is:

either $W_cL/4 = 335 \times 6.5/4 = 544$ kN

or $2W_c(L/2 - c/4)^2/L$
$= 2 \times 335(6.5/2 - 3.5/4)^2/6.5$
$= 581$ kNm (664 kNm when acting alone)

Moment due to dead load = 21.8 × 6.5/8 = 18 kNm
Max. ultimate moment M_x = 581 + 18 = **599 kNm (682 kNm** when acting alone)
Although the dead load maximum BM occurs at mid-span, and the wheel maximum occurs a distance $c/4$ away, it it usual to assume the value of M_x to be the sum of the maxima as shown.
Moment due to surge load = 2 × 9.1(6.5/2 − 3.5/4)²/6.5 = 15.8 kNm
Max. ultimate moment M_y = **15.8 kNm**
Vertical shear force due to wheel loads is:

$W_c(2 - c/L) = 335(2 - 3.5/6.5)$
$= 490$ kN (560 kN when acting alone)

Vertical shear force due to dead load = 21.8/2 = 11 kN
Max. ultimate shear force F_x = 490 + 11
= **501 kN (571 kN** when acting alone)
Lateral shear force due to surge load = 9.1(2 − 3.5/6.5) = 13.3 kN

Max. ultimate shear force F_y = **13.3 kN**
Max. ultimate reaction $R_x = 490 + 21.8 = 512$ kN
$R_y = 13.3$ kN

(e) Shear capacity

clause 4.2.3 *Table 1.2*

Design strength for chosen section with flange 23.6 mm thick:
$p_y = 265$ N/mm^2
Shear capacity $P_{vx} = 0.6 p_y A_v$
$= 0.6 \times 265 \times 617.5 \times 14.1 = 1380$ kN
$F_x/P_{vx} = \mathbf{0.41}$
Shear capacity $P_{vy} = 0.6 \times 265 \times (307 \times 23.6 + 280 \times 20) = 2030$ kN
$F_y/P_{vy} = \mathbf{0.01}$

(f) Moment capacity

clause 3.5.5

The chosen section (Fig. 5.6) is a plastic section with:

b/T (internal) $= 280/20 = 14$
b/T (outstand) $= 153.5/23.6 = 6.5$

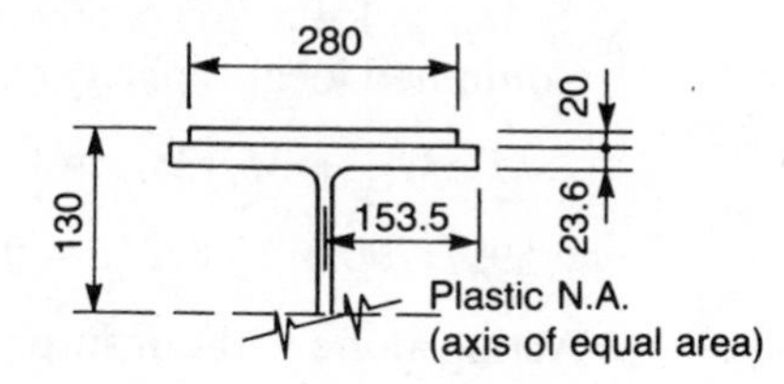

Fig. 5.6

For the built-up section chosen, the designer may need to calculate the plastic modulus (S_x)(6,7), if this is not available in published tables. The properties may be obtained from formulae given in Appendix A.
Area of plate $A_p = 280 \times 20 = 5600$ mm^2
Total area $A = 228 + 5600 \times 10^{-2} = 284$ cm^2
Plastic section properties $d_p = 5600/(2 \times 14.1) = 198.6$ mm
(i.e. 130 mm below the top face)
$S_x = 5520 + 14.1 \times 198.6^2 \times 10^{-3} + 5600(617.5/2 + 20/2 - 198.6)10^{-3}$
$= 6750$ cm^3
S_y (for top flange only) $= (23.6 \times 307^2/4 + 20 \times 280^2/4)10^{-3}$
$= 948$ cm^3
Elastic section properties $d_e = 5600(617.5 + 20)/2(284 \times 10^2)$
$= 62.8$ mm
$I_x = 152\,000 + 228 \times 62.8^2 \times 10^{-2}$
$+ 5600(617.5/2 + 20/2 - 62.8)^2 \times 10^{-4}$
$= 198\,000$ cm^4
$Z_x = 198\,000/(617.5/2 + 62.8)\,10^{-1} = 5320$ cm^3
For tension flange about y axis $I_{tf} = 23.6 \times 307^3/(12 \times 10^4) = 5690$ cm^4
For compression flange about y axis $I_{cf} = 5690 + 20 \times 280^3/(12 \times 10^4)$
$= 9350$ cm^4
Z_y (for top flange only) $= 2I_{cf}/B$
$= 2 \times 9350/(307 \times 10^{-1}) = 609$ cm^3

I_y (for combined section) $= 11\,400 + 20 \times 280^3/(12\times10^4) = 15\,100$ cm^4
$r_y = \sqrt{(I_y/A)}$
$= \sqrt{(15\,100/284)} = 7.28$ cm

Torsional index may be calculated using the approximate method in BS appendix B.2.5.1(c).

$h_s = 617.5 + 20 - 23.6/2 - 43.6/2 = 604$ mm
$\Sigma bt + h_w t_w = 28\,400 + 570.3 \times 14.1 = 36\,400$ mm^2
$\Sigma bt^3 + h_w t_w{}^3 = 307 \times 41.8^3 + 307 \times 23.6^3 + 570.3 \times 14.1^3 \times 2$
$= 29.65 \times 10^6$ mm^4
$x = 604\sqrt{(36400/[29.65 \times 10^6])} = 21.2$

Moment capacity $M_{cx} = p_y S_x$
$= 265 \times 6750 \times 10^{-3} = 1790$ kNm

To prevent plasticity at working load (see Section 3.4), $M_{cx} \not> 1.4 p_y Z_x$ where factored load/unfactored load = 1.4

$M_{cx} \not> 1.4 \times 265 \times 5320 \times 10^{-3} = 1970$ kNm
$M_{cy} = p_y S_y$ where S_y is for the top flange only
$M_{cy} = 265 \times 948 \times 10^{-3} = 251$ kNm

But $M_{cy} \not> 1.4 p_y Z_y$ using constant 1.4 as noted above and Z_y for top flange only

$M_{cy} \not> 1.4 \times 265 \times 609 \times 10^{-3} = 226$ kNm

Combined local capacity check:

$M_x/M_{cx} + M_y/M_{cy} \not> 1$

$599/1790 + 15.8/226 = \mathbf{0.40}$

Acting alone without surge

$M_x/M_{cx} = 683/1790 = \mathbf{0.38}$

Hence section chosen is satisfactory.

(g) Buckling resistance

The buckling resistance may be found in the same way as in Section 3.8(f), but allowing for the destabilizing effect on the crane load.

clause 4.11.3 Hence $m = n = 1.0$

BS table 13 No restraint is provided between the ends of the girder. At the supports the diaphragm gives partial restraint against torsion, but the compression flange is not restrained (Fig 5.7).

BS table 9 $L_E = 1.2(L + 2D)$
$= 1.2(6.5 + 2 \times 0.637) = 9.33$ mm

Slenderness $\lambda = L_E/r_y$
$= 9330/72.8 = 128$
$\lambda/x = 128/21.2 = 6.0$

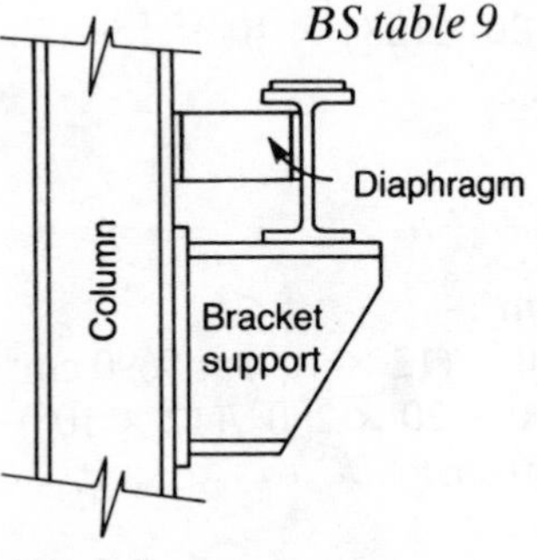

Fig. 5.7

BS table 14	$N = I_{cf}/(I_{cf} + I_{tf})$
	$= 9350/(9350 + 5690) = 0.62$
	$v = 0.73$
BS table 13	$n = 1.0$
clause 4.3.7.5	$u = 1.0$ (conservatively)
	$\lambda_{LT} = nuv\lambda$
	$= 1.0 \times 1.0 \times 0.73 \times 128 = 93$
BS table 12	Bending strength $p_b = 128\ \text{N/mm}^2$
	Buckling resistance $M_b = p_b S_x$
	$= 128 \times 6750 \times 10^{-3} = 864$ kNm
clause 4.11.3	Equivalent uniform moment factor $m = 1.0$
	Overall buckling check:

$$m_x M/M_b + m_y M_y/M_{cy} \not> 1$$

$$599/864 + 15.8/(265 \times 609 \times 10^{-3}) = \mathbf{0.79}$$

Acting alone without surge:

$$M_x/M_b = 683/864 = \mathbf{0.79}$$

Hence the section is satisfactory.

(h) Web buckling

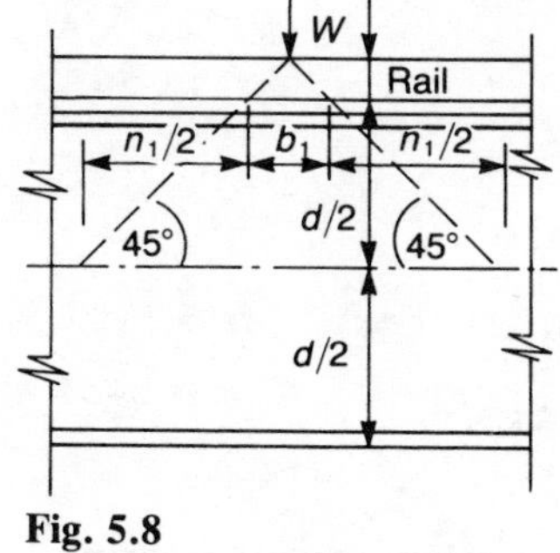

Fig. 5.8

At points of concentrated load (wheel loads or reactions) the web of the girder must be checked for local buckling[8] (See Fig. 5.8.). If necessary load carrying stiffeners must be introduced to prevent local buckling of the web.

Dispersion length **under wheel:**

clause 4.11.5	$b_1 = 2 \times 75 = 150$ mm
	$n_1 = 617 + 20 = 637$ mm
clause 4.5.2.1	Web slenderness $\lambda = 2.5d/t$
	$= 2.5 \times 537/14.1 = 95$
BS table 27c	Compressive strength $p_c = 131\ \text{N/mm}^2$
	Buckling resistance $P_w = (b_1 + n_1)tp_c$
	$= (150 + 637)14.1 \times 0.131 = \mathbf{1450\ kN}$

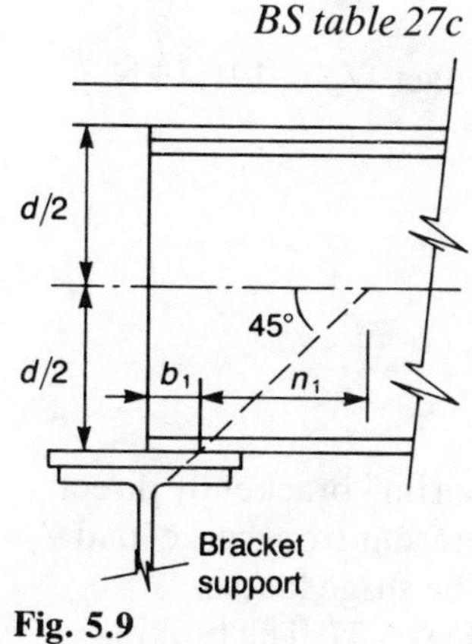

Fig. 5.9

Max. wheel load = 383 kN

Hence buckling resistance satisfactory.

Assume stiff bearing length **at support** (Fig. 5.9):

$b_1 = 70$ mm

$n_1 = 319$ mm

$P_w = (70 + 319)14.1 \times 0.131 = \mathbf{718\ kN}$

Support reaction = 571 kN $< P_w$

(i) Web bearing

At the same points the web of the girder must be checked for local crushing(8) (See Fig. 5.10.). If necessary bearing stiffeners must be introduced to prevent local crushing of the web.

clause 4.11.5 **Load dispersion under wheel** = 75 + 2(43.6 + 16.5) = 195 mm
Bearing capacity u.c. – P_{crip} = 195 × 14.1 × 0.265 = **729 kN**
Wheel load = 383 kN

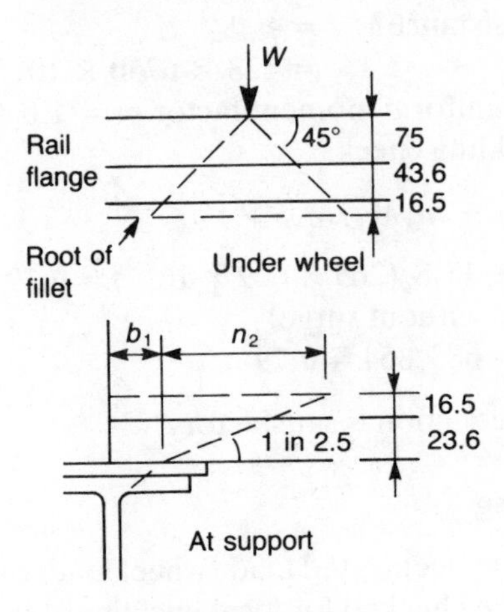

Fig. 5.10

Load dispersion at support:

$$n_2 = (23.6 + 16.5)2.5 = 100 \text{ mm}$$

$$\text{u.c.} - P_{crip} = (b_1 + n_2)tp_y$$

$$= (70 + 100)14.1 \times 0.265 = \mathbf{635\ kN}$$

Support reaction = 571 kN
Hence bearing capacity is satisfactory. Note that adequate stiff bearing is needed at the supports to prevent the web bearing capacity being exceeded.

(j) Deflection

Serviceability vertical wheel load excluding impact W_c = 191.4 kN
Max. deflection for position given(9):

$\delta_c = W_c L^3(3a/4L - a^3/L^3)/6EI$

where $a = (L - c)/2 = 1.5$ m

$\delta_c = 3.5$ mm

Deflection limit = 6500/600 = 10.8 mm

(k) Connection

The vertical forces are transmitted to the supporting bracket by direct bearing (Fig. 5.11). Horizontal reactions are present from surge load (13.3 kN) and horizontal braking (15.3 kN). The surge load is transmitted to the column by the diaphragm (Fig. 5.7). The braking

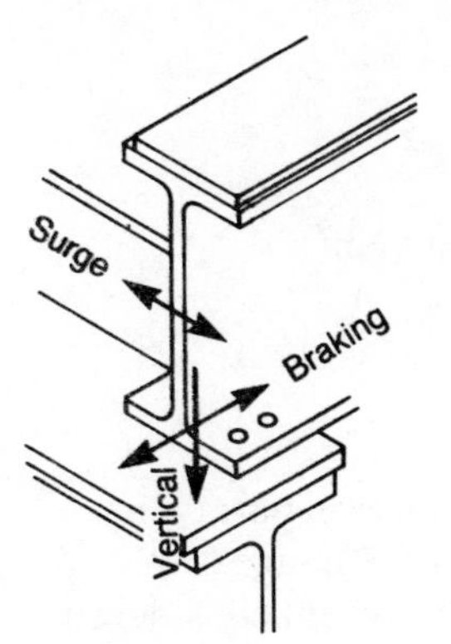

Fig. 5.11

force will be transmitted by nominal bolts, provide, say, two no. M20 (grade 4.6).

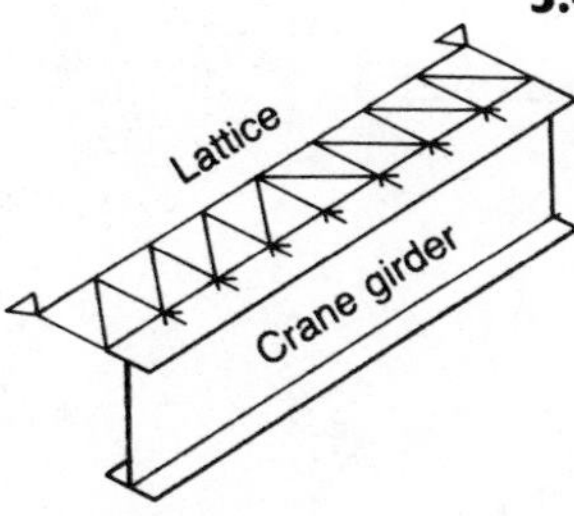

Fig. 5.12

5.4 EXAMPLE 10. CRANE GIRDER WITH LATERAL RESTRAINT

The design in Section 5.3 may be repeated but including a lattice restraint to the compression flange. In practice such a lattice girder may have been provided specifically for this purpose, or to support access platforms or walkways (Fig. 5.12). In modern UK practice it is rarely economic to include such a restraint in order to reduce the beam size, except in heavy industrial buildings.

(a) Dimensions

As Section 5.3

(b) Loading

As Section 5.3 (making no correction for the changed self weight)

(c) Wheel loads

As Section 5.3

(d) BM and SF

As Section 5.3, i.e.
Max. M_x = 599 kNm (683 kNm when acting alone)
Max. M_y = 15.8 kNm

(e) Shear capacity

A smaller UB may be chosen as lateral restraint is provided. Assume a **533 × 210 × 122 UB** with no plate added. Note that in the design in Section 5.3 web bearing and buckling were critical, hence a revised section having a similar web thickness is chosen.

clause 4.2.3 Shear capacity $P_{vx} = 0.6 p_y A_v$
$= 0.6 \times 0.265 \times 544.6 \times 12.8 = 1108$ kN
$F_x/P_{vx} = 571/1108 = \mathbf{0.52}$

(f) Moment capacity

The chosen UB is a plastic section ($b/T = 5.0$)
Moment capacity $M_{cx} = p_y S_x$
$= 265 \times 3200 \times 10^{-3} = 849$ kNm
But $M_{cx} \not> 1.2 p_y Z_y$
$= 1.2 \times 265 \times 2800 \times 10^{-3} = 890$ kNm
$M_x/M_{cx} = 683/849 = \mathbf{0.80}$
Section is satisfactory.

clause 4.3.2 The effect of the lateral moment M_y is considered later in part (j) in combination with a restraint force equal to 2.5% of the flange force. Note that the value of 1% given in clause 4.3.2 is at present considered to be too small.

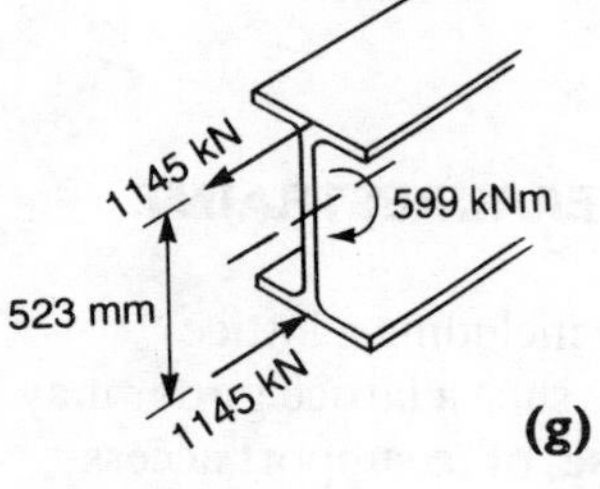

Fig. 5.13

Flange force may be estimated as: 599/0.523 = **1145 kN** (1306 kN when acting alone), see Fig. 5.13.
Restraint force = **28.6 kN** (32.6 kN when acting alone)

(g) Web buckling/bearing

Stiff bearing length at support:

$b_1 = 85$ mm (assumed see note below)

$n_1 = 544.6/2 = 272.3$ mm

$\lambda = 2.5 \times 477/12.8 = 93$

BS table 27c $p_c = 134$ N/mm^2

Buckling $P_w = (85 + 272.3)12.8 \times 0.134 = \mathbf{613}$ **kN**
Support reaction = 571 kN
Flange dispersion at support $n_2 = (21.3 + 12.7)2.5 = 85$ mm
Bearing u.c. − $P_{crip} = (85 + 85)12.8 \times 0.265 = \mathbf{577}$ **kN**
Note that adequate stiff bearing (85 mm) is needed at the supports to give the bearing strength.

(h) Deflection

Wheel load = 191.4 kN
Calculation as in Section 5.3(j) gives $\delta = 9.0$ mm
Limit = 6500/600 = 10.5 mm

(i) Connection

As Section 5.3 but the lattice transmits horizontal forces to the column; hence the diaphragm is not needed.

(j) Lattice

The lattice girder (Fig. 5.14) which provides to the crane girder top flange is loaded by:

(i) the restraint force of 28.6 kN (32.6 kN when acting alone) which is considered to be distributed between the nodes of the lattice;
(ii) the surge load of 9.1 kN per wheel.

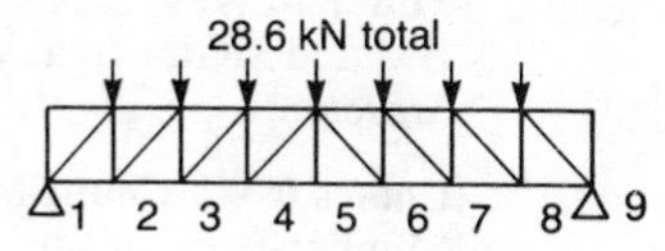

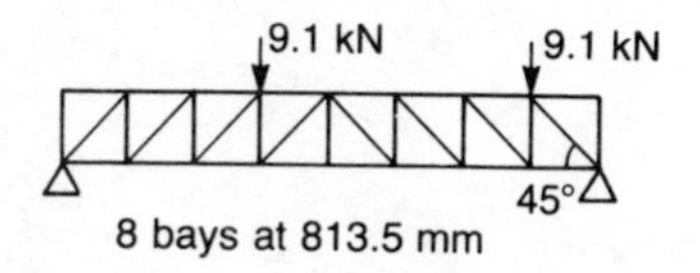

Fig. 5.14

The truss can be analysed by graphics, calculation, or computer, giving member forces as tabulated below (kN):

Panel	Top chord	Bottom chord	Diagonal	Post
1	0	21.2	30.0	0
2	21.2	38.3	24.2	17.1
3	38.3	51.3	18.4	13.0
4	51.3	51.0	0.3	0.3
5	46.6	51.0	6.1	0
6	38.2	46.6	11.9	4.3
7	25.7	38.2	17.7	8.4
8	0	25.7	36.3	12.5
9				0

The applied forces may act in either direction and hence the member forces may be either tension or compression. Designing the chord members for the maximum compression of 51.3 kN:

Use a **45 × 45 × 6 equal angle**

$\lambda = L_E/r_{rr} = 1.0 \times 813/8.67 = 94$

BS table 27c $p_c = 135\ \text{N/mm}^2$

Compression resistance $P_c = A_g\, p_c$

$= 5.09 \times 135 \times 10^{-1} = \mathbf{69\ kN}$

Designing the diagonals for a compression of 36.3 kN using the same single angle:

Effective length $L_E = 1.0 \times 813/\cos 45° = 1150$ mm

$\lambda = 1150/8.67 = 133$

BS table 27c $p_c = 83\ \text{N/mm}^2$

$P_c = 5.09 \times 83 \times 10^{-1} = \mathbf{42\ kN}$

The basic lattice member is therefore a 45 × 45 × 6 equal angle and might be fabricated as a welded truss. For a more detailed consideration of truss design reference should be made to Chapters 6 and 11.

STUDY REFERENCES

Topic	*Reference*
1. Crane loading	BS 6399 *Loading for buildings* Part 1: *Dead and imposed loads* (1984)
2. Crane types	BS 2573 *Permissible stresses in cranes and design and design rules* Part 1: *Structures* (1977)

3. Influence lines	**Marshall W.T.** & **Nelson H.M.** (1977) Maximum BM and SF due to a system of point loads, *Structures,* pp. 35–41. Longman
4. Influence lines	**Coates R.C., Coutie M.G.** & **Kong F.K.** (1988) Mueller-Breslau's principle. Model analysis *Structural Analysis,* pp. 127–131. Van Nostrand Reinhold
5. Influence lines	**Wang C.K.** (1983) Influence lines for statically determinate beams, *Intermediate Structural Analysis,* pp. 459–67. McGraw-Hill
6. Plastic modulus	**Marshall W.T.** & **Nelson H.M.** (1977) Plastic bending, *Structures,* pp. 104–106. Longman
7. Plastic modulus	**Horne M.R.** & **Morris L.J.** (1981) *Plastic design of low-rise frames.* Collins
8. Web buckling and bearing	*Steelwork Design,* vol. 3, Commentary on BS 5950: Part 1. Steel Construction Institute (to be published)
9. Deflection formulae	(1972) Simply supported beams, *Steel Designers' Manual,* pp. 17–38. Crosby Lockwood Staples

6

TRUSSES

Trusses and lattice girders are fabricated from the various steel sections available, joined together by welding or by bolting usually via gusset (connecting) plates. Generally they act in one plane and are usually designed as pin-jointed frames, although some main members may be designed as continuous. Where members lie in three dimensions the truss is known as a space frame. Trusses and lattice girders are particularly suited to long spans, as they can be made to any overall depth, and are commonly used in bridge construction. In buildings they have particular application for roof structures, and for members supporting heavy loads (columns from floors above) and for members requiring longer spans.

The use of a greater overall depth leads to a large saving in weight of steel compared with a universal beam. This saving of material cost can offset the extra fabrication costs in certain cases.

6.1 TYPES OF TRUSS AND THEIR USE

A selection of roof trusses is shown in Fig. 6.1, where the roof slopes and spans dictate the shape of the truss and the layout of the members. Hipped trusses are used for small spans, economically up to 6 m, the lattice girders for medium spans, and the mansard for large spans. Such trusses are lightly loaded by snow and wind load, together with a small allowance for services. It is unusual for lifting facilities to be supported from the roof trusses. The resulting member and connection sizes are therefore relatively small.

Heavy trusses may be used in multi-storey buildings where column loads from the floors over need to be carried. Examples of these are shown in Fig. 6.2. Trusses of this type can carry very heavy loads, and are similar in layout and member size to bridge structures.

A common method of providing stability to a building, whether single or multi-storey, is to use an arrangement of bracing members. These are

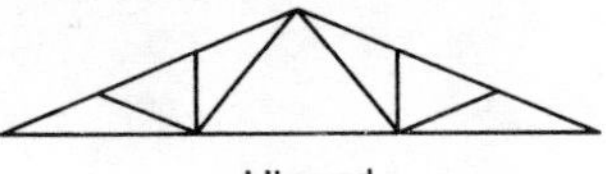

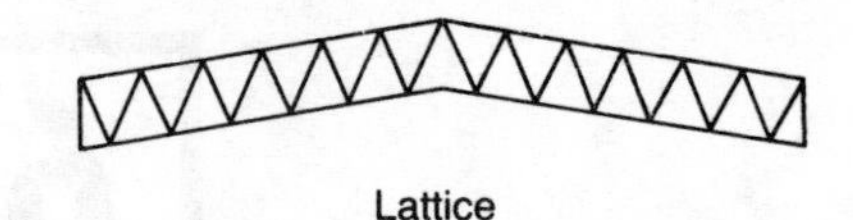

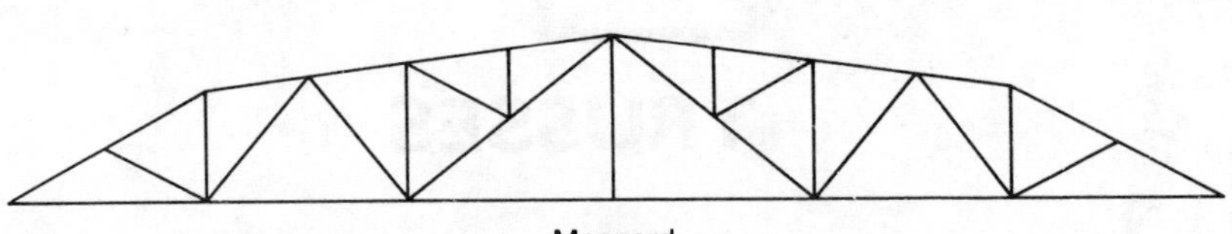

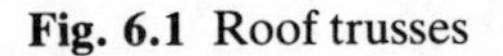

Fig. 6.1 Roof trusses

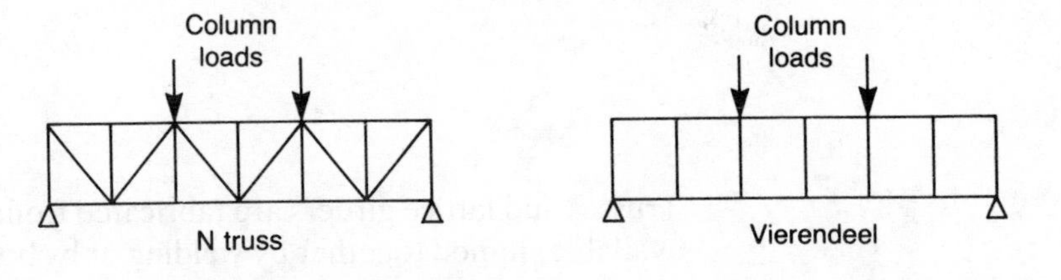

Fig. 6.2 Support trusses

essentially formed into a truss and carry the horizontal loads, such as wind, to the foundations by acting as horizontal and vertical frameworks. Examples are shown in Fig. 6.3. Bracing is considered in more detail in Chapter 10, and is graphically illustrated in Chapters 11 and 12 in terms of the overall stability of single-storey buildings.

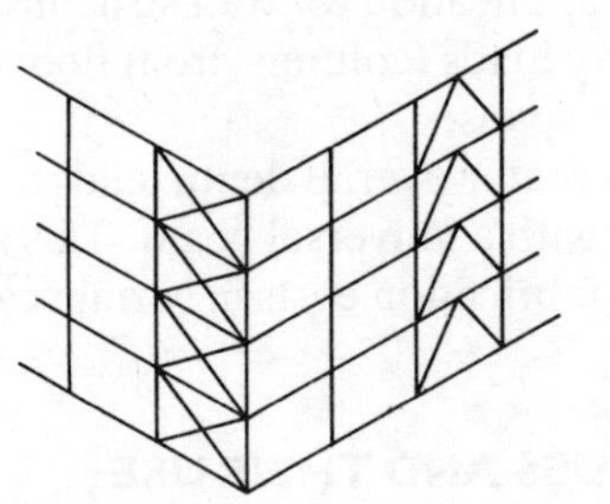

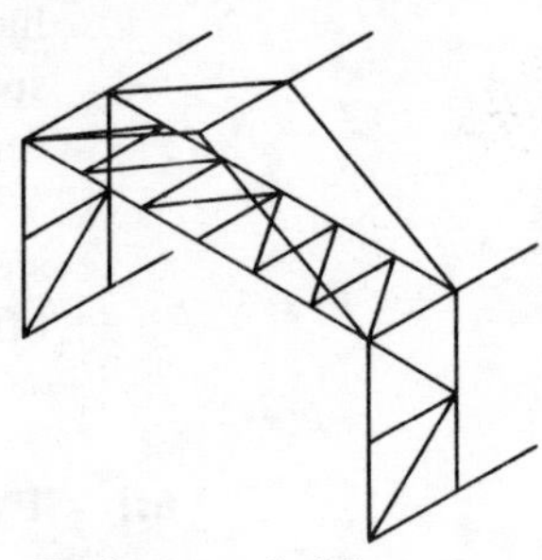

Fig. 6.3 Bracing

6.2 LOADING AND ANALYSIS

Loading will consist of dead, imposed and wind loads as described in Chapter 2. Combinations of loads giving maximum effects in individual members must be considered (see Section 2.4) and safety factors γ_f must be included (Section 1.7).

The loads will usually be transferred to the truss by other members such as purlins (Fig. 1.2) or by beams in the case of a floor truss. A wind bracing will be loaded by the gable posts, or by side members such as the eaves beam. It is ideal if the loads can be transferred to the truss at the node points, but commonly (as shown in Fig. 1.2) this is not possible. In roof truss design the purlin positions may not be known initially, and allowing for the possibility of purlin changes during future re-roofing, a random position for loads is often allowed.

The analysis therefore involves several stages:

(a) Analysis of the truss assuming pin-joints (except Vierendeel trusses) and loading at the nodes.
This may proceed using manual methods – joint resolution, method of sections and graphical means are all suitable[1,2] – or computer techniques. Several analyses may be needed where different arrangements of dead, imposed and wind loading must be considered.
(b) Analysis of the load bearing member such as the rafter as a continuous beam supported at the nodes and loaded by the purlins. In cases where the load positions are uncertain the rafter moment may be taken as $WL/6$ (clause 4.10c), where W is the purlin load and L is the node to node length perpendicular to W.
(c) Assessment of stresses due to eccentricity of the connections. Ideally the centroidal axes of members should meet at the nodes. Where this is not possible the members and connections should be designed for the moments due to the eccentricity if significant.
(d) Assessments of the effects of joint rigidity and deflections. Secondary stresses become important in some trusses having short thick members, but may be neglected where more slender members are used (clause 4.10).

The overall analysis of the truss will therefore involve the summation of two or more effects. Analyses (a) and (b) must always be considered, while (c) and (d) may be avoided by meeting certain conditions.

6.3 SLENDERNESS OF MEMBERS

The slenderness λ of a compression member (a strut) is given by:

$$\lambda = L_E/r$$

where L_E is the effective length of the strut about the appropriate axis
r is the radius of gyration about the appropriate axis

The requirements of clauses 4.7.2 and 4.7.10 define the effective lengths of the chord and internal members of trusses and are illustrated in Figs. 6.4 and 6.5. These requirements take into account the effect of the nodes (joints) which divide the top chord into a number of in-plane effective lengths. In the lateral (out-of-plane) direction the purlins restrain the top chord. The radius of gyration appropriate to a given strut depends on the possible axis of buckling, and these are shown in Figs 6.4 and 6.5.

The effective lengths of discontinuous struts are increased where single bolt connections are used resulting in reduced compressive strengths. Hence single bolted connections usually result in less economic trusses.

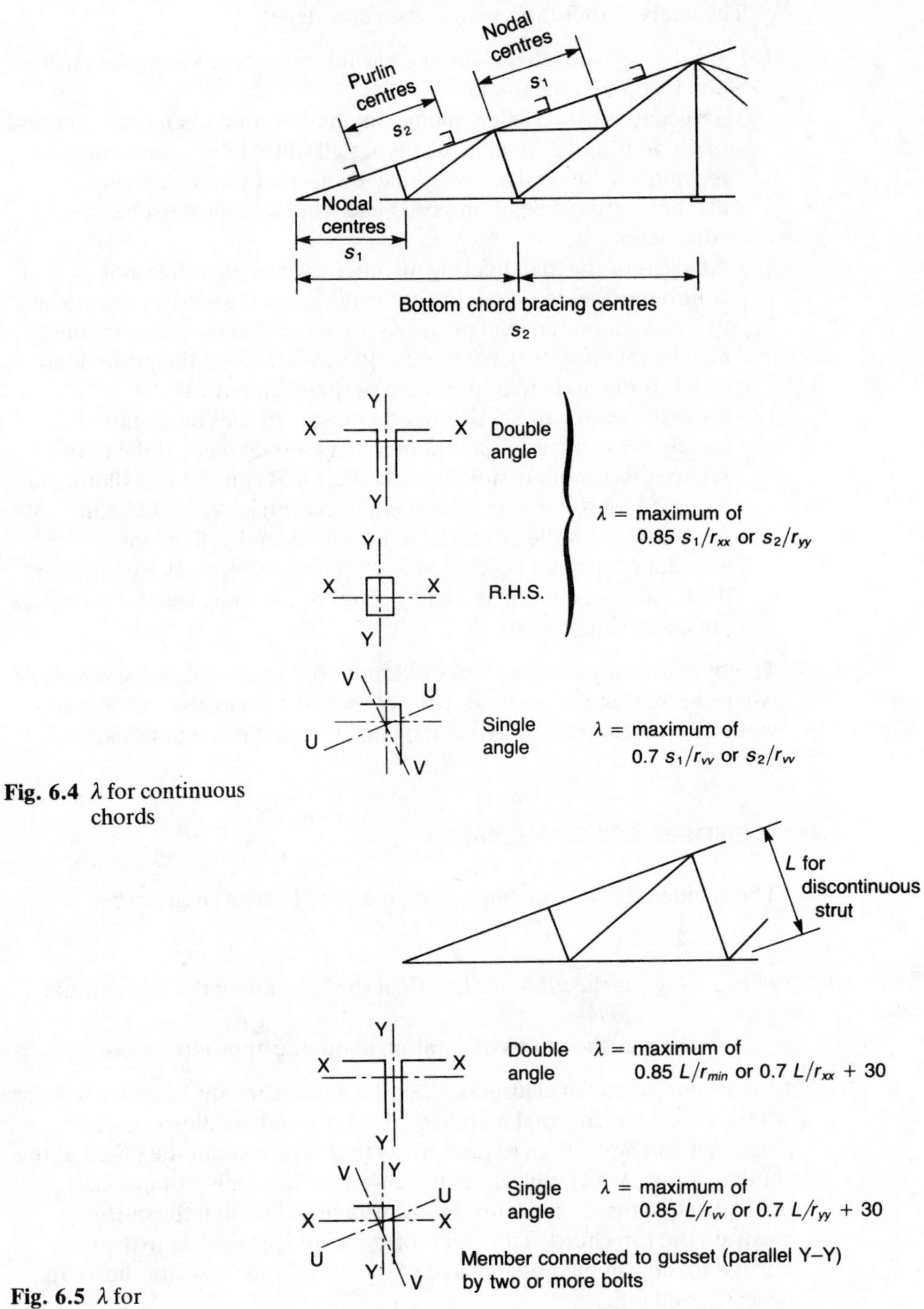

Fig. 6.4 λ for continuous chords

Fig. 6.5 λ for discontinuous struts

Where double angles are used as shown in Figs 6.4 and 6.5 it is necessary to reduce the slenderness of the individual component (single angle) by interconnecting the angles at points between the joints. These connections are usually a single bolt (minimum 16 mm diameter), with a packing between the angles equal to the gusset thickness, and are commonly placed at third or quarter points along the member.

Slenderness $\lambda_b = \sqrt{(\lambda_m{}^2 + \lambda_c{}^2)}$

where $\lambda_m = s_2/r_{yy}$

$\lambda_c = s_2/3r_{vv}$ for members having connections which divide s_2 into three equal parts

λ_c should not exceed 50.

These requirements are given in clauses 4.7.13.1(d) and 4.7.9(c).

Slenderness for any strut should not exceed 180 for general members resisting loads other than wind loads, or 250 for members resisting self weight and wind load only (clause 4.7.3.2). For a member normally acting as a tie, but subject to reversal of stress due to wind, the slenderness should not exceed 350. In addition very small sections should be avoided, so that damage during transport and erection does not occur, e.g. a minimum size of angle would be 50 × 50 × 6 generally.

6.4 COMPRESSION RESISTANCE

The compression resistance of struts is discussed also in Chapter 7. The compressive strength p_c depends on the slenderness λ and the design strength p_y. Tests on axially loaded, pin-ended struts show that their behaviour can be represented by a number of curves which relate to the type of section and the axis of buckling. These curves are dependent on material strength and initial imperfections, which affect the inelastic behaviour and the inelastic buckling load. For design the value of p_c is obtained from one of four strut curves or tables (BS tables 27a to 27d). The appropriate table is chosen by reference to section type and thickness, and to the axis of buckling (BS table 25).

The compression resistance P_c is

either $P_c = A_g p'_c$ for slender sections (see Section 1.5)
or $P_c = A_g p_c$ for all other sections

where A_g is the gross sectional area
p'_c is the compressive strength based on a reduced design strength (clause 3.6).

6.5 TENSION CAPACITY

The tension capacity P_t of a member is:

$$P_t = A_e p_y$$

where A_e is the effective sectional area as defined in clause 3.3.3.

Where a member is connected eccentrically to its axis then allowance should be made for the resulting moment. Alternatively such eccentric effects may be neglected by using a lower value of the effective area A_e. For a single angle connected through one leg:

$$A_e = a_1 + 3a_1a_2/(3a_1 + a_2)$$

where a_1 is the net sectional area of the connected leg
a_2 is the sectional area of the unconnected leg

Full details of these reduced effective areas are given in clause 4.6.3.

6.6 CONNECTIONS

Connections are required to join one member to another (internal joints), and to connect the truss to the rest of the building (external joints). Three main types of connection are used:

- bolting to a gusset plate
- welding to a gusset plate
- welding member to member

Examples of these are shown in Fig. 6.6. The choice of connection type is often made by the fabricator, and will depend on his available equipment, with welding becoming more economical the larger the number of truss and member repetitions.

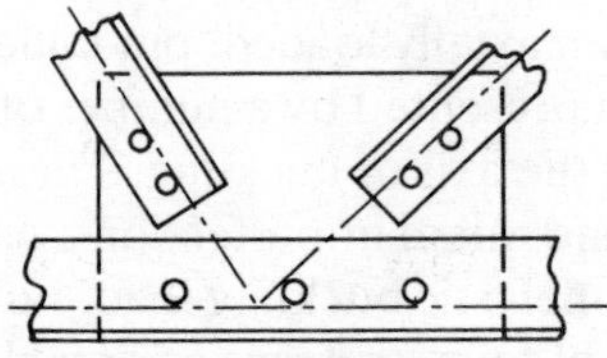

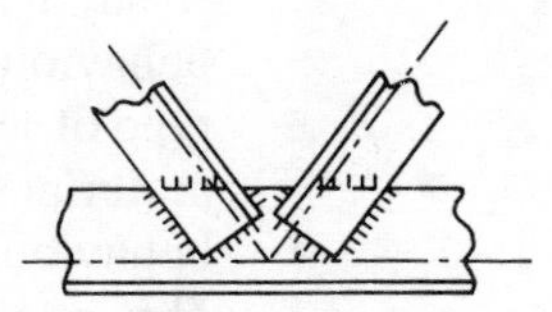

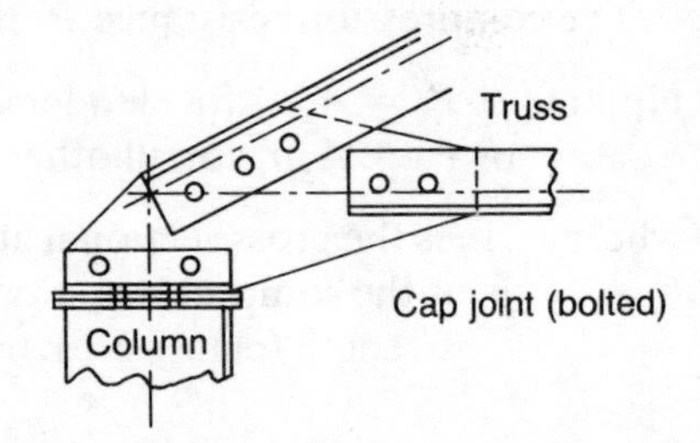

Fig. 6.6 Connection details

Ideally members should be connected so that centroidal axes (or bolt centre lines in the case of angles or tees) meet at a point (Fig. 6.6). If this cannot be achieved then both members and connection must be designed for the eccentricity. In many cases the gusset plate will not lie in the plane of the member centroidal axes, but stresses due to this eccentricity are ignored in construction using angles, channels and tees (clause 4.7.6c).

Design of the bolts or welds follows conventional methods[3,4]. Bolts (grade 4.6 or 8.8) must be designed for both shear stress and bearing stress (see Section 3.7(g)). Friction grip fasteners may be used[5], but would not usually be economic unless bolt slip were unacceptable. Design stresses in the gusset may be checked as a short beam with a combined axial load. Such design is not realistic, however, and it is sufficient to check the direct stress only at the end of the member (Fig. 6.7), on an area $b \times t$ as shown.

Minimum bolt size is usually 16 mm, and minimum gusset plate thickness is 6 mm (internal) or 8 mm (exposed).

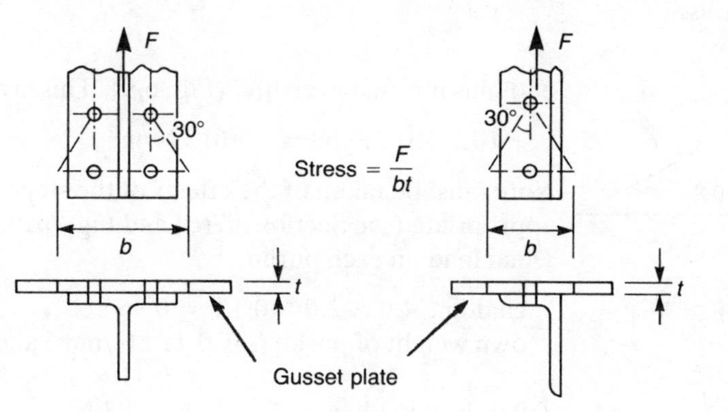

Fig. 6.7 Gusset plate stress

6.7 EXAMPLE 11. ROOF TRUSS WITH SLOPING RAFTER

(a) Dimensions

(See Fig. 6.8)

Span of truss	16.0 m
Rise of truss	3.2 m
Roof slope	21.8°
Truss spacing	4.0 m
Rafter length	8.62 m

(b) Loading

Cladding + insulation	0.12 kN/m²
Roof truss self weight (estimated)	8 kN
Snow + services	0.75 kN/m²
Wind pressure q	0.68 kN/m²

Using internal and external pressure coefficients given in reference (6), the worst case for wind loading on the roof slope is found to be 'wind on

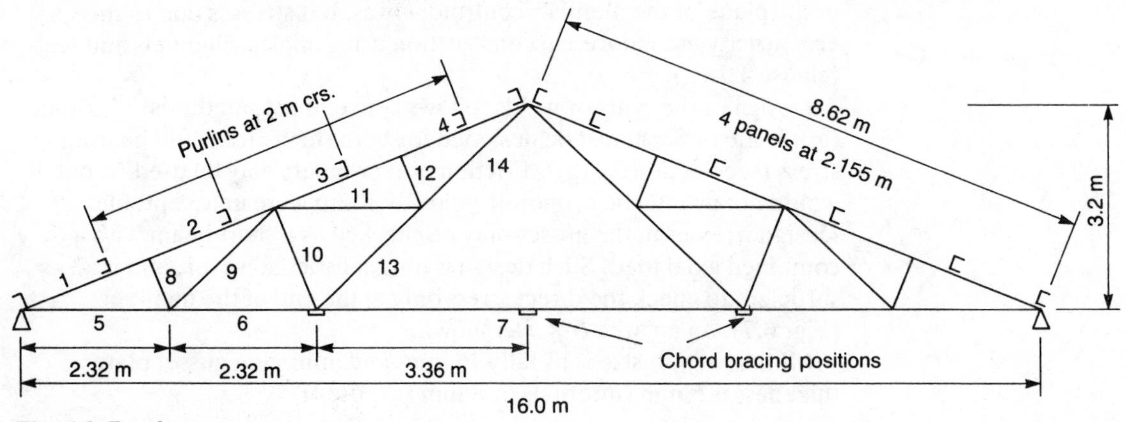

Fig. 6.8 Roof truss

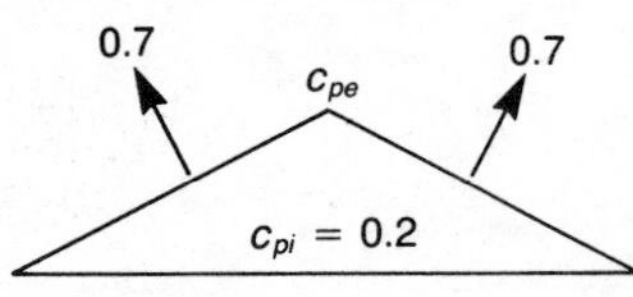

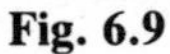
Fig. 6.9

end plus internal pressure' (Fig. 6.9). This gives an outward pressure of:

$$-(0.7 + 0.2)0.68 = -0.61 \text{ kN/m}^2$$

Note must be taken of the effects of the slope on values of each load as appropriate (see Section 2.7(d) and Fig. 6.10.)
Dead load on each purlin:

cladding $4.0 \times 2.0 \times 0.12 = 0.96$ kN
own weight of purlin (say 0.11 kN/mm) and truss
$= 0.11 \times 4.0 + 8.0/10 = 1.24$ kN

Total dead load $= 2.20$ kN

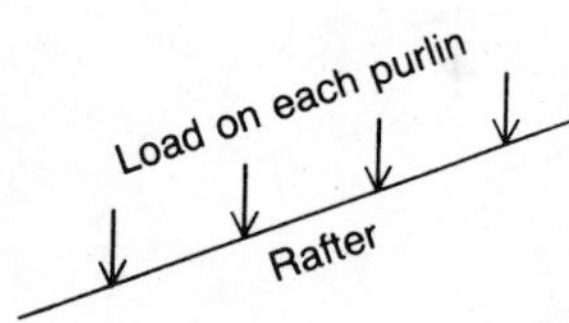

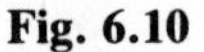
Fig. 6.10

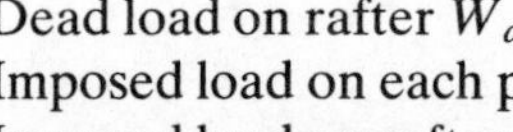

Dead load on rafter W_d $= 2.20 \times 8.62/2.0 = 9.48$ kN
Imposed load on each purlin $= 4.0 \times 2.0 \cos 21.8° \times 0.75 = 5.56$ kN
Imposed load on rafter W_i $= 5.56 \times 8.62/2.0 = 24.0$ kN
Wind load on each purlin $= 4.0 \times 2.0 \times 0.61 = 4.88$ kN (suction)
Wind load on rafter W_w $= 4.88 \times 8.62/2.0 = 21.0$ kN (suction)

(c) Truss forces (dead)

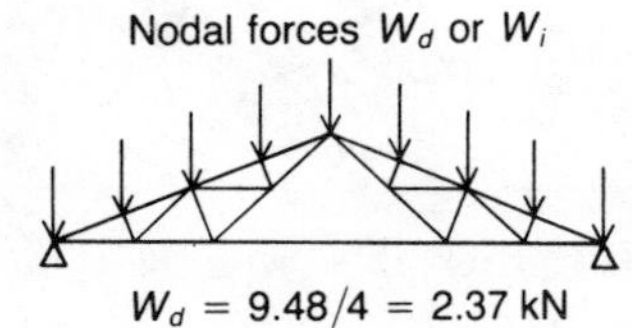

Fig. 6.11

Truss analysis is carried out by placing concentrated loads at the nodes of the truss, i.e. dividing the rafter load proportional to the nodal centres (Fig. 6.11). Analysis by manual or computer techniques gives forces as in the table (due to symmetry half of the truss only is recorded).

(d) Truss forces (imposed)

The forces are arranged as in (c) but have values of 6.00 kN instead of 2.37 kN. Member forces are given in the table.

(e) Truss forces (wind)

Wind forces on the truss are as shown (Fig. 6.12) and member forces are again analysed and the results given in the table.

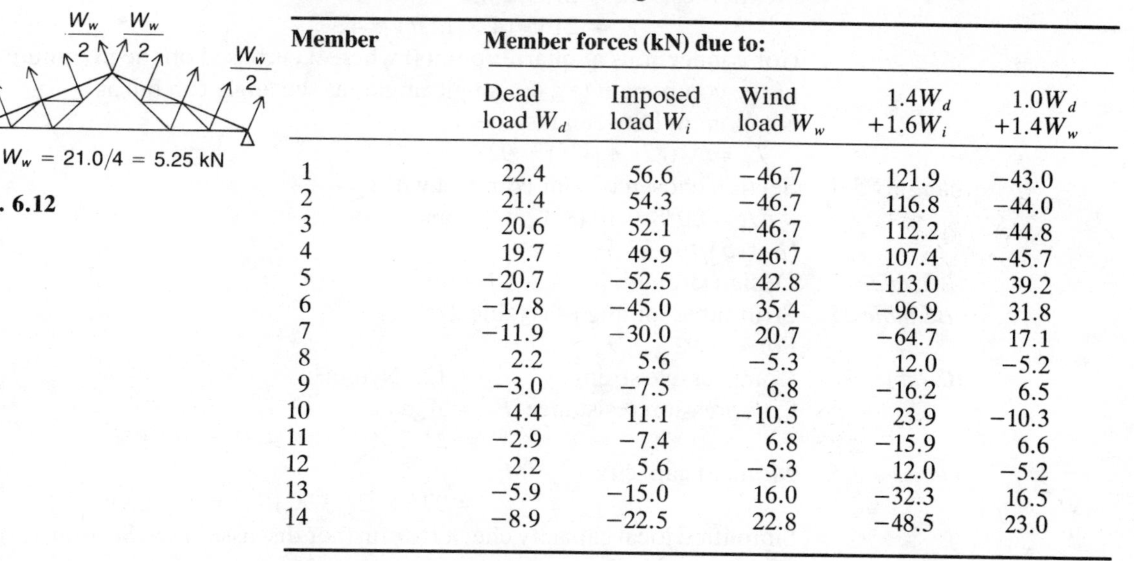

Fig. 6.12

Member	Member forces (kN) due to:				
	Dead load W_d	Imposed load W_i	Wind load W_w	$1.4W_d + 1.6W_i$	$1.0W_d + 1.4W_w$
1	22.4	56.6	−46.7	121.9	−43.0
2	21.4	54.3	−46.7	116.8	−44.0
3	20.6	52.1	−46.7	112.2	−44.8
4	19.7	49.9	−46.7	107.4	−45.7
5	−20.7	−52.5	42.8	−113.0	39.2
6	−17.8	−45.0	35.4	−96.9	31.8
7	−11.9	−30.0	20.7	−64.7	17.1
8	2.2	5.6	−5.3	12.0	−5.2
9	−3.0	−7.5	6.8	−16.2	6.5
10	4.4	11.1	−10.5	23.9	−10.3
11	−2.9	−7.4	6.8	−15.9	6.6
12	2.2	5.6	−5.3	12.0	−5.2
13	−5.9	−15.0	16.0	−32.3	16.5
14	−8.9	−22.5	22.8	−48.5	23.0

Compression force is positive.
As the wind load is suction, the load combinations $1.4W_d + 1.4W_w$ and $1.2(W_d + W_i + W_w)$ may be ignored.

(f) BM in rafter

Assuming purlin positions are not known:
Purlin load $W = 1.4(0.96 + 0.44) + 1.6 \times 5.56 = 10.9$ kN

clause 4.10(c)

$$\text{BM} = WL/6$$
$$= 10.9 \times 2.0 \cos 21.8°/6 = \mathbf{3.36\ kNm}$$

(g) Rafter design

Max. compression F = 121.9 kN
Max. tension = 45.7 kN
Nodal distance = 8.62/4 = 2.16 m
Use 2 – **80 × 60 × 8 unequal angles** spaced 8 mm apart to allow for gusset plates with spacing washers at quarter points, i.e. 0.54 mm centres (Fig. 6.13).

$r_x = 25.0$ mm

Slenderness $\lambda_x = 0.85 \times 2155/25.0 = 73$

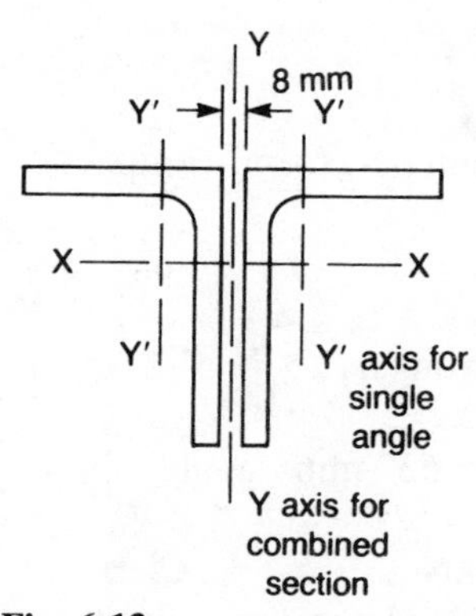

Fig. 6.13

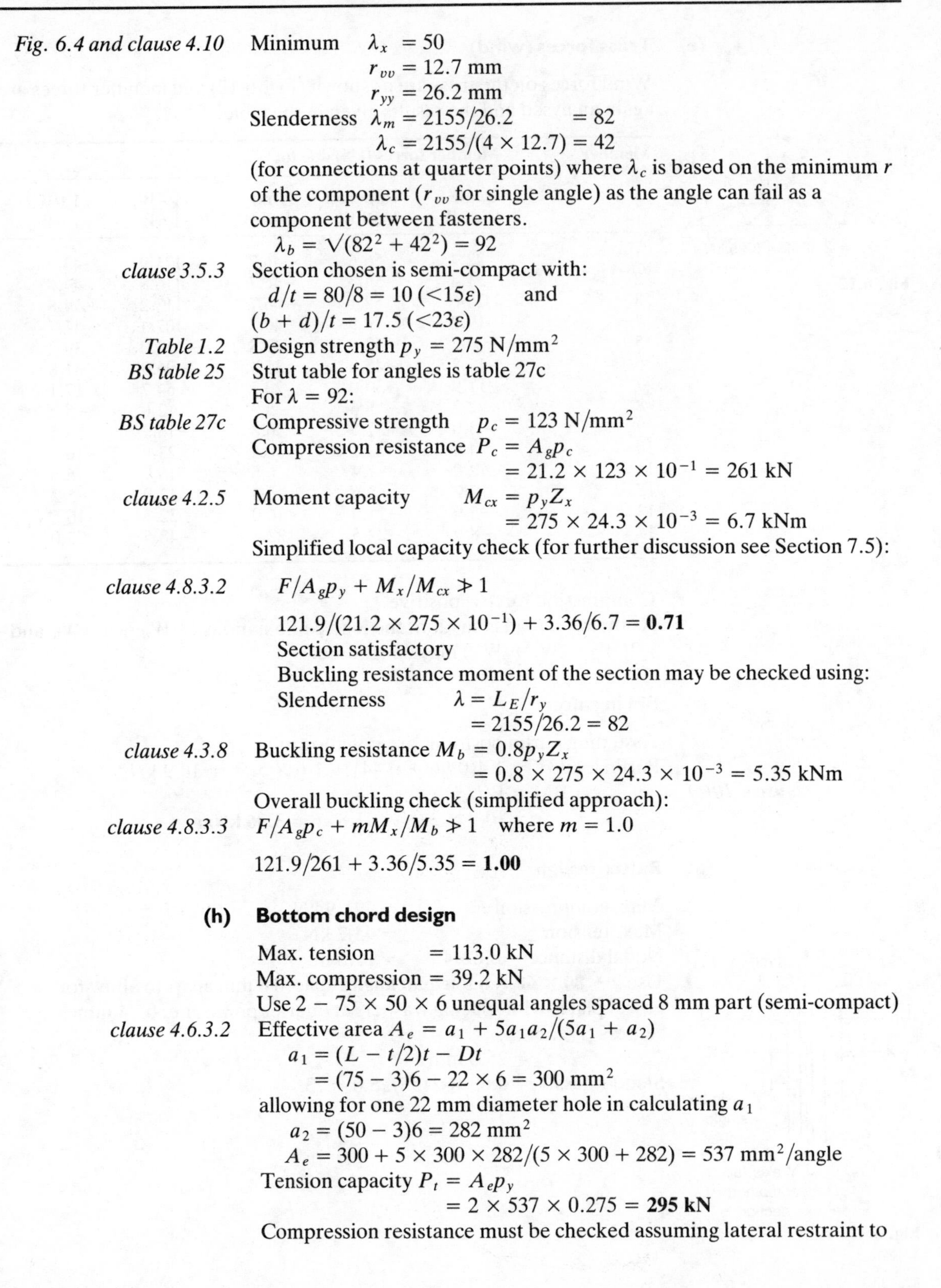

Fig. 6.4 and clause 4.10 Minimum $\lambda_x = 50$

$r_{vv} = 12.7$ mm

$r_{yy} = 26.2$ mm

Slenderness $\lambda_m = 2155/26.2 = 82$

$\lambda_c = 2155/(4 \times 12.7) = 42$

(for connections at quarter points) where λ_c is based on the minimum r of the component (r_{vv} for single angle) as the angle can fail as a component between fasteners.

$\lambda_b = \sqrt{(82^2 + 42^2)} = 92$

clause 3.5.3 Section chosen is semi-compact with:

$d/t = 80/8 = 10\ (<15\varepsilon)$ and

$(b + d)/t = 17.5\ (<23\varepsilon)$

Table 1.2 Design strength $p_y = 275$ N/mm^2

BS table 25 Strut table for angles is table 27c

For $\lambda = 92$:

BS table 27c Compressive strength $p_c = 123$ N/mm^2

Compression resistance $P_c = A_g p_c$

$= 21.2 \times 123 \times 10^{-1} = 261$ kN

clause 4.2.5 Moment capacity $M_{cx} = p_y Z_x$

$= 275 \times 24.3 \times 10^{-3} = 6.7$ kNm

Simplified local capacity check (for further discussion see Section 7.5):

clause 4.8.3.2 $F/A_g p_y + M_x/M_{cx} \not> 1$

$121.9/(21.2 \times 275 \times 10^{-1}) + 3.36/6.7 = \mathbf{0.71}$

Section satisfactory

Buckling resistance moment of the section may be checked using:

Slenderness $\lambda = L_E/r_y$

$= 2155/26.2 = 82$

clause 4.3.8 Buckling resistance $M_b = 0.8 p_y Z_x$

$= 0.8 \times 275 \times 24.3 \times 10^{-3} = 5.35$ kNm

Overall buckling check (simplified approach):

clause 4.8.3.3 $F/A_g p_c + mM_x/M_b \not> 1$ where $m = 1.0$

$121.9/261 + 3.36/5.35 = \mathbf{1.00}$

(h) Bottom chord design

Max. tension = 113.0 kN

Max. compression = 39.2 kN

Use 2 – 75 × 50 × 6 unequal angles spaced 8 mm part (semi-compact)

clause 4.6.3.2 Effective area $A_e = a_1 + 5a_1a_2/(5a_1 + a_2)$

$a_1 = (L - t/2)t - Dt$

$= (75 - 3)6 - 22 \times 6 = 300$ mm^2

allowing for one 22 mm diameter hole in calculating a_1

$a_2 = (50 - 3)6 = 282$ mm^2

$A_e = 300 + 5 \times 300 \times 282/(5 \times 300 + 282) = 537$ mm^2/angle

Tension capacity $P_t = A_e p_y$

$= 2 \times 537 \times 0.275 = \mathbf{295\ kN}$

Compression resistance must be checked assuming lateral restraint to the bottom chord at the chord bracing connections, maximum spacing 4.64 m (Fig. 6.8).

$r_{yy} = 21.4$ mm
$\lambda_m = 4640/21.4 = 217$
$\lambda_c = 580/10.8 = 54$
(for connections at quarter points, i.e. 0.58 m centres)
$\lambda_b = \sqrt{(217^2 + 54^2)} = 224$

clause 4.7.3.2b Max. slenderness = 250

BS table 27c Compression strength $p_c = 34$ N/mm^2

Compression resistance $P_c = A_g p_c$
$= 14.4 \times 34 \times 10^{-1} =$ **50.4 kN**

Section satisfactory

(i) Strut (member 10)

Max. compression F = 23.9 kN
Max. tension = 10.3 kN
Use **60 × 60 × 6 equal angle**

$r_{vv} = 11.7$ mm
$r_{yy} = 18.2$ mm
$L = 1.72$ m

Fig. 6.5 and clause 4.7.10.2a $\lambda = 0.85L/r_{vv} = 125$
or $\lambda = 0.7L/r_{yy} + 30 = 96$

BS table 27c Compressive strength $p_c = 91$ N/mm^2 for $\lambda = 125$

Compression resistance $P_c = A_g p_c$
$= 6.91 \times 91 \times 10^{-1} =$ **63 kN**

Section satisfactory

(j) Connection

Check the strength of the connection joining members 6, 7, 10 and 13. Assume a 8 mm thick gusset plate and 20 mm bolts (grade 4.6) (Fig. 6.14).
Max. force (member 13) = 32.3 kN
Force change between members 6 and 7 is 32.2 kN. If members 6 and 7 were to be joined at this connection (to change the size or reduce fabrication lengths) then max. force (member 6) would be 96.9 kN.

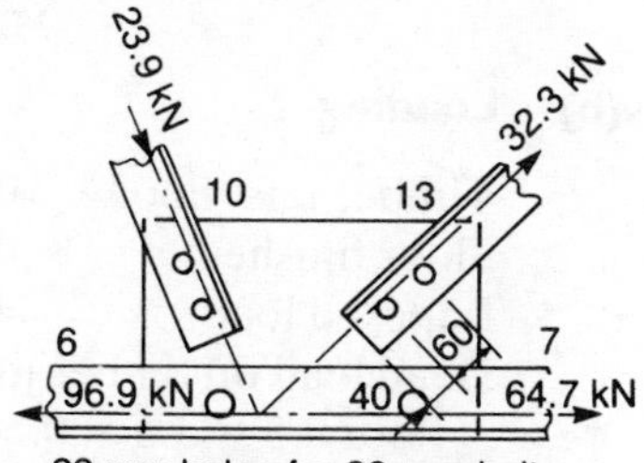

Fig. 6.14

clause 6.3.2 Shear capacity of bolts (double shear) $P_s = p_s A_s$
$= 160 \times 2 \times 0.245 = 78$ kN

clause 6.3.3.2 Bearing capacity of bolts $P_{bb} = dtp_{bb}$
$= 20 \times 6 \times 0.435 \times 2 = 104$ kN

clause 6.3.3.3 Bearing capacity of angles $P_{bs} = dtp_{bs}$
$= 20 \times 6 \times 0.450 \times 2 = 108$ kN

but $P_{bs} \not> etp_{bs}/2$
$= 40 \times 6 \times 0.450/2 = 54$ kN/bolt

Clearly a bolt size of as low as 16 mm would be possible.
Gusset plate stress can be based on an effective width (Fig. 6.7) of
$(60/\cos 30°) - 22 = 47$ mm
Plate stress $= 32.3 \times 10^3/(6 \times 47) = 115$ N/mm^2
(Maximum is $p_y = 275$ N/mm^2)

The remaining truss members may be designed in the same way, with the compression force usually being the main design criterion. It should be noted that it is good practice to limit the number of different member sizes being used, probably not more than four in total. The deflection of a roof truss is not usually critical to the design, and the effects of sag under load may be offset by pre-cambering the truss during fabrication by say 50 mm or 100 mm. If the deflection is required then the Williot-Mohr graphical method may be used[7,8], or the virtual work method[9,10], or appropriate computer programs. It should be noted that deflection due to bolt slip can be significant compared with dead load elastic deflections.

6.8 EXAMPLE 12. LATTICE GIRDER

(a) Dimensions

See Fig. 6.15: Lattice girder fabricated from tubular sections, span 8.0 m; spacing 3.5 m.

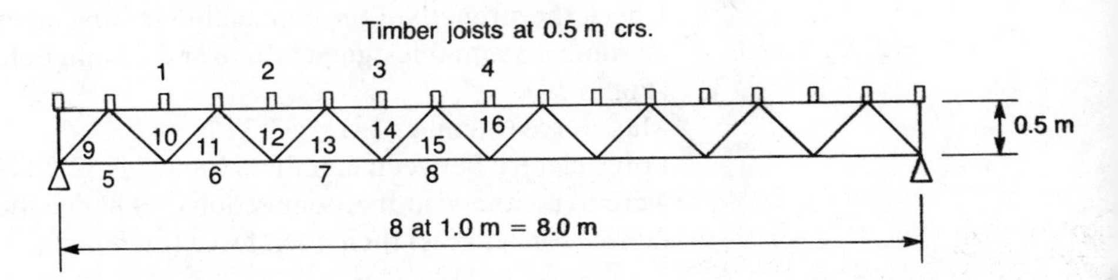

Fig. 6.15 Lattice girder

(b) Loading

Timber joist floor	0.5 kN/m^2
Floor finishes	0.2 kN/m^2
Imposed load	4.0 kN/m^2

Dead load on timber joist:

floor $0.5 \times 3.5 \times 0.5$	= 0.88 kN
finishes $0.2 \times 3.5 \times 0.5$	= 0.35 kN
weight of truss (estimated)	= 0.37 kN
total load W_d	= 1.60 kN

Imposed load on timber joist:
$W_i = 4.0 \times 3.5 \times 0.5 = 7.00$ kN
Load on girder from each joist $= 1.4W_d + 1.6W_i$
$= 1.4 \times 1.60 + 1.6 \times 7.00 = 13.4$ kN
Concentrated load at nodes $=$ **26.8 kN**

(c) Truss forces

Analysis of the truss[1,2] under the nodal loads gives member forces (kN) in the table.

Top chord		Bottom chord		Diagonals	
Member	Force	Member	Force	Member	Force
1	188	5	−107	9	152
2	322	6	−268	10	−144
3	402	7	−375	11	114
4	429	8	−429	12	−76
				13	76
				14	−38
				15	38
				16	0

Compression force is positive.

(d) BM in top chord

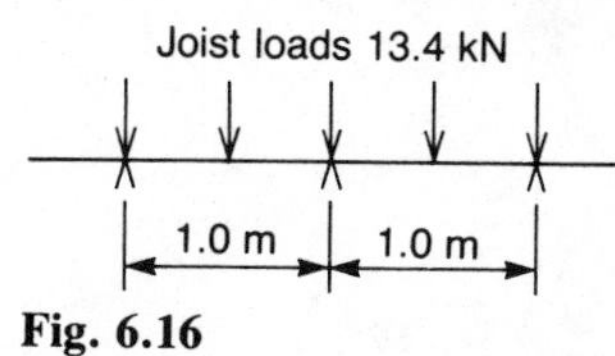

Fig. 6.16

(See Fig. 6.16).
BM in continuous top chord $= WL/8$
$= 13.4 \times 1.0/8 =$ **1.68 kNm**

(e) Top chord

Max. compression $F = 429$ kN
Max. BM $M_x = 1.68$ kNm
Use **90 × 90 × 6.3 RHS**
$r_x = 34.1$ mm

Fig. 6.4 Slenderness $\lambda = 0.85 \times 1000/34.1 = 25$
BS table 25 Strut table for RHS is table 27a
BS table 27a Compressive strength $p_c = 270$ N/mm²
Compression resistance $P_c = A_g p_c$
$= 20.9 \times 270 \times 10^{-1} = 564$ kN
Section chosen is plastic ($b/T = 14$)
Moment capacity $M_c = p_y S$
$= 275 \times 65.3 \times 10^{-3} = 17.96$ kNm
clause 4.8.3.2 Local capacity check (simplified):

$F/A_g p_y + M_x/M_{cx} \ngtr 1$

$429/(20.9 \times 275 \times 10^{-1}) + 1.68/17.96 =$ **0.84**
Section satisfactory

Lateral torsional buckling need not be considered if compression flange is positively connected to the timber joist floor, nor for box sections sections (clause B.2.6.1).

(f) Bottom chord

Max. tension $F = 429$ kN
Use 90 × 90 × 5 RHS
Tension capacity $P_t = A_e p_y$
$= 16.9 \times 275 \times 10^{-1} =$ **465 kN**
Section is satisfactory

(g) Diagonal

(See Fig. 6.17.)
Max. compression $F = 152$ kN
Max. tension $= 114$ kN
Use 50 × 50 × 3.2 RHS
$r_x = 19.1$ mm
Slenderness $\lambda = 0.7L/r_x$
$= 0.7 \times 500/19.1 \cos 45° = 26$

BS table 27a Compressive strength $p_c = 269$ N/mm^2
Compression resistance $P_c = A_g p_c$
$= 5.94 \times 269 \times 10^{-1} =$ **160 kN**
Tension resistance $P_t = 5.94 \times 275 \times 10^{-1} =$ **163 kN**
Section satisfactory

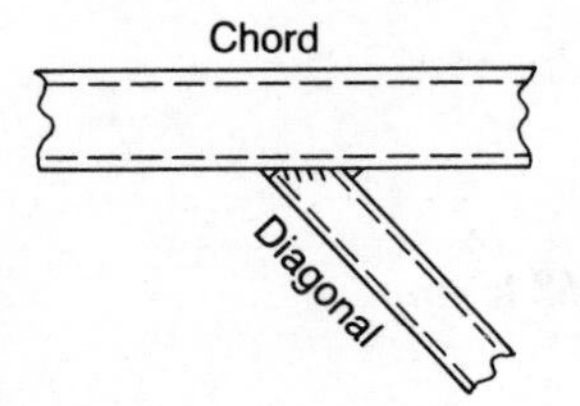

Fig. 6.17

(h) Connections

All welded joints with continuous 4 mm welds
Weld length $= 2(50 + 50/\cos 45°) = 241$ mm

clause 6.6.5.1 Design strength $p_w = 215$ N/mm^2
Weld resistance $= 0.7 \times 4 \times 241 \times 215 = 145$ kN
The maximum forces which may be transmitted between hollow steel sections are more complex and detailed references may be consulted if required(11).

As in Section 6.7 the number of section sizes would be limited in practice, with the allowable variation in member sizes depending on the degree of repetition expected.

STUDY REFERENCES

Topic	*Reference*
1. Truss analysis	**Marshall W.T. & Nelson H.M.** (1977) Analysis of statically determinate structures, *Structures,* pp. 14–24. Longman
2. Truss analysis	**Coates R.C., Coutie M.G. & Kong F.K.** (1988) Analysis of plane trusses, *Structural Analysis,* pp. 41–51. Van Nostrand Reinhold
3. Bolt design	**Pask J.W.** (1982) Fasteners, *Manual on Connections,* pp. 8–12. BCSA Ltd
4. Welding	**Pask J.W.** (1982) Welding, *Manual on Connections,* pp. 13–18. BCSA
5. Friction grip bolts	**Pask J.W.** (1982) Fasteners, *Manual on Connections,* pp. 8–12. BCSA Ltd
6. Wind pressure coefficients	BS 6399 *Loading for buildings* Part 2: *Wind loads* (to be published; presently CP3 Ch. V part 2)
7. Williot-Mohr diagrams	**Marshall W.T. & Nelson H.M.** (1977) Williot displacement diagram, *Structures,* pp. 140–47. Longman
8. Williot-Mohr diagrams	**Wang C.K.** (1983) The graphical method, *Intermediate Structural Analysis,* pp. 93–9. Pergamon
9. Virtual work	**Marshall W.T. & Nelson H.M.** (1977) Principle of virtual work, *Structures,* pp. 147–54. Longman
10. Virtual work	**Coates R.C., Coutie M.G. & Kong F.K.** (1988) Applications of the principle of virtual work, *Structural Analysis,* pp. 134–75. Van Nostrand Reinhold
11. Connection of RHS	CIDECT (1985) Welded joints, *Construction with hollow steel sections,* pp. 129–42. British Steel Corporation Tubes Division

7

SIMPLE AND COMPOUND COLUMNS

Columns, sometimes known as stanchions, are vertical steel members in both single and multi-storey frames. They are principally designed to carry axial loads in compression, but can also be subjected to moments due to eccentricities or lateral loads, or as a result of being part of a rigid frame, i.e. continuity moments. In some structures, particularly single-storey frames and top lengths in multi-storey frames, the moments may have greater effect in the design than the axial compression; and under some loading combinations axial tension may occur (see Section 2.7(d)) and Table 11.7 in Chapter 11).

Where columns are principally compression members their behaviour and design are similar to those of struts (Section 6.4). The loads carried are usually larger than those in typical truss members, and the simplifications adopted in truss design (Section 6.2) should not be applied.

7.1 TYPES OF COLUMN

Typical cross-sections used in column design are shown in Fig. 7.1. While any section may be used, the problem of instability under axial compression results in preferred sections of circular or square types. Buckling is likely to occur about the axis of lower bending resistance, and the use of sections having one axis of very low bending resistance is usually uneconomic. However, as the range of hollow structural sections is limited, and making connections is more difficult, the H and I sections are frequently used.

Columns may also be built up from smaller sections or plates, being either welded or bolted together as in Fig. 7.2. In some cases the column may take the form of a lattice girder, which will be particularly economic where high moments occur, such as tall building wind frames, masts and cranes. The laced column and the battened column shown in

Fig. 7.3 are used in cases where overturning moments due to wind or eccentricity (crane gantry loading) are high.

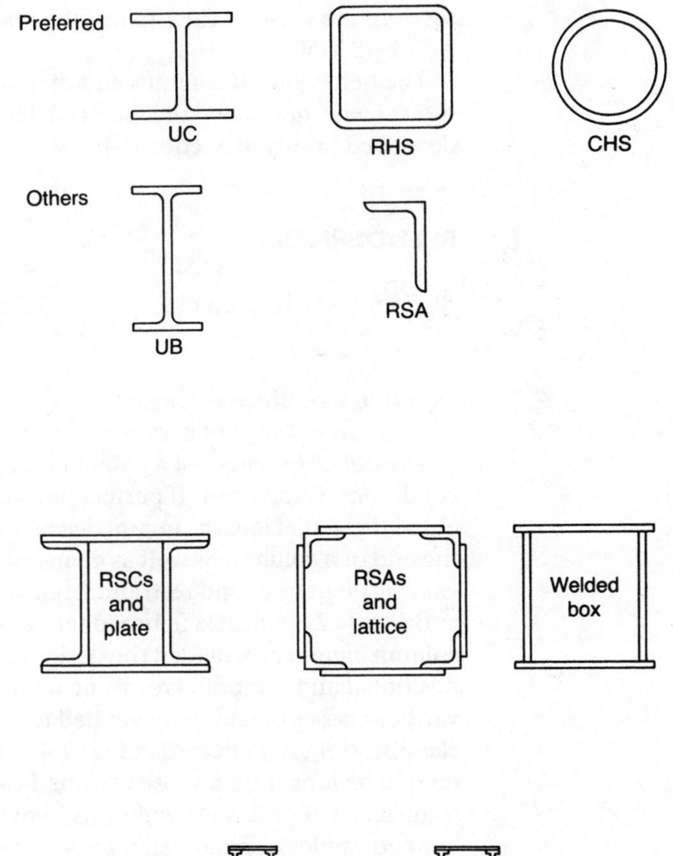

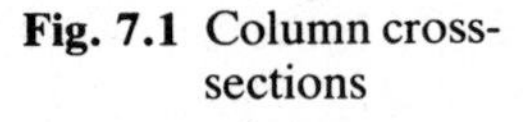

Fig. 7.1 Column cross-sections

Fig. 7.2 Compound sections

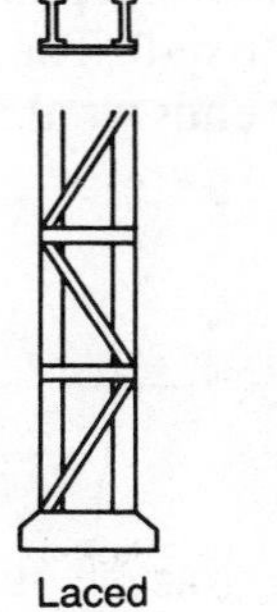

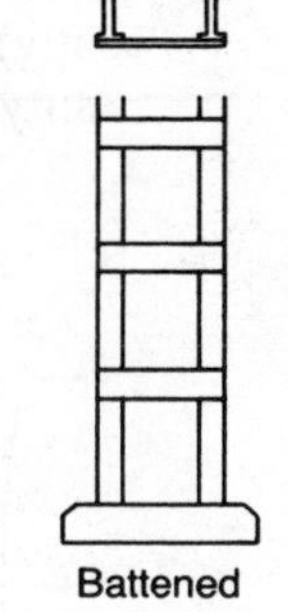

Fig. 7.3 Column types

7.2 AXIAL COMPRESSION

Columns with idealized end connections may be considered as failing in an Euler-type buckling mode[1,2]. However, practical columns usually fail by inelastic buckling and do not conform to the Euler theory assumptions[3,4], particularly with respect to elastic behaviour. Only extremely slender columns remain linearly elastic up to failure. Local buckling of thin flanges rarely occurs in practice when normal rolled sections are used, as their flange thicknesses usually conform to clause 3.5 of BS 5950.

The behaviour of columns and their ultimate strength is assessed by their slenderness λ and the material design strength p_y, which are described briefly in Section 6.4.

7.3 SLENDERNESS

Slenderness λ is given by:

$$\lambda = L_E/r$$

where L_E = effective length
r = radius of gyration

The effective length of a column is dependent on the restraint conditions at each end. If perfect pin connections existed then L_E would equal the actual length. In practice beam type members connected to the end of a column, as well as giving positional restraint, provide varying degrees of end restraint from virtual full fixity to near pinned.

BS table 24 indicates in broad terms the nominal effective length of a column member, provided the designer can define the amount of positional and rotational restraint acting on the column in question. The various classes of end fixity alluded to in BS table 24, represent a crude classification, as indicated in Fig. 7.4. The nominal effective lengths tend to be longer than those obtained using the Euler values as they take cognizance of practical conditions; that is, there is no such condition as 'pinned' (unless a pin is deliberately manufactured, which would be costly) or 'fixed'. Therefore, taking the case having both ends 'fixed', in reality the ends would have some flexibility, hence a value of 0.7, rather

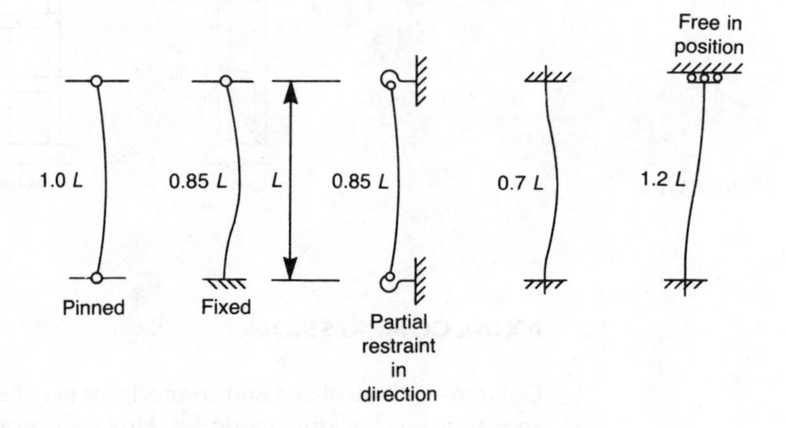

Fig. 7.4 End fixity/ effective length

than the idealized Euler value of 0.5, is used. Although these values in BS table 24 are adequate for column members in multi-storey buildings designed by the 'simple design' method, a more accurate assessment is provided by section 5.7 and appendix E of BS 5950: Part 1 for columns in 'rigid' frames (continuous construction).

The restraint provided may be different about the two column axes, and in practice in steelwork frames this will generally be the case. The effective lengths in the two planes will therefore generally be different, and so will be the slenderness (λ_x and λ_y). For axially loaded columns the compressive strength p_c is selected from BS tables 27a to 27d for both values of slenderness, and the lower strength is used in the design irrespective of axis.

The effective length of a column in a single-storey frame is difficult to assess from considerations of fixity as in Fig. 7.4 and hence BS 5950: Part 1 gives these special consideration. Clause 4.7.2c and appendix D1 should be used for these cases.

It is essential to have a realistic assessment of effective length, but it should be realized that this assessment is not precise and is open to debate, and consequently it is not reasonable to expect highly accurate strength predictions in column design. Interpolation between the values in BS table 24 might produce more accurate estimates, but such interpolation must reflect the actual restraint conditions.

7.4 BENDING AND ECCENTRICITY

In addition to axial compression, columns will usually be subjected to moments due to horizontal loading and eccentricity of connections carrying vertical loads. The types of load that can occur are summarized in Fig. 7.5, in which:

W_a is the vertical load from a roof truss, taken as applied concentrically (clause 4.7.6a(2))
W_b are horizontal loads due to wind, applied by side rails
W_c and W_{hc} are the crane gantry loads, applied through a bracket at a known eccentricity
W_d is the self weight of the column/sheeting
W_e is the resultant force carried through the truss bottom chord.

In single-storey column and truss structures force W_e occurs whenever the columns carry unequal horizontal loads or unequal moments. Values of this resultant force for different arrangements of horizontal loading are shown in Fig. 7.6.

In cases such as most beam/column connections where eccentricity is not known precisely, clause 4.7.6 states that a value of 100 mm from the column face (flange or web as appropriate) should be used.

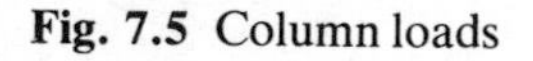

Fig. 7.5 Column loads

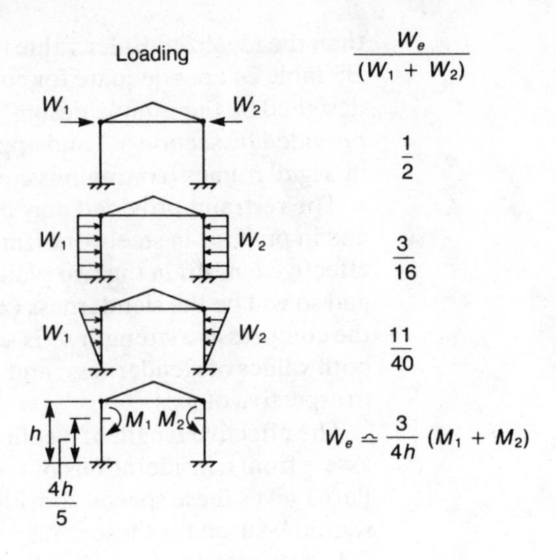

Fig. 7.6 Resultant force in roof truss bottom chord

7.5 LOCAL CAPACITY

At any section in a column the sum of effects of axial load F and moments M_x and M_y must not exceed the local capacity. This is considered to be satisfactory if the combination relationship satisfies the following interaction equation:

(a) For **plastic** and **compact** sections:
UB, UC and joist sections

$$(M_x/M_{rx})^2 + M_y/M_{ry} \not> 1$$

RHS, CHS and solid sections

$$(M_x/M_{rx})^{5/3} + (M_y/M_{ry})^{5/3} \not> 1$$

Channel, angle and all other sections

$$M_x/M_{rx} + M_y/M_{ry} \not> 1$$

where M_{rx} and M_{ry} are reduced moments in the presence of axial load

(b) For **semi-compact** and **slender** sections, and as a simplified method for compact sections in (a) above:

$$F/A_g p_y + M_x/M_{cx} + M_y/M_{cy} \not> 1$$

where A_g is the gross sectional area
M_{cx} and M_{cy} are the moment capacities defined in Section 3.4

7.6 OVERALL BUCKLING

The failure of columns, whether carrying moments combined with an axial load or not, will commonly involve member buckling. The assessment of overall buckling resistance involves the same factors and procedures as given in Section 3.2, but with an additional term for axial load. Using a simplified approach (clause 4.8.3.3.1) the overall buckling check is considered to be satisfactory if the combination relationship is satisfied:

$$F/A_g p_c + m_x M_x/M_b + m_y M_y/p_y Z_y \ngtr 1$$

where p_c is the compressive strength (Section 6.4)
m_x, m_y are the equivalent uniform moment factors (Section 3.2)
M_b is the buckling resistance moment (Section 3.5)
Z_y is the minor axis elastic section modulus (compression face)

A more exact approach (clause 4.8.3.3.2) may be carried out using M_{ax} and M_{ay}, the maximum buckling moments about major and minor axes in the presence of axial load. In this approach

$$m_x M_x/M_{ax} + m_y M_y/M_{ay} \ngtr 1$$

7.7 EXAMPLE 13. COLUMN FOR INDUSTRIAL BUILDING

(a) Dimensions

(See Fig. 7.7.)

Overall height	12.5 m
Height of crane rail	10.0 m
Side rail spacing	2.5 m

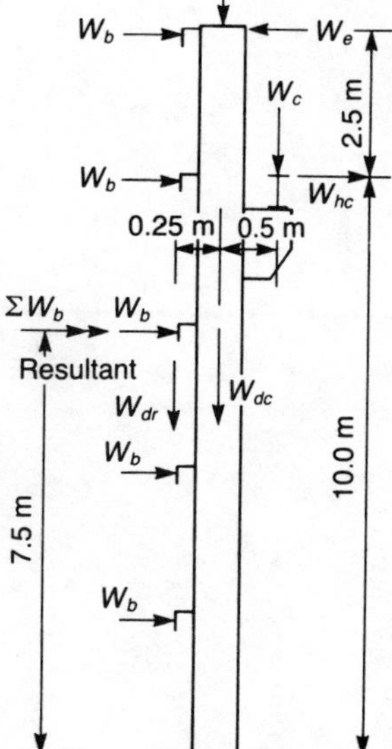

Fig. 7.7

(b) Loading

Roof truss reactions:

Dead load	50 kN
Imposed load	78 kN
Wind load	90 kN (suction)

Crane girder reactions (including dynamic effects where appropriate):

Self weight of girder	20 kN
Crane load (nearside)	220 kN
Crane load (far side)	100 kN
Horizontal surge	6 kN

Wind on side of building:

Side rail load (assumed constant over column)	11 kN
Cladding dead load	16 kN
Column self weight	10 kN

(c) Combined force and BM

Combinations of loads must be considered (including γ_f factors) to give rise to:

(i) maximum BM + maximum force
(ii) maximum BM + minimum force
(iii) maximum force + any BM (if different from (i) above)

clause 2.4.1.1 Possible combinations of loads, dead W_d, imposed W_i, wind W_w and crane W_c are:

$1.4W_d + 1.6W_i$
$1.4W_d + 1.4W_w$
$1.4W_d + 1.6W_c$
$1.2W_d + 1.2W_i + 1.2W_w$
$1.2W_d + 1.6W_i + 1.6W_c$
$1.2W_d + 1.2W_w + 1.2W_c$
$1.2W_d + 1.2W_i + 1.2W_w + 1.2W_c$

The value of $1.6W_c$ may be reduced to $1.4W_c$ where vertical and horizontal crane loads are both included.

Max $W_a = 1.4 \times 50 + 1.6 \times 78 = 194.8$ kN
Max. $W_b = 1.4 \times 11 = 15.4$ kN
Min. $W_b = 0$
Max. $W_c = 1.6 \times 220 + 1.4 \times 20 = 380.0$ kN

But W_c acting with W_{hc} (combination (c.i) below)
$= 1.4 \times 220 + 1.2 \times 20 = 332.0$ kN

Max. W_{hc} acting with $W_c = 1.4 \times 6 = 8.4$ kN
Max. $W_{dc} = 1.4 \times 10 = 14$ kN
Max. $W_{dr} = 1.4 \times 16 = 22.4$ kN

The resultant force W_e depends on the difference in wind loading on the column being designed and that on the similar column on the far side of the building (Fig. 7.6). Assuming the wind load on the far column is zero:

Max. $W_e = 3/16 \times 5 \times 15.4 = 14.4$ kN

There is also a similar small resultant force due to the asymmetric crane bracket loading but this is neglected in this design.

(c.i) Max. BM + Max. force

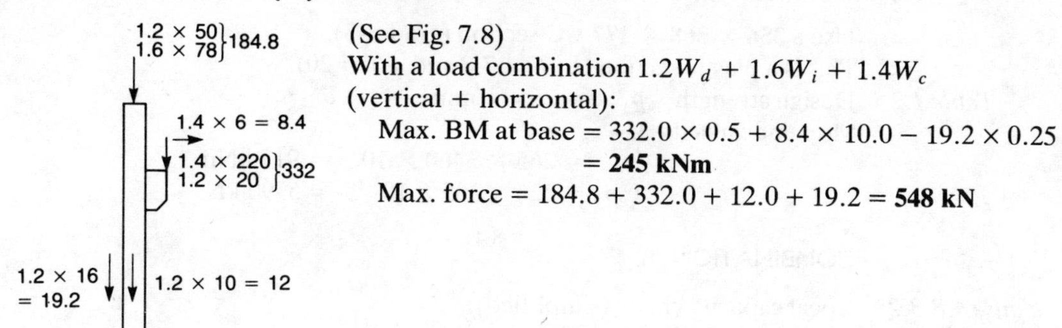

(See Fig. 7.8)
With a load combination $1.2W_d + 1.6W_i + 1.4W_c$ (vertical + horizontal):

Max. BM at base = $332.0 \times 0.5 + 8.4 \times 10.0 - 19.2 \times 0.25$
= **245 kNm**

Max. force = $184.8 + 332.0 + 12.0 + 19.2 =$ **548 kN**

Fig. 7.8

(c.ii) Max. BM + associated force

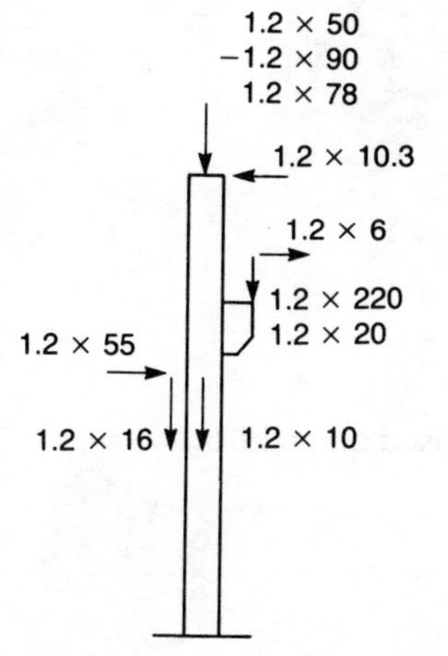

(See Fig. 7.9)
With a load combination $1.2W_d + 1.2W_i + 1.2W_c + 1.2W_w$

Max. BM = $288.0 \times 0.5 + 7.2 \times 10.0 - 19.2 \times 0.25$
$+ 66.0 \times 7.5 - 12.4 \times 12.5$
= **551 kNm**

Associated force = $45.6 + 288.0 + 12.0 + 19.2 =$ **365 kN**

Fig. 7.9

(c.iii) Max. force

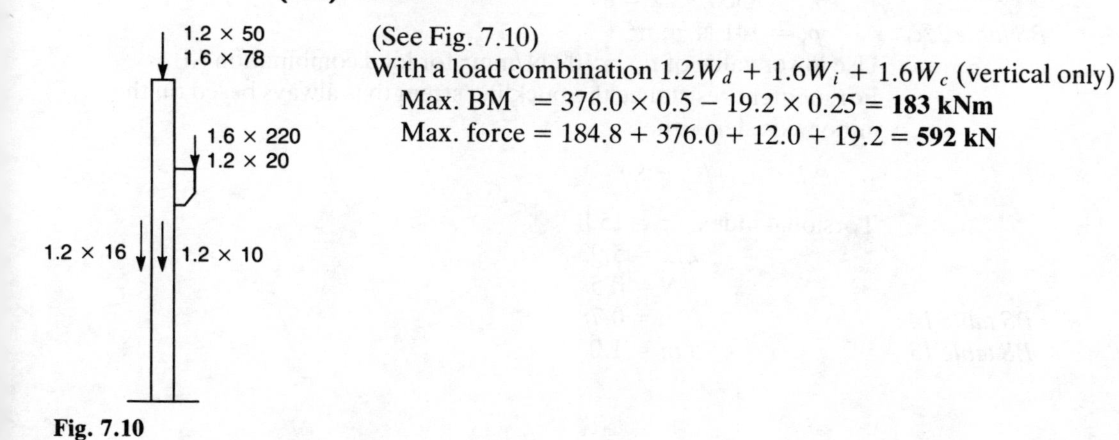

(See Fig. 7.10)
With a load combination $1.2W_d + 1.6W_i + 1.6W_c$ (vertical only)

Max. BM = $376.0 \times 0.5 - 19.2 \times 0.25 =$ **183 kNm**

Max. force = $184.8 + 376.0 + 12.0 + 19.2 =$ **592 kN**

Fig. 7.10

(d) Local capacity

Use a **356 × 368 × 177 UC** section (grade 43).
This is a plastic section ($b/T = 7.8$ and $d/t = 20$).

Table 1.2 Design strength $p_y = 265$ N/mm^2
Moment capacity $M_{cx} = p_y S_x$
$= 265 \times 3460 \times 10^{-3} = 917$ kNm
$A_g p_y = 226 \times 265 \times 10^{-1} = 5990$ kN

COMBINATION (i)

clause 4.8.3.2 Local capacity check (simplified):

$$F/A_g p_y + M_x/M_{cx} \not> 1$$

$548/5990 + 245/917 =$ **0.36**

COMBINATION (ii)

Local capacity check:

$365/5990 + 551/917 =$ **0.66**

COMBINATION (iii)

Local capacity check:

$592/5990 + 183/917 =$ **0.30**

(e) Overall buckling

For compressive strength the slenderness λ is based on an effective length:

BS appendix D (Fig. 19) $L_{Ex} = 1.5 \times 12.5 = 18.75$ m
or $L_{Ey} = 0.85 \times 10.0 = 8.5$ m
$\lambda_x = L_{Ex}/r_x$
$= 18750/159 = 118$

BS table 27b $p_c = 109$ N/mm^2
$\lambda_y = 8500/95.2 = 89$

BS table 27c $p_c = 141$ N/mm^2

Use lower value of $p_c = 109$ N/mm^2 for load combination (ii)
For moment resistance the buckling strength is always based on the minor axis slenderness:

$$\lambda = L_{Ey}/r_y = 89$$

Torsional index $x = 15.0$
$\lambda/x = 5.93$
$N = 0.5$

BS table 14 $v = 0.78$

BS table 13 $m = 1.0$

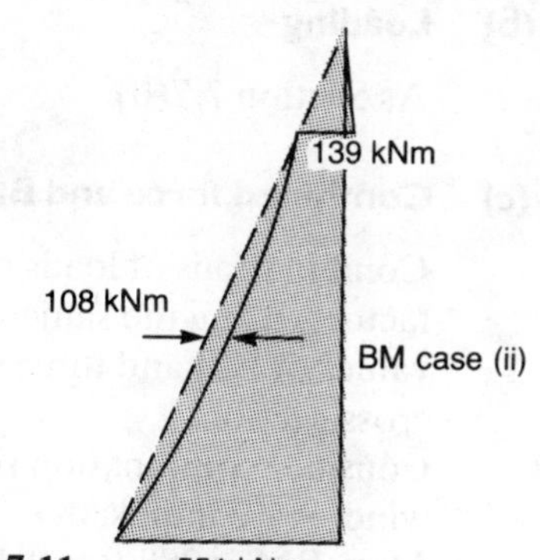

Fig. 7.11

BS tables 16, 17 as the column is loaded along its length by side rails

$\beta = 139/551 = 0.2$ (Fig. 7.11)

$\gamma = M/M_o \ = 551/108 = -5$ approx.

$n = 0.70$

clause 4.3.7.5 $u = 0.844$

$\lambda_{LT} = nuv\lambda$

$= 0.70 \times 0.844 \times 0.78 \times 89 = 41$

BS table 11 Buckling strength $p_b = 252\ \text{N/mm}^2$

Buckling resistance moment $M_b = p_b S_x$

$= 252 \times 3460 \times 10^{-3} = 872$ kNm

Simplified overall buckling check:

$F/A_g p_c + m_x M_x/M_b \ngtr 1$

$365/(226 \times 109 \times 10^{-1}) + 1.0 \times 551/872 = \mathbf{0.78}$

For load combination (i) $v = 0.78$ as before

BS table 13 As the column is not loaded along its length for this combination:

$n = 1.0$

BS table 16 $m = 0.57$ (for $\beta = 0$)

$u = 0.844$

$\lambda_{LT} = 1.0 \times 0.844 \times 0.78 \times 89 = 59$

BS table 11 Buckling strength $p_b = 209\ \text{N/mm}^2$

Buckling resistance $M_b = 209 \times 3460 \times 10^{-3} = 723$ kNm

Simplified overall buckling check:

$548/(226 \times 109 \times 10^{-1}) + 0.57 \times 245/723 = \mathbf{0.42}$

Section is satisfactory

7.8 EXAMPLE 14. LACED COLUMN FOR INDUSTRIAL BUILDING

(a) Dimensions

Principal dimensions are as Section 7.7(a) but the cross-section is revised as shown (Fig. 7.12).

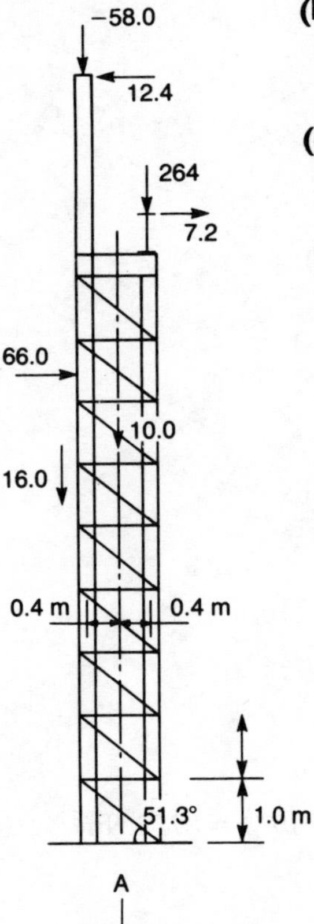

Fig. 7.12

(b) Loading

As Section 7.7(b)

(c) Combined force and BM

Combinations of loads may be considered as before (including γ_f factors) using the same values of loads W_a to W_e inclusive. Maximum values of BM and force will vary slightly due to the revised cross-section.
Consider combination (ii) only with 1.2 dead + 1.2 crane + 1.2 wind + 1.2 imposed:

$$\text{Max. BM} = 288.0 \times 0.4 + 7.2 \times 10.0 + 66.0 \times 7.5 - 12.4 \times 12.5 - 19.2 \times 0.65 - 45.6 \times 0.4 = \mathbf{497\ kNm}$$

$$\text{Associated force} = 45.6 + 288.0 + 12.0 + 19.2 = \mathbf{365\ kN}$$

(d) Local capacity

clause 4.7.8 The design of a laced column may be carried out assuming it to be a single integral member, and then checking the design of the lacings.
Use 2 − **305 × 127 × 37 UB** sections.
These are semi-compact section ($b/T = 14.2$)
Design strength $p_y = 275\ \text{N/mm}^2$
Second moment of area of combined section in cm units:

$$I_a = 2(337 + 47.5 \times 40^2) = 152\,670\ \text{cm}^4$$
$$Z_a = 152\,670/55 = 2780\ \text{cm}^3$$

clause 4.2.5 Moment capacity $M_{cx} = p_y Z_a = 275 \times 2780 \times 10^{-3} = 763\ \text{kNm}$
Local capacity check:

$$F/A_g p_y + M_x/M_{cx} \not> 1$$

$$365/(2 \times 47.5 \times 275 \times 10^{-1}) + 497/763 = \mathbf{0.79}$$

(e) Overall buckling

For compressive strength the overall slenderness λ is based on an effective length:

BS appendix D (Fig. 20) $L_{Ea} = 1.5 \times 10.0 = 15.0\ \text{m}$

or $L_{Eb} = 0.85 \times 10.0 = 8.5$ m

$r_a = \sqrt{(I_a/A)}$

$= \sqrt{(152\,670/[2 \times 47.5])} = 40.1$ cm

$\lambda_a = L_{Ea}/r_a$

$= 1500/40.1 = 37$

$\lambda_b = L_{Eb}/r_b$

$= 850/12.3 = 69$

Local slenderness $\lambda_c = L/r_y$

$= 1000/26.7 = 38$

clause 4.7.8g Maximum value $\lambda_c = 50$

Overall slenderness $\lambda_b \nless 1.4\lambda_c$

$= 1.4 \times 38 = 53$

BS table 27a Compressive strength p_c based on the highest value of λ, i.e. 69

$p_c = 225$ N/mm^2

Bending about the A–A axis (Fig. 7.12) can be assumed to produce axial forces in the UB sections.

Axial force = moment/centroidal distance between UBs

= 497/0.8 = 621 kN (tension and compression)

Max. compression in one UB = 621 + 365/2 = 804 kN

Compression resistance $P_c = A_g p_c$

$= 47.5 \times 225 \times 10^{-1} =$ **1069 kN**

Section is satisfactory

(f) Lacings

Max. force in lowest diagonal

(66.0 + 7.2 − 12.4)/cos 51.3° = 97.4 kN (compression)

clause 4.7.8i In addition the lacings should carry a transverse force equal to 2.5% of the axial column force, i.e:

$0.025 \times 365 = 9.1$ kN

Note that the value of 1% in clause 4.7.8i is at present considered to be too small.

Lacing force = 9.1/cos 51.3° = 13.3 kN

Total force = 97.4 + 13.3 = 111 kN

i.e. 55.5 kN on each side of column

clause 4.7.8h Effective length of lacing = $\sqrt{(0.80^2 + 1.00^2)} = 1.28$ m

Use **60 × 60 × 6 equal angle**

$\lambda = 1280/11.7 = 109$

BS table 27c Compressive strength $p_c = 111$ N/mm^2

Compression resistance $P_c = A_g p_c$

$= 6.91 \times 111 \times 10^{-1} =$ **76.7 kN**

Section is satisfactory

STUDY REFERENCES

Topic	*Reference*
1. Euler load	**Marshall W.T. & Nelson H.M.** (1977) Compression members and stability problems, *Structures*, pp. 329–64. Longman
2. Euler load	**Coates R.C., Coutie M.G. & Kong F.K.** (1988) Instability of struts and frameworks, *Structural Analysis*, pp. 304–361. Van Nostrand Reinhold
3. Column behaviour	*Steelwork Design*, vol 3, Commentary on BS 5950: Part 1. Steel Construction Institute (to be published)
4. Column behaviour	**Kirby P.A. & Nethercot D.A.** (1979) In-plane instability of columns, *Design for Structural Stability*, Collins

8

COLUMN BASES & BRACKETS

The transfer of force between one element of a structure and the next requires particular care by the designer. A good detail may result from long experience of the use of structural steelwork, and many examples are available for students to copy or adapt[1,2] However, equally good details may be developed with less experience providing the following basic principles are adhered to:

- The forces to be carried must be defined and transferred between different elements of the connection, i.e. a realistic load path must be assumed within the connection.
- Simplicity of detail usually produces the most effective and robust engineering solution, provided strength and stiffness requirements are satisfied.
- Any detail must be practicable and cost-effective, both from the point of view of the steelwork fabricator and from that of the site erector.

Column bases and brackets are connections which carry forces and reactions to a column element. Bases transfer reactions from the foundation to the column, while brackets may be used to transfer loads from crane girders or similar members. In addition columns may receive loads via beam connections (see Section 3.7(h)) or cap plates (see Section 6.6).

8.1 COLUMN BASES

Two main forms of column base are used, and these are shown in Fig. 8.1. Welded or bolted construction can be used, or a combination of both, the choice being dependent on whether or not the base is attached to the column during fabrication, or later during site erection. In general the simpler slab base is used in small and medium construction when axial load dominates. The gusseted base is used in heavy construction with larger column loads and where a certain amount of fixity is required.

Construction requirements and details are given in reference (3).

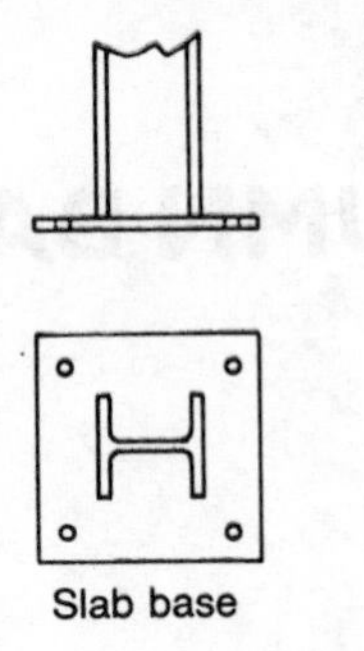

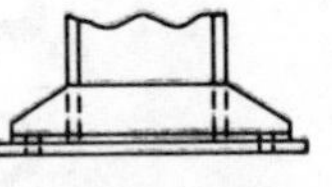

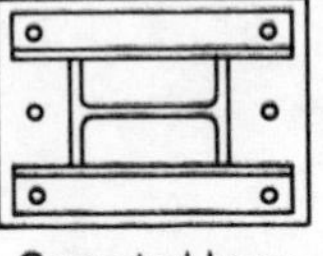

Fig. 8.1 Column bases

8.2 DESIGN OF COLUMN BASES

Column bases transmit forces and moments from the column to its foundation. The forces will be axial loads, shear forces and moments about either axis, or any combination of them. Shear force is in reality probably transmitted by friction between the base plate and the foundation concrete, but it is common in design for this shear force to be resisted totally by the holding down bolts.

The common design case deals with axial load and moment about one axis. If the ratio of moment/axial load is less than base length/6, then a positive bearing pressure exists over the whole base and may be calculated from equilibrium alone. Nominal holding bolts are provided in this case, and at least two are in fact provided to locate the base plate accurately. If the ratio exceeds base length/6, holding down bolts are required to provide a tensile force. Both arrangements are shown in Fig. 8.2.

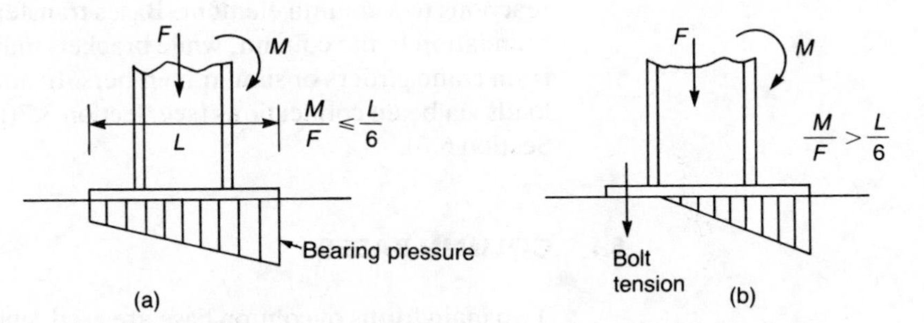

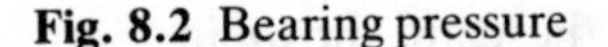
Fig. 8.2 Bearing pressure

Where tension does occur in the holding down bolts a number of methods of design are possible:

1. It is assumed that the bearing pressure has a linear distribution to a maximum value of $0.4f_{cu}$, where f_{cu} is the concrete cube strength.

This basis is suggested in clause 4.13.1, and analysis of the bearing pressure and bolt stresses may follow reinforced concrete theory (Fig. 8.3).

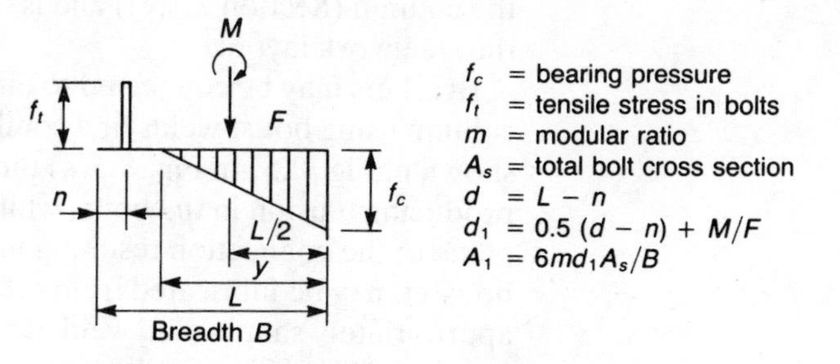

y is solution of $y^3 - 3(d - d_1)\,y^2 + A_1 y - A_1 d = 0$

$$f_c = 6d_1F/[By(3d - y)]$$
$$f_t = mf_c\,(d/y - 1)$$

Fig. 8.3 Reinforced concrete theory

2. An alternative approximate analysis is sometimes used which assumes that the permissible stresses for steel tension and concrete bearing are reached together. This method is shown in Fig. 8.4.

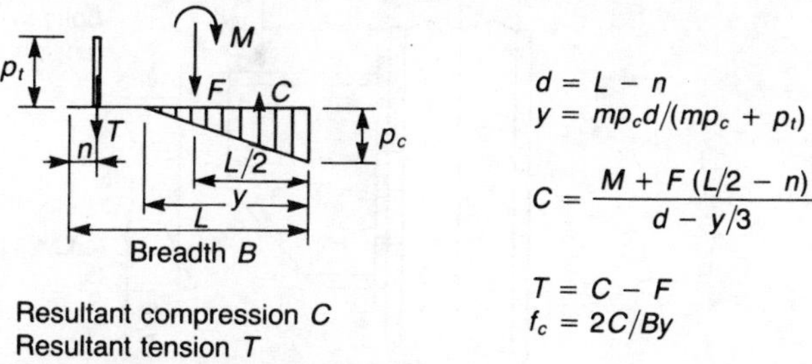

$$d = L - n$$
$$y = mp_cd/(mp_c + p_t)$$
$$C = \frac{M + F\,(L/2 - n)}{d - y/3}$$
$$T = C - F$$
$$f_c = 2C/By$$

Resultant compression C
Resultant tension T
Permissible compressive stress p_c
Permissible tensile stress p_t

Fig. 8.4 Approximate analysis

3. A rectangular pressure distribution may also be used, which leads to a slightly different plate thickness and bolt sizes. The analysis is based on reinforced concrete theory for ultimate limit state.

In general the first method is used to obtain bearing pressures and bolts stresses. Thickness of base plate is obtained using the steel strength p_{yp} from Table 1.2 of Chapter 1, but not more than 270 N/mm^2.

Maximum moment in plate ≯ $1.2p_{yp}Z$ (clause 4.13.2.3) where Z is the elastic modulus of the plate section.

For the case of concentric forces only, base plate thickness may be obtained from clause 4.13.2.2.

8.3 BRACKETS

Brackets are used as an alternative to cleated connections (Section 3.7(h)) only where the latter are unsuitable. A typical case of this situation is the crane girder support on a column (Section 5.3(k)). The eccentric connection by the bracket is necessary for the chosen structural arrangement. It does, however, generate large moments in the column (Section 7.7(c)) and is therefore only used where essential to the steelwork layout.

Brackets may be connected to either the web or the flange of the column using bolts, welds or a combination of the two. Examples are shown in Fig. 8.5. In Fig. 8.5(a) the moment acts out-of-plane producing tension in the bolts, while in Fig. 8.5(b) the moment is in the plane of the connection resulting in a shear effect in the bolts. The bracket may be fabricated from offcuts of rolled sections, or from plates appropriately shaped and welded together. Connection to the column may be made during fabrication, or alternatively the brackets may be attached during site erection.

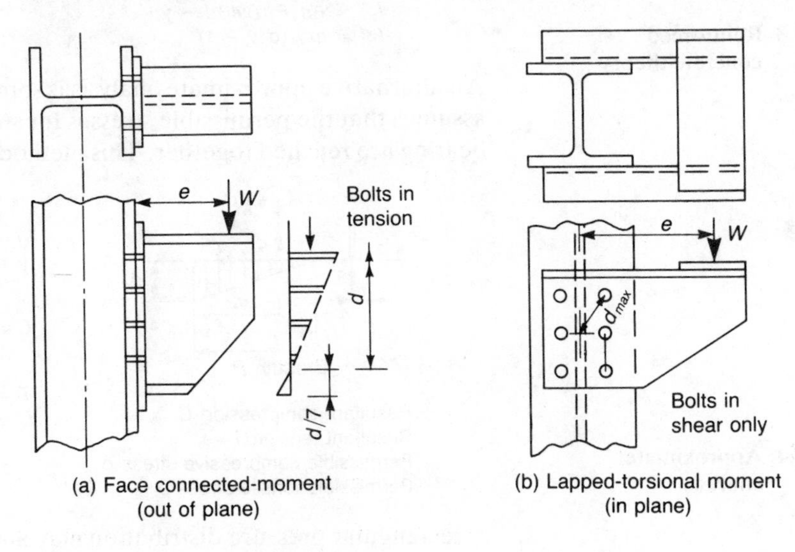

Fig. 8.5 Brackets

8.4 DESIGN OF BRACKETS

Brackets are subjected both to a vertical shear load and to a moment due to the eccentricity of the vertical load. Note that the moment will vary with the point in the bracket under consideration. A bracket may also be subjected to horizontal loads, but these are usually of a secondary nature, or may be covered by a special detail (Fig. 5.11).

Vertical loads are supported by welds or by bolts acting in shear. Ideally the moments also should be carried by bolts in shear (or by welds), but some bracket arrangements shown in Fig. 8.5a will give rise to bolt tension.

Vertical load W is divided between the bolts or weld group uniformly so that:

Bolt shear = W/N kN

or Weld shear = W/L_w kN/mm

where N is the number of bolts
L_w is the total weld length (mm)

For a weld design strength of 215 N/mm² (BS table 36), weld capacities (kN/mm) may be calculated for each weld size. Note that the weld size is in fact the leg length, and the design dimension is the throat distance, i.e. throat size = leg length/√2. Values of weld capacity are given in reference (4).

Where the moment produces bolt shear only (Fig. 8.5) then the shear on each bolt is given approximately by:

$$\text{Bolt shear} = Wed_{max}/\Sigma d^2$$

where d is the bolt distance from the bolt group centroid

For a weld group the approximation is:

$$\text{Weld shear} = Wed_{max}/I'_p$$

where $I'_p = I'_x + I'_y$ for the weld group (Fig. 8.11)

In some cases the moment produces bolt tension and in these cases bolt force is considered proportional to distance from a neutral axis. The neutral axis may be taken as $d/7$ in depth[(2)], and as a result:

$$\text{Bolt tension} = Wed_{max}/\Sigma d^2$$

where d is the bolt distance from the neutral axis

For a weld group:

$$\text{Weld shear} = Wed_{max}/I'_x$$

where I'_x is the equivalent second moment of area of the welds about the weld group centroid

The effects of vertical load and moment due to eccentricity must be added for either individual bolts, or for points in a weld run. Clearly points of maximum force or stress need to be checked, which occur at positions furthest from the group centroid or neutral axis.

Where the moment produces shear in a bolt, vectorial addition may be used. In cases where the moment produces tension a combined check may be used (clause 6.3.6.3):

$$F_s/P_s + F_t/P_t \ngtr 1.4 \text{ for ordinary bolts}$$

(note that neither part of the interaction equation may exceed 1.0)

where F_s is the applied shear
P_s is shear capacity
F_t is applied tension
P_t is tension capacity

For a weld group all combinations of vertical load and moment produce shear in the weld, and vectorial addition is used as necessary.

8.5 EXAMPLE 15. DESIGN OF SLAB BASE

(a) Dimensions

305 × 305 × 137 UC column

(b) Loading

All loads include appropriate values of γ_f

Case (i) Max. vertical load 1400 kN

Case (ii) Largest moment under max. load conditions: 60 kNm and 850 kN

Case (iii) Largest moment under min. load conditions: 85 kNm and 450 kN

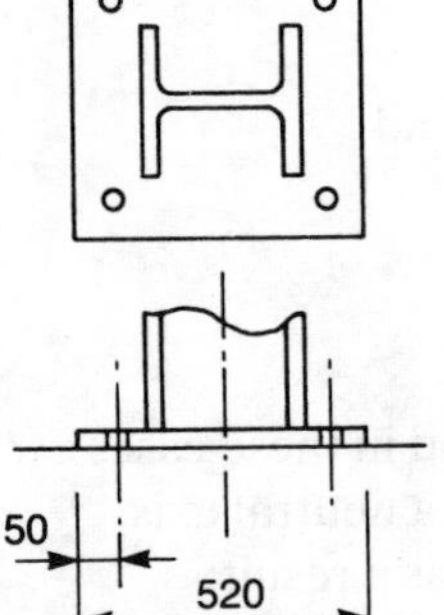
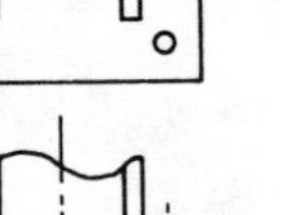
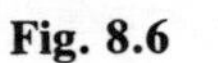

Fig. 8.6

(c) Bearing pressure

(See Fig. 8.6.)

Assume base **520 × 520 × plate**

Assume 4 bolts (grade 4.6) 20 mm diameter

Tension bolt area $A_s = 2 \times 245 = 490\ \text{mm}^2$

$d = 520 - 50 = 470$ mm

CASE (i) LOADING

Pressure $= 1400 \times 10^3/520 \times 520 = \mathbf{5.18\ N/mm^2}$

Assuming concrete cube strength $f_{cu} = 30\ \text{N/mm}^2$:

clause 4.13.1 permissible pressure $= 0.4 \times 30 = 12\ \text{N/mm}^2$

CASE (ii) LOADING

$M/F = 60/850 = 71$ mm

$L/6 = 520/6 = 87$ mm

$M/F < L/6$

Base area $A = 520 \times 520\ \text{mm}^2 = 2700\ \text{cm}^2$

Base modulus $Z = 520 \times 520^2/6\ \text{mm}^3 = 23\,400\ \text{cm}^3$

Pressure $= F/A + M/Z = \mathbf{6.11\ N/mm^2}$

CASE (iii) LOADING

$M/F = 85/450 = 189$ mm

Fig. 8.3 $d_1 = 0.5(470 - 50) + 85 \times 10^3/450 = 399$ mm

$A_1 = 6 \times 399 \times 15 \times 490/520 = 33.8 \times 10^3\ \text{mm}^2$

y is solution of:

$y^3 - 3(d - d_1)y^2 + A_1y - A_1d = 0$

$y^3 - 3(470 - 399)y^2 + 33.8 \times 10^3y - 33.8 \times 10^3 \times 470 = 0$

Hence $y = 288$ mm

Pressure $f_c = 6d_1F/By(3d - y)$

$= 6 \times 399 \times 450 \times 10^3/520 \times 288(3 \times 470 - 288)$

$= \mathbf{6.41\ N/mm^2}$

Bearing pressure satisfactory ($< 12\ \text{N/mm}^2$)

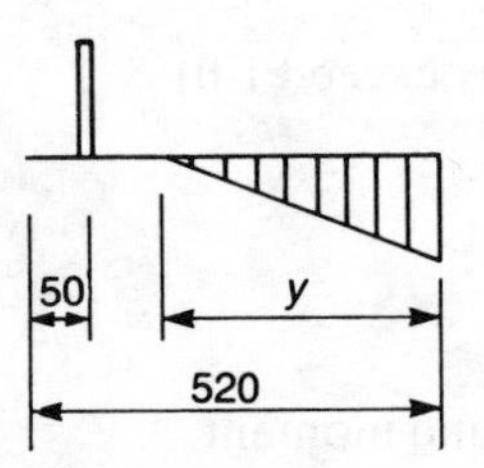

Fig. 8.7

(d) Bolt capacity

Bolt stress $f_t = mf_c(d/y - 1)$
$= 15 \times 6.41(470/288 - 1) = 61\ \text{N/mm}^2$

Force/bolt $= 61 \times 245 \times 10^{-3} = 14.9\ \text{kN}$

clause 6.3.6.1 Bolt capacity $P_t = p_t A_t$
$= 195 \times 245 \times 10^{-3} =$ **47.8 kN**

Bolts satisfactory

(e) Plate thickness

(See Fig. 8.8.)

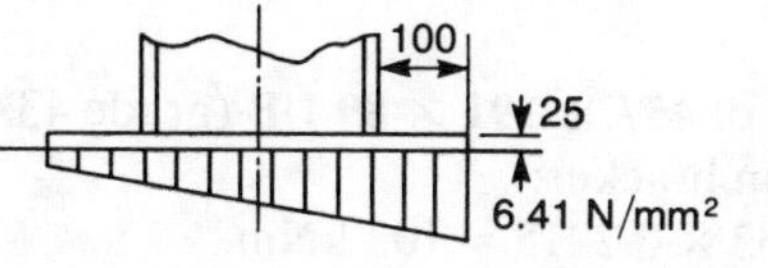

Fig. 8.8

Max. bearing pressure from case (iii) loading $= 6.41\ \text{N/mm}^2$
Max. BM (assuming constant pressure)
$= 6.41 \times 520 \times 100^2/2 = 16.7\ \text{kNm}$

Some reduction of BM may be found by using the trapezium pressure distribution:

Try 25 mm thick plate:

Table 1.2 $p_{yp} = 265\ \text{N/mm}^2$

Plate modulus $Z = 520 \times 25^2/6 = 54.2 \times 10^3\ \text{mm}^3$

clause 4.13.2.3 Moment capacity $= 1.2 p_{yp} Z$
$= 1.2 \times 265 \times 54.2 \times 10^{-3} =$ **17.2 kNm**

Plate satisfactory

For larger loads and/or moments a gusseted base may be required, particularly if the thickness of a slab base would otherwise exceed 50 mm. The design is the same as given above, but in Section 8.5(e) the plate modulus Z is based on the combined effect of plate and gussets. At thicknesses greater than 25 mm steel grades other than 43A may be needed to avoid the possibility of brittle fracture (BS table 4).

(f) Column/base plate weld

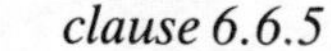

Fig. 8.9

The weld is commonly designed to carry the maximum moment, ignoring the effect of vertical load. All compression is taken in direct bearing (Fig. 8.9).

Max. tension in flange $= M/(D - T)$
$= 85 \times 10^3/300 = 283\ \text{kN}$

For one flange weld length $= 2 \times 308 = 616\ \text{mm}$

Weld shear $= 283/616 = 0.459\ \text{kN/mm}$

clause 6.6.5 Use **6 mm weld**, capacity[4] $= 0.903\ \text{kN/mm}$

Weld satisfactory

8.6 EXAMPLE 16. DESIGN OF CRANE GIRDER BRACKET (FACE)

(a) Dimensions

(See Fig. 8.10.)
Column 610 × 229 × 140 UB
Crane girder eccentricity 550 mm

(b) Loading

Max. rail reaction 462 kN
(including appropriate values of γ_f)
Crane surge load carried by diaphragm restraint

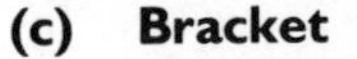

(c) Bracket

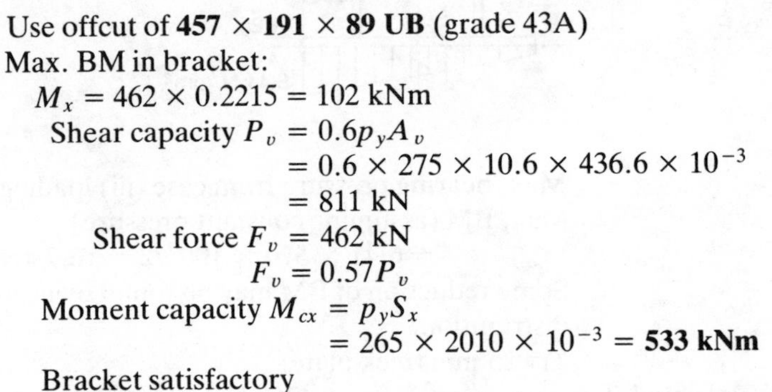

Use offcut of **457 × 191 × 89 UB** (grade 43A)
Max. BM in bracket:

$M_x = 462 \times 0.2215 = 102$ kNm
Shear capacity $P_v = 0.6 p_y A_v$
$= 0.6 \times 275 \times 10.6 \times 436.6 \times 10^{-3}$
$= 811$ kN
Shear force $F_v = 462$ kN
$F_v = 0.57 P_v$
Moment capacity $M_{cx} = p_y S_x$
$= 265 \times 2010 \times 10^{-3} =$ **533 kNm**

Fig. 8.10

Bracket satisfactory

(d) End plate weld

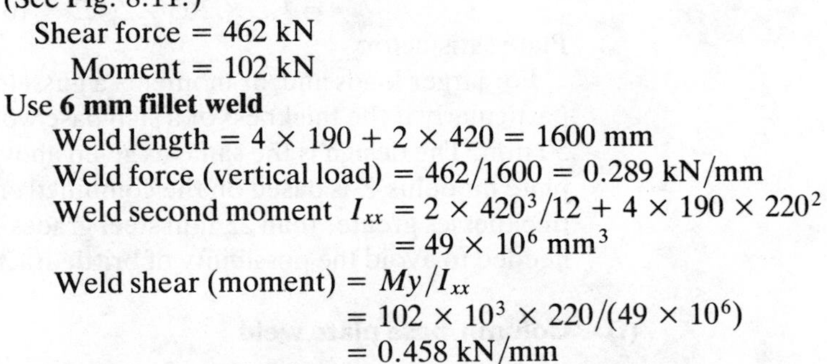

(See Fig. 8.11.)
Shear force = 462 kN
Moment = 102 kN

clause 6.6.5.3 Use **6 mm fillet weld**

Weld length = 4 × 190 + 2 × 420 = 1600 mm
Weld force (vertical load) = 462/1600 = 0.289 kN/mm
Weld second moment $I_{xx} = 2 \times 420^3/12 + 4 \times 190 \times 220^2$
$= 49 \times 10^6$ mm³
Weld shear (moment) $= My/I_{xx}$
$= 102 \times 10^3 \times 220/(49 \times 10^6)$
$= 0.458$ kN/mm

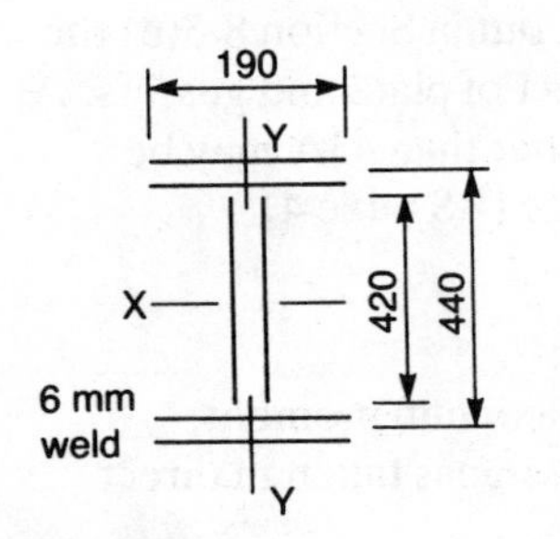

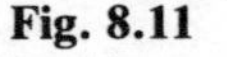
Fig. 8.11

Note that in this case the vertical shear and the shear due to moment act perpendicular to each other, and the resultant shear is obtained by vectorial addition.

Resultant $= \sqrt{(0.289^2 + 0.458^2)} =$ **0.542 kN/mm**
Weld capacity = 0.903 kN/mm
Weld satisfactory

(e) Connection bolts

Shear force = 462 kN
Out-of-plane moment = 462 × 0.2415 = 112 kNm
Use **8 no. 22 mm bolts** (grade 8.8)

Shear/bolt $F_s = 462/8 = 57.8$ kN

clause 6.3.2 Shear capacity $P_s = p_s A_s$
$= 375 \times 303 = 114$ kN

clause 6.3.3 Bearing capacity of plate $P_{bs} = dtp_{bs}$
$= 22 \times 20 \times 450 = 198$ kN

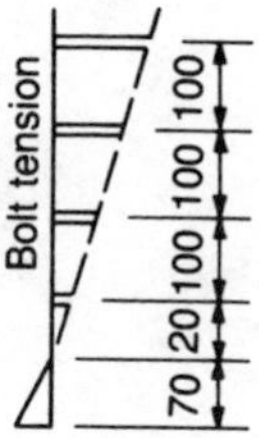

Tensile force (Fig. 8.12):
$$F_t = Md_{max}/\Sigma d^2$$
$$= 112 \times 10^3 \times 320/[(20^2 + 120^2 + 220^2 + 320^2)2] = 108 \text{ kN}$$

Fig. 8.12

clause 6.3.6 Tension capacity $P_t = p_t A_t$
$= 0.450 \times 303 = 136$ kN

Combined check:
$$F_s/P_s + F_t/P_t \ngtr 1.4$$
$$57.8/114 + 108/136 = 0.51 + 0.79 = \mathbf{1.30}$$
Bolts satisfactory

8.7 EXAMPLE 17. DESIGN OF CRANE GIRDER BRACKET (LAPPED)

(a) Dimensions

(See Fig. 8.13.)
Column 305 × 305 × 158 UC
Crane girder eccentricity 550 mm

(b) Loading

As Section 8.6(b)
Max. vertical reaction 462 kN

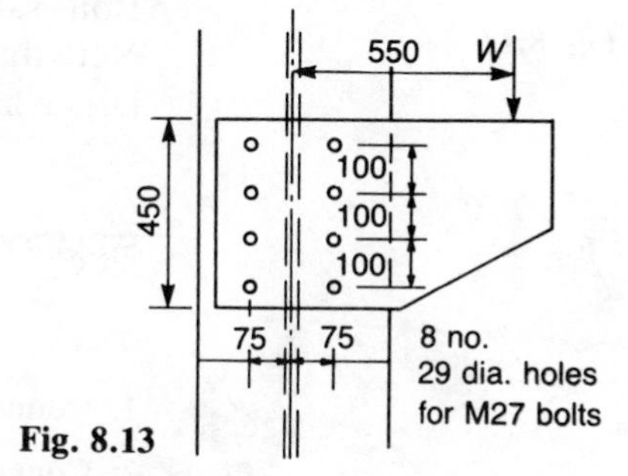

Fig. 8.13

(c) Bracket

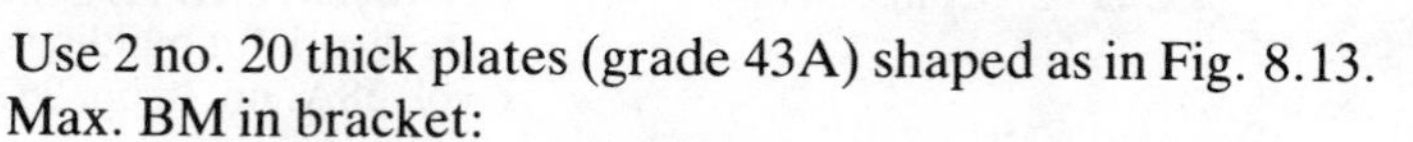
Use 2 no. 20 thick plates (grade 43A) shaped as in Fig. 8.13.
Max. BM in bracket:
$$M_x = 462 \times 0.550 = 254 \text{ kNm}$$

clause 4.2.3(c) Shear area $A_v = 0.9(450 - 4 \times 29)20 \times 2 = 12\,020$ mm²

Shear capacity $P_v = 0.6 p_y A_v$
$= 0.6 \times 265 \times 12\,020 \times 10^{-3} = 1910$ kN

Hence $F_v/P_v = 0.24$

Second moment of area of plate (cm units)
$= 2 \times 20 \times 450^3/12 = 30\,380$ cm⁴

Minus bolt holes:
$4 \times 20 \times 29 \times 50^2 = 580$ cm⁴
$4 \times 20 \times 29 \times 150^2 = 5\,220$ cm⁴

Net $I = 24\,580 \text{ cm}^4$
Modulus $Z = 24\,580/22.5 = 1090 \text{ cm}^3$

For brackets of this type it may be assumed that the bolts or welds provide lateral restraint to the compression zones. The moment capacity should be taken as:

$M_{cx} = p_y Z_x = 265 \times 1090 \times 10^{-3} = \mathbf{289\ kNm}$

Bracket satisfactory

(d) Column bolts

(See Fig. 8.14.)

Shear force = 462 kN
Moment = 462 × 0.550 = 254 kNm

Use **8 no. 27 mm bolts** (grade 8.8) on each face

Shear/bolt due to vertical load = 462/8 × 2 = 28.9 kN
Shear/bolt due to moment $= Md_{max}/\Sigma d^2$
$= 254 \times 10^3 \times 168/[8(90^2 + 168^2)]$
$= 147$ kN
Vector sum of shear = **162 kN/bolt**
Shear capacity $P_s = p_s A_s$
$= 375 \times 459 =$ **172 kN**
Bearing capacity of plate $P_{bs} = dtp_{bs}$
$= 27 \times 20 \times 450 = 243$ kN

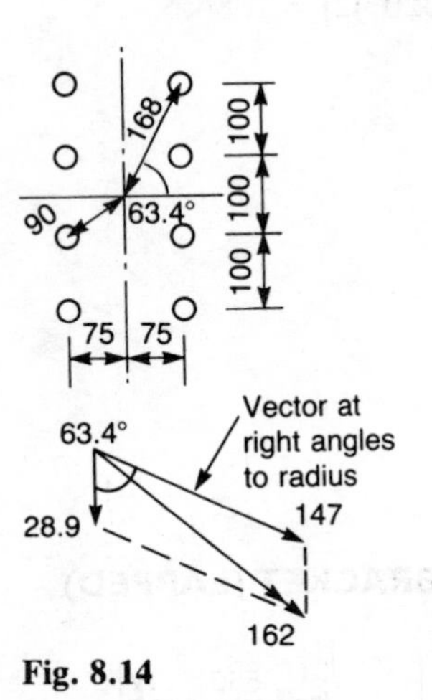

Fig. 8.14

Bolts satisfactory

Note that the lapped bracket requires twice the number of bolts of a larger size compared with the face bracket.

STUDY REFERENCES

Topic	*Reference*
1. Connections	**Pask J.W.** (1982) *Manual on Connections.* BCSA
2. Connections	**Needham F.H.** (1980) Connections in structural steelwork for buildings, *The Structural Engineer,* vol. 58A, no. 9, pp. 267–77
3. Column bases	(1980) *Holding Down Systems for Steel Stanchions.* Concrete Society/BCSA/Steel Construction Institute
4. Weld capacity	(1985) Strength of fillet welds, *Steelwork Design,* vol. 1 Section properties, member capacities, p. 205. Steel Construction Institute

9

COMPOSITE BEAMS & SLABS

The term composite can be used of any structural medium in which two or more materials interact to provide the required strength and stiffness. In steelwork construction the term refers to cross-sections which combine steel sections with concrete in such a way that the two act together as one unit. Typical cross-sections of beams and slabs are shown in Fig. 9.1.

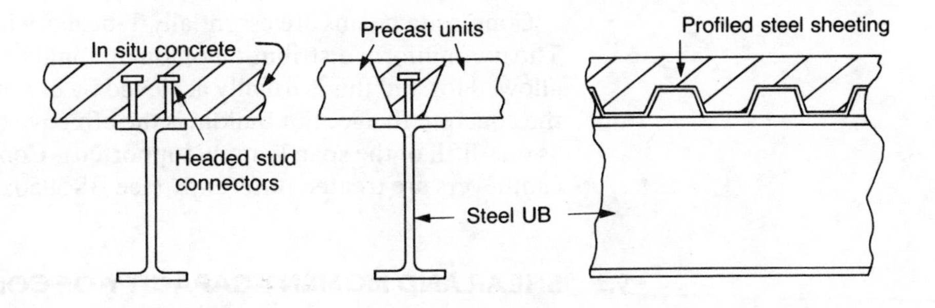

Fig. 9.1 Composite sections

The performance of composite beams is similar to that of reinforced concrete beams[1], but there are two main differences. Firstly, the steel section has a significant depth and its second moment of area may not be ignored, unlike that of the steel bar reinforcement. Secondly, the concrete to reinforcement bond, which is essential for reinforced concrete action, is absent in composite beams generally and must be provided by shear connection. Design methods for composite beams therefore follow those methods for reinforced concrete with modifications as indicated. Due to the presence of the concrete slab, problems of steel compression flange instability and local buckling of the steel member are not usually relevant in simply supported members except during erection.

Recommendations for design in composite construction are not included in Part 1 of BS 5950 but are included in:

Part 3: *Design in composite construction* (to be published)
Part 4: *Design of floors with profiled steel sheeting* (1982)

The basis of design used in this chapter is given in Section 9.7.

9.1 COMPOSITE BEAMS

The advantages of composite beams compared with normal steelwork beams are the increased moment capacity and stiffness, or alternatively the reduced steel sizes for the same moment capacity. Apart from a saving in material, the reduced construction depth can be worth while in multi-storey frames. The main disadvantage of composite construction is the need to provide shear connectors to ensure interaction of the parts.

As in all beam design, shear capacity and moment capacity of a composite section must be shown to be adequate. But in addition the strength of the shear connection must be shown to be satisfactory, both with regard to connector failure and also local shear failure of the surrounding concrete (see Section 9.4). For full interaction of the steel and concrete, sufficient shear connection must be provided to ensure that the ultimate moment capacity of the section can be reached. Lower levels of connection will result in partial interaction which is not covered in this chapter[(2)].

Composite beams are essentially T-beams with wide concrete flanges. The non-uniform distribution of longitudinal bending stress must be allowed for and this is usually achieved by use of an effective breadth for the concrete flange. For buildings the effective breadth b_e may be taken as one-fifth of the span (simply supported). Continuous beams and cantilevers are treated differently (see BS 5950: Part 3).

9.2 SHEAR AND MOMENT CAPACITY OF COMPOSITE BEAMS

The shear capacity of a composite beam is based on the resistance of the web of the steel section alone. Calculation of the shear capacity P_v is given in Section 3.7(d):

$$P_v = 0.6 p_y A_v$$

Moment capacity is based on assumed ultimate stress conditions shown in Figs 9.2 and 9.3. When the neutral axis lies in the concrete slab (Fig. 9.2) the value of x_p may be found by equilibrium of the tension and compression forces. The moment capacity M_c is given by:

$$M_c = A_b p_y (d_c + D/2 - x_p/2)$$

When the neutral axis lies in the steel section (Fig. 9.3) the value of A_{bc} may be found by equilibrium. The centroid of the compression steel A_{bc} must be located, and moment capacity M_c is given by:

$$M_c = A_b p_y (D/2 + d_c/2) - 2A_{bc} p_y (d_{sc} - d_c/2)$$

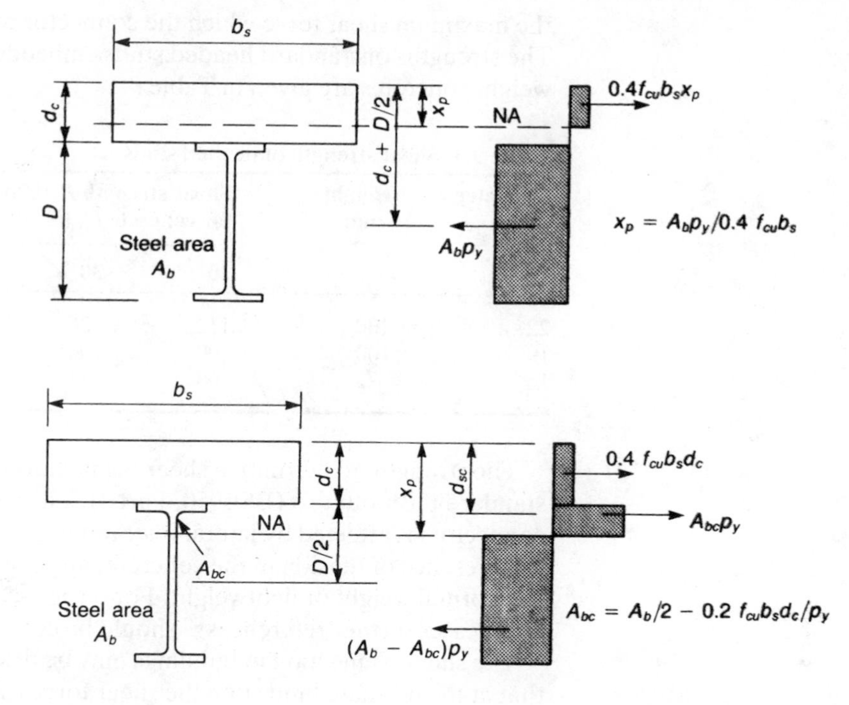

Fig. 9.2 Moment capacity (NA in slab)

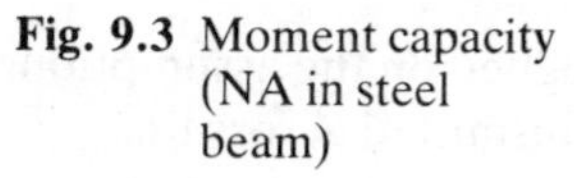

Fig. 9.3 Moment capacity (NA in steel beam)

9.3 SHEAR CONNECTORS

Many forms of shear connector have been used, of which two are shown in Fig. 9.4, but the preferred type is the headed stud. This combines ease of fixing with economy. Shear connectors must perform the primary function of transferring shear at the steel/concrete interface (equivalent to bond), and hence control slip between the two parts. In addition they have the secondary function of carrying tension between the parts and hence controlling separation.

The relationship between shear force and slip for a given connector is important in design where partial interaction is expected. For the design in this section, where full interaction is assumed, a knowledge of only

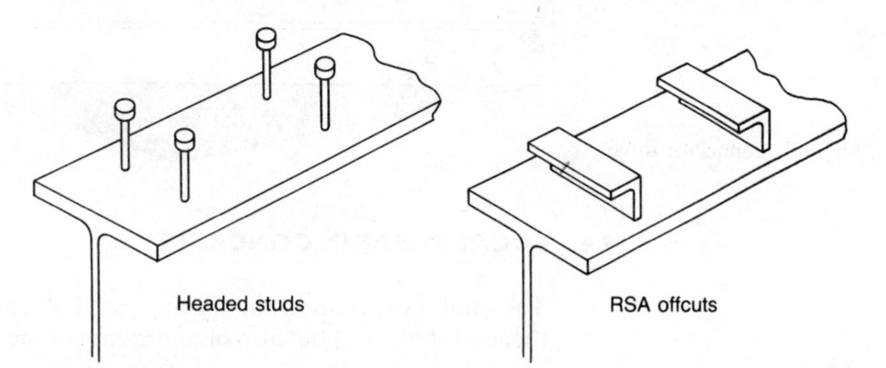

Fig. 9.4 Shear connectors

the maximum shear force which the connector can sustain is required. The strengths of standard headed studs embedded in different normal weight concretes are given in Table 9.1.

Table 9.1 Shear strength of headed studs

Diameter (mm)	Height (mm)	Shear strength P_k (kN) for concrete f_{cu} (N/mm²)			
		20	30	40	50
22	100	112	126	139	153
19	100	90	100	109	119
16	75	66	74	82	90

The strength of alternative shear connectors can be found by use of a standard push-out test (BS 5950: Part 3). The performance of all shear connectors is affected by lateral restraint of the surrounding concrete, the presence of tension in the concrete, and the type of concrete used, i.e. normal weight or lightweight. For design of composite beams in these cases further references[2] should be consulted.

The shear connection in buildings may be designed on the assumption that at the ultimate limit state the shear force transmitted across the interface is distributed evenly between the connectors. The shear force is based on the moment capacity of the section and connector force P_c as shown in Fig. 9.5.

$$P_c = F_c/N_{sc}$$

where $F_c = 0.4f_{cu}b_s x_p$ (when NA in concrete)
or $F_c = 0.4f_{cu}b_s d_c$ (when NA in steel)

The connector force P_c must be checked:

$$P_c \ngtr 0.75P_k$$

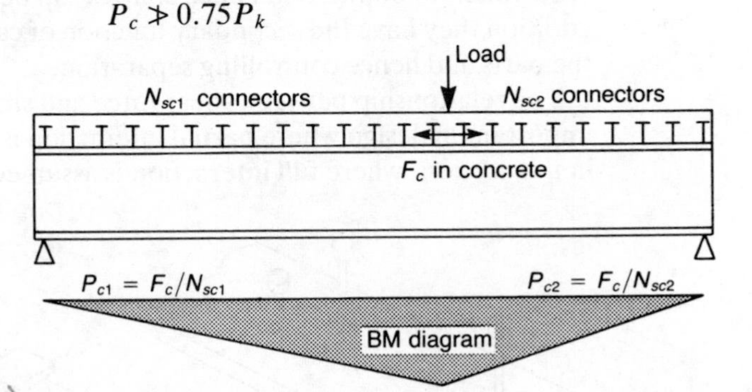

Fig. 9.5 Connector force

9.4 LOCAL SHEAR IN CONCRETE

The total shear connection depends not only on the shear connector (headed stud, etc.) but also on the ability of the surrounding concrete to transmit the shear stresses. Longitudinal shear failure is possible on the planes shown in Fig. 9.6. Transverse reinforcement should be provided with strength greater than the applied shear per unit length q, such that:

$$q \ngtr 0.15 L_s f_{cu}$$

and $$q \ngtr 0.9 L_s + 0.7 A_e p_{ry}$$

where A_e is either $(A_{rt} + A_{rb})$ or $2A_{rb}$, depending on the shear path
p_{ry} is design strength of the reinforcement
f_{cu} is the concrete cube strength
L_s is either (connector width + twice stud height) or (twice slab depth)

Note that the constant 0.9 has units N/mm^2

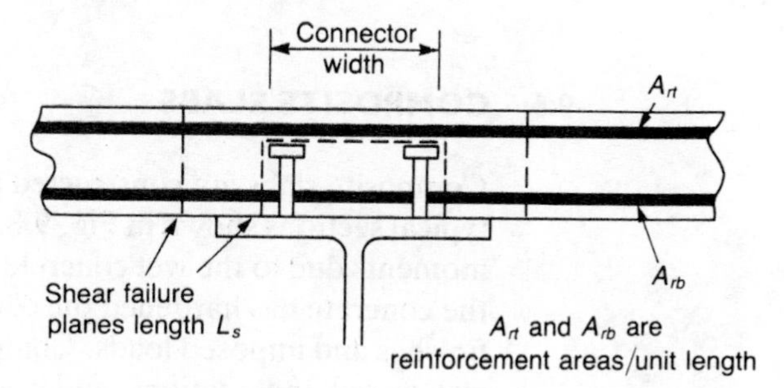

Fig. 9.6 Shear in concrete

9.5 DEFLECTIONS

As in steel beam design, deflection must be calculated at the serviceability limit state, i.e. with unfactored loads. The presence of concrete in the section means that the two different elastic moduli (steel and concrete) must be included, which is usually achieved by use of the transformed (or equivalent) cross-section [3,4]. The elastic modulus for concrete is usually modified to allow for creep. Under sustained loading the elastic modulus is about half that under short-term loading. The modular ratio m $(= E_s/E_c)$ is therefore:

f_{cu} (N/mm^2)	**Modular ratio (m)** Short-term	Sustained
20	8.2	16.4
30	7.3	14.6
40	6.6	13.2
50	6.0	12.0

The values of neutral axis depth x_e and equivalent second moment of area I_{bc} are shown in Fig. 9.7. This allows deflections to be calculated using normal elastic formulae with a value for E_s of 205 kN/mm^2.

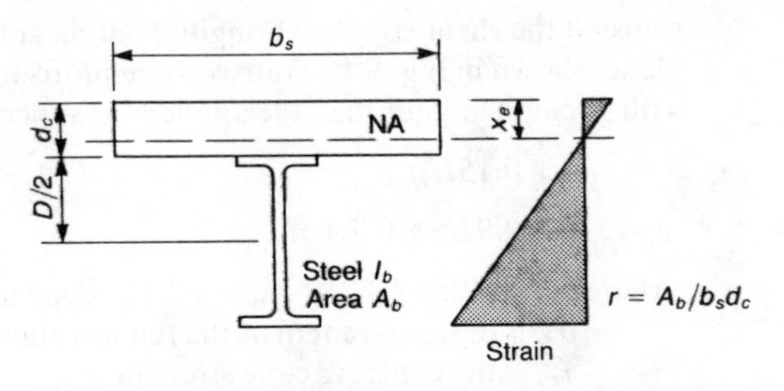

$$x_e = [d_c/2 + mr(D/2 + d_c)]/(1 + mr)$$
$$I_{bc} = A_b(D + d_c)^2/4(1 + mr) + b_s d_c^3/12m + I_b$$

Fig. 9.7 Transformed section

9.6 COMPOSITE SLABS

Composite slabs are constructed from profiled steel sheeting with two typical sections shown in Fig. 9.8. The sheeting alone must resist the moments due to the wet concrete and other construction loads. When the concrete has hardened the composite section resists moments due to finishes and imposed loads. Composite action is achieved by bond as well as web indentations, and in some cases by end anchorage where the connectors for composite beams are welded through the sheeting.

Fig. 9.8 Profiled sheeting

In most cases design is controlled by the construction condition rather than the performance as a composite section. In general the failure of the slab as a composite section takes place due to incomplete interaction, i.e. slip on the steel/concrete interface. For these reasons design of composite slabs with profiled sheeting has evolved from testing. Details of the test information are available from manufacturers and CIRIA[(4)]. The effects of the sheeting profile on connector performance and on beam behaviour are also given in the CIRIA publication[(4)].

9.7 EXAMPLE 18. COMPOSITE BEAM IN BUILDING

The design follows that given in Section 3.7 for a non-composite beam. The notation follows that of the CIRIA publication[(4)], to which reference should be made until Part 3 of BS 5950 is published.

(a) Dimensions

(See Fig. 3.2.)
Span 7.4 m simply supported
Beams at 6.0 m centres
Concrete slab 250 mm thick (f_{cu} = 30 N/mm^2) spanning in two directions
Finishing screed 40 mm thick

(b) Loading

As Section 3.7(b) allowing the same self weight of beam

Dead load W_d = 180 kN
Imposed load W_i = 132 kN

(c) BM and SF

Ultimate moment M_x = 573 kNm
Ultimate shear force F_x = 237 kN

(d) Shear capacity

Assume the beam to be **406 × 140 × 46 UB**

Shear capacity $P_v = 0.6p_yA_v$
$= 0.6 \times 0.275 \times 402.3 \times 6.9 = 458$ kN
Shear force $F_x/P_v = 0.52$

(e) Moment capacity

Use effective breadth b_s as $L/5$, i.e. 1.48 m
For neutral axis in the concrete slab, see Fig. 9.2.

$x_p = A_bp_y/0.4b_sf_{cu}$
$= 5900 \times 275/(0.4 \times 1480 \times 30) = 91$ mm

In slab 250 mm thick, see Fig. 9.9.

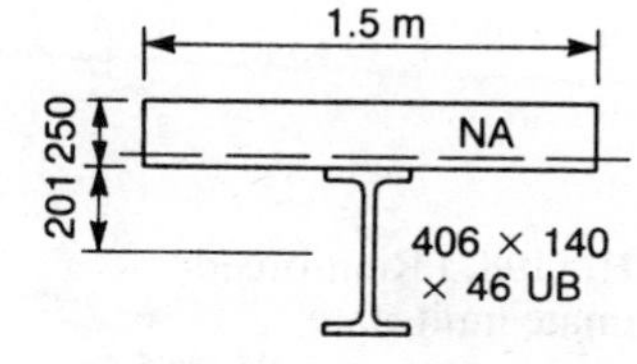

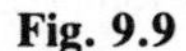

Fig. 9.9

Moment capacity $M_{pc} = A_bp_y(d_c + D/2 - x_p/2)$
$= 5900 \times 275(250 + 402.3/2 - 91/2)10^{-6}$
$=$ **658 kNm**
$M_x/M_{pc} = 0.87$

Section satisfactory

(f) Shear connectors

Force in concrete at mid-span:

$F_c = 0.4f_{cu}b_sx_p$
$= 0.4 \times 30 \times 1480 \times 91 \times 10^{-3} = 1620$ kN

Use 19 mm diameter by 100 mm high headed stud connectors

Table 9.1 P_k = 100 kN
$N_{sc} = 1620/100 \times 0.75 =$ **22 studs**

These are distributed evenly in each half span
Spacing = 3700/22 = **170 mm**

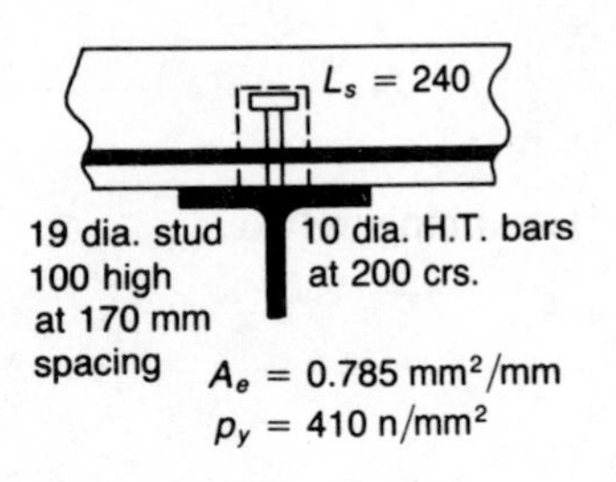

Fig. 9.10

(See Fig. 9.6 and 9.10)

Length of shear path $L_s = 40 + 2 \times 100 = 240$ mm

Shear per unit length $q = F_c/(L/2)$

$= 1620/3700 = 438$ N/mm

Longitudinal shear capacity $\not> 0.15 L_s f_{cu}$

$= 0.15 \times 240 \times 30 = 1080$ N/mm

and $\not> 0.9 L_s + 0.7 A_e p_y$

$= 0.9 \times 240 + 0.7 \times 0.785 \times 410$

$= 441$ N/mm

Local shear satisfactory

(g) Deflection

Using unfactored imposed loads as in Section 3.7(f) $W = 132$ kN

The properties of the transformed sections(4) (Fig. 9.7.) are:

(Fig. 9.7) $r = A_b/b_s d_c$

$= 5900/1500 \times 250 = 0.0157$

section 9.5 $m = 14.6$

$x_e = [250/2 + 14.6 \times 0.0157(201 + 250)]/(1 + 14.6 \times 0.0157)$

$= 186$ mm

$I_{bc} = 80\,000$ cm^4

Deflection $= WL^3/60EI_{bc}$

$= 132 \times 7400^3/(60 \times 205 \times 80\,000 \times 10^4) =$ **5.6 mm**

Deflection limit $= 7400/360 = 20.6$ mm

Comparing the section used (406 × 140 × 46 UB) with that required in non-composite (533 × 210 × 92 UB) gives a clear indication of the weight saving achieved in composite construction. However, as discussed in Section 9.1, some other costs must be taken into account in any cost comparison.

STUDY REFERENCES

Topic	*Reference*
1. Reinforced concrete	**Kong, F.K. & Evans, R.H.** (1987) Reinforced concrete beams – the ultimate limit state, *Reinforced and Prestressed Concrete,* pp. 85–155. Van Nostrand Reinhold
2. Composite Construction	**Johnson, R.P.** (1982) Simply supported composite beams and slabs, *Composite Structures of Steel and Concrete,* pp. 40–100. Granada Publishing
3. Transformed cross-section	**Kong, F.K. & Evans, R.H.** (1987) Elastic theory, *Reinforced and Prestressed Concrete,* pp. 157–167. Van Nostrand Reinhold
4. Composite slabs	**Lawson, R.M.** (1983) *Composite Beams and Slabs with Profiled Steel Sheeting,* CIRIA Report 99 London

BRACING

10.1 LOADING RESISTED BY BRACING

Bracing members, or braced bay frames, consist usually of simple steel sections such as flats, angles, channels or hollow sections arranged to form a truss (Section 6.1). The members are often arranged, using cross-bracing, so that design may be on a tension only basis.

A bracing will carry loading which is usually horizontal, derived from a number of sources:

- wind, crane and machinery loads acting horizontally on a structure;
- earthquake loads derived as an equivalent static horizontal load;
- notional loads to ensure sway stability;
- beam or column bracing forces as a proportion of the longitudinal force;
- loads present during the temporary construction stage.

In addition bracing, whether permanent or temporary, is usually necessary for steelwork erectors to line and level properly the steel framework during construction.

10.2 SWAY STABILITY

It is important that all structures should have adequate stiffness against sway. Such stiffness is generally present where the frame is designed to resist horizontal forces due to the wind loading. To ensure a minimum sway provision notional forces are suggested in clause 2.4.2.3. applied horizontally:

1% of $\gamma_f W_d$

or 0.5% of $\gamma_f(W_d + W_i)$ if greater

acting in conjunction with $1.4W_d + 1.3W_i$ vertically

This requirement is in place of the horizontal wind or other loads and in practice forms a minimum provision.

10.3 MULTI-STOREY BRACING

In multi-storey frames horizontal forces may be resisted by:

- rigidly jointing the framework with connections capable of resisting the applied moments and analysing the frame accordingly;
- providing stiff concrete shear walls usually at stair and lift wells, and designing these to absorb all the horizontal loads;
- arranging braced bay frames of steel members forming trusses as shown in Fig. 6.3.

In all but the first case the steel beams may be designed as simply supported.

The arrangement of steel bracing or wind towers of concrete walls requires care to ensure economy and simplicity. Alternative arrangements are shown in Fig. 10.1. Symmetrical arrangements are preferred as they avoid torsion on the braced frames.

The vertical bracing must be used in conjunction with suitable horizontal framing. Wind loads are transmitted by the cladding of the building on to the floors, and then to the vertical braced bays or towers. Design should ensure that adequate horizontal frames exist at floor levels to carry these loads to the vertical bracing. Where concrete floors are provided no further provision may be required but in open frame industrial buildings horizontal bracing is also needed (Fig. 10.1).

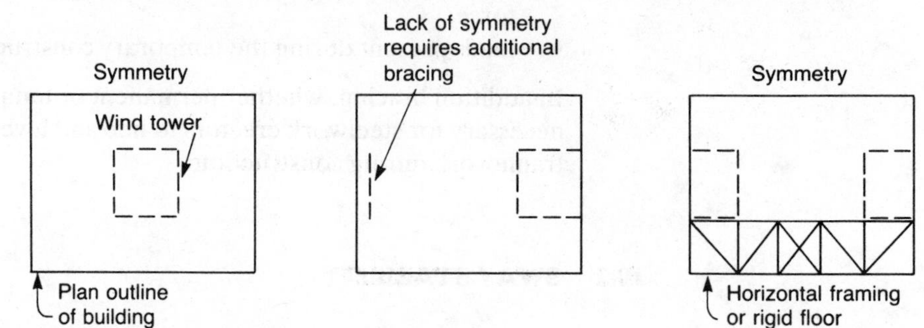

Fig. 10.1 Wind towers and bracing

Braced bay frames may take a number of different forms as shown in Fig. 10.2. Cross-bracing, whilst it allows a tension only design, creates difficulty where door or window openings are required. The alternatives shown may be used to accommodate openings, but will involve compression in the bracing members. In the design of such members slenderness must be kept as low as possible by use of structural hollow sections, and by reducing effective lengths as far as practicable.

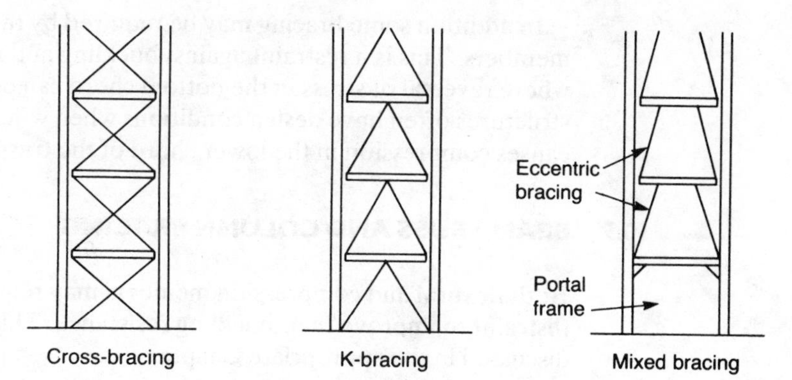

Fig. 10.2 Braced bay frames

10.4 SINGLE-STOREY BRACING

The principal loading which requires the provision of bracing in a single-storey building is that due to wind. In addition the longitudinal crane forces will require braced bay support. The horizontal (wind and crane surge) loads transverse to the building are supported by portal frame action, or column cantilever action, and no further bracing is needed in this direction.

Longitudinal forces do, however, require support by a braced bay frame as shown in Fig. 6.3. The wind forces arise from pressures or suctions on the gable end and frictional drag on the cladding of both the roof and sides of a building (see Section 11.4.3). Gable wind girders are needed therefore at each end of the building, and may be provided at the level of the rafters (low-pitch) or at the level of the eaves, as shown in Fig. 10.3. The gable wind girders are supported by vertical side bracing as shown, which is also used to support the longitudinal crane forces. The gable posts themselves are designed to span vertically carrying the wind load between the base and the gable wind girder.

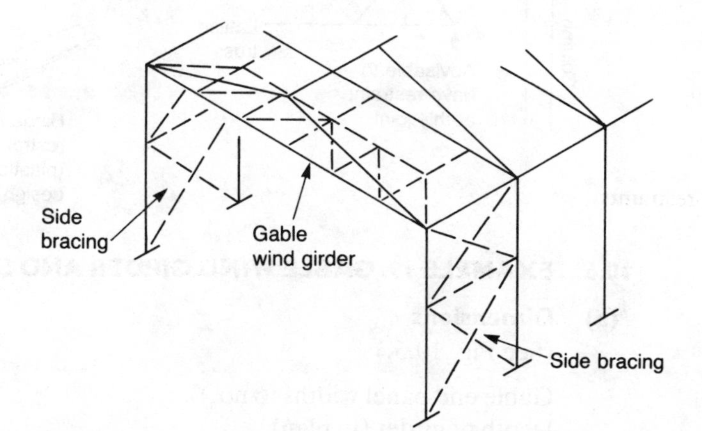

Fig. 10.3 Gable wind girder

In addition some bracing may be required by the truss lower chord members. This is a restraint against buckling and is needed in cases where reversal of stress in the bottom chord can occur. Lightweight roof structures often have design conditions when wind suction on the roof causes compression in the lower chord of the truss.

10.5 BEAM TRUSS AND COLUMN BRACING

Both flexural and compression members may require lateral bracing or restraint to improve their buckling resistance. This provision has been discussed in the appropriate Chapters:

Beams in buildings	—	Chapter 3
Crane girders	—	Chapter 5
Trusses	—	Chapter 6
Columns	—	Chapter 7

In each case the effective length of the portion of the member in compression may be reduced by providing single members or frameworks capable of resisting the lateral buckling forces. The values of these lateral forces have been assessed from test data and are given in the appropriate clauses of BS 5950. In some cases, e.g. crane girders, the buckling force is combined with other lateral forces in the design of the bracing.

The designer should always be aware of the need for bracing in unusual positions, and should examine all compression members, and compression flanges, to ensure that adequate lateral restraint exists and is satisfactory. Examples of restraints needed in lattice frameworks and portal frames (plastic design) are given in Fig. 10.4.

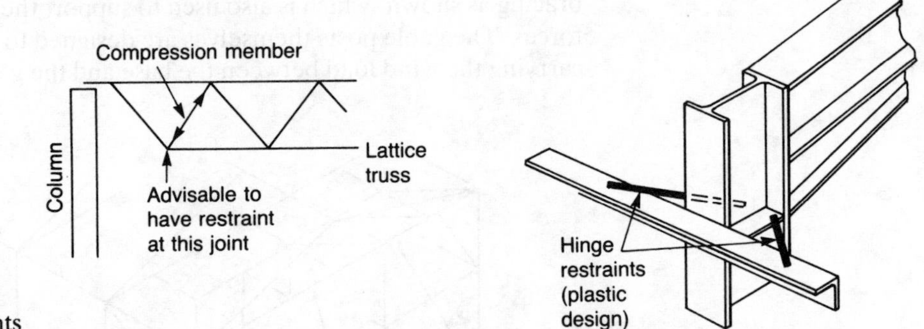

Fig. 10.4 Special restraints

10.6 EXAMPLE 19. GABLE WIND GIRDER AND SIDE BRACING

(a) Dimensions

(See Fig. 10.5.)

Gable end panel widths (6 no.)	5.0 m each
Depth of girder (in plan)	3.0 m
Side bay width	6.0 m
Eaves height	12.5 m

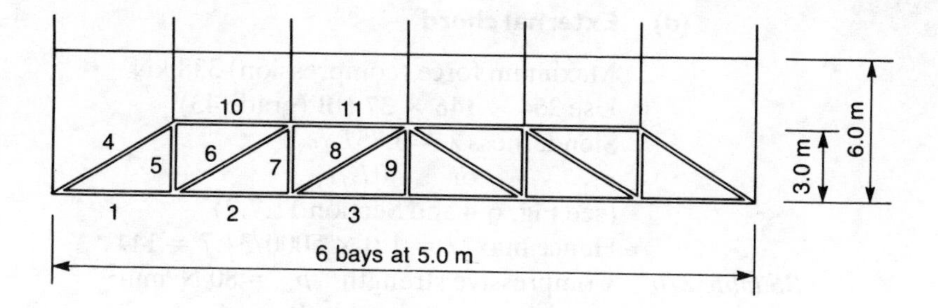

Fig. 10.5 Wind girder dimensions

(b) Loading

Reactions (excluding γ_f) from gable stanchions (spanning vertically) vary as shown (Fig. 10.6).
Wind pressure or suction (in brackets) results in two sets of reactions. These values are derived knowing C_{pe} and C_{pi} and are given in Section 2.3.
Longitudinal load from crane (Section 5.1) = 12.0 kN.

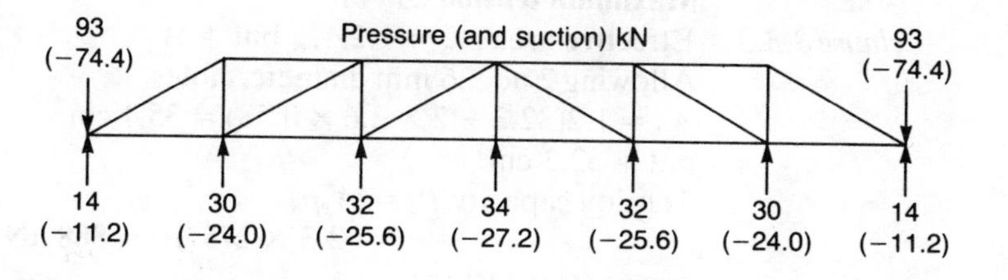

Fig. 10.6 Wind girder loading

(c) Member forces

Member forces may be obtained by any of the methods of analysis (Section 6.2(a)) and the pressure and suction cases are shown in the table; the loads incorporate the factor $\gamma_f = 1.4$.
Compression is positive.

Member	Factored member force (kN)	
	Pressure	Suction
1	184	−147
2	299	−239
3	338	−270
4	−215	172
5	111	−89
6	−133	106
7	69	−55
8	−46	37
9	48	−38
10	−184	147
11	−299	239

(d) External chord

Maximum force (compression) 338 kN
Use **254 × 146 × 37 UB** (grade 43)
Slenderness $\lambda = 0.85L/r_x$
or $1.0L/r_y$
(see Fig. 6.4 and Section 11.7.2)
Hence max. $\lambda = 1.0 \times 5000/34.7 = 144$

BS table 27b Compressive strength $p_c = 80 \text{ N/mm}^2$
Compression resistance $P_c = A_g p_c$
$= 47.5 \times 80/10 =$ **380 kN**
Section satisfactory

(e) Internal chord

Maximum compression 239 kN
Use **203 × 133 × 30 UB** (grade 43)
Slenderness $\lambda = 1.0 \times 5000/31.8 = 157$

BS table 27b Compressive strength $p_c = 68 \text{ N/mm}^2$
Compression resistance $P_c = 38.0 \times 68/10 =$ **258 kN**
Maximum tension 299 kN

clause 3.3.3 Effective area $A_e = 1.2A_{net}$ but $\not> A_g$
Allowing 2 no. 26 mm diameter holes
$A_e = 1.2(32.3 - 2 \times 2.6 \times 0.58) = 35.1 \text{ cm}^2$
but $\not> 32.3 \text{ cm}^2$
Tension capacity $P_t = A_e p_y$
$= 32.3 \times 275/10 =$ **888 kN**
Section satisfactory

(f) Diagonals/struts

Maximum compression (diagonal member 4) = 172 kN
Use **203 × 133 × 30 UB** (grade 43)
Slenderness $\lambda = 1.0 \times 5830/31.8 = 183$

BS table 27b $p_c = 52 \text{ N/mm}^2$
Compression resistance $P_c = 38.0 \times 52/10 =$ **198 kN**
Maximum compression (strut member 5) = 111 kN
$\lambda = 1.0 \times 3000/31.8 = 94$
Use same section

(g) Side bracing

Reaction from wind girder 93 kN
Crane load 12 kN
Maximum design load = 1.4 × 93 = 130 kN

section 2.7(f) or 1.2 × 93 + 1.2 × 12 = 126 kN
(see Fig. 10.7.)

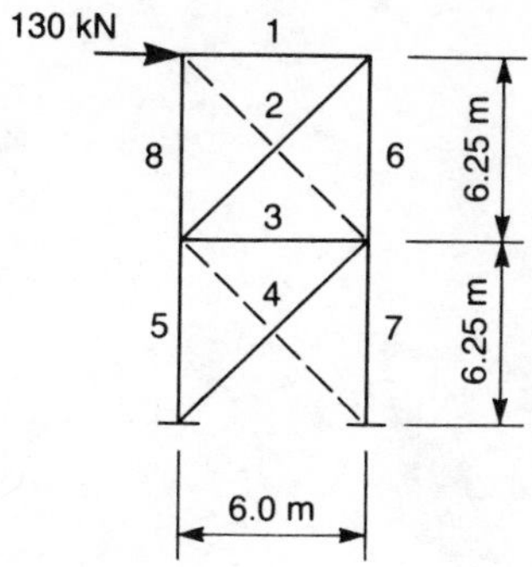

130 kN 1 2 8 6 3 4 5 7 6.25 m 6.25 m 6.0 m

Fig. 10.7

Factored member forces (kN)

1	130
2	−188
3	130
4	−188
5	−135
6	135
7	270
8	0

Maximum tension in diagonals 2 and 4 (assuming cross bracing to avoid compression) = 188 kN

Use **100 × 65 × 7 angle**

clause 4.6.3.1 Effective area $A_e = a_1 + 3a_1a_2/(3a_1 + a_2)$

$a_1 = (100 - 7/2)7 - 22 \times 7 = 522\ \text{mm}^2$

$a_2 = (65 - 7/2)7 = 431\ \text{mm}^2$

allowing one 22 mm diameter hole in connected leg (100 mm)

$A_e = 522 + 3 \times 522 \times 431/(3 \times 522 + 431) = 860\ \text{mm}^2$

Tension capacity $P_t = A_e p_y$

$= 860 \times 275 \times 10^{-3} =$ **237 kN**

Maximum compression in strut 3 = 130 kN

Use **203 × 133 × 25 kg UB** (grade 43)

$\lambda = 1.0 \times 6000/31.0 = 194$

BS table 27b $p_c = 46\ \text{N/mm}^2$

Compression resistance $p_c = 32.3 \times 46/10 =$ **149 kN**

Forces in the eaves girder 1 and main frame members 5, 6 and 7 should be considered in the design of these members when appropriate. The values of these forces will need to be adjusted for λ_f used in the combination of forces for each member.

10.7 EXAMPLE 20. MULTI-STOREY WIND BRACING

(a) Dimensions

(See Fig. 10.8.)
7 storeys at 3.5 m high
Bay width 4.0 m
Cross bracing with K-bracing on alternate floors to allow door openings

(b) Loading

Wind loading transmitted to bracing by concrete floor slabs at each level

(c) Member forces

Member forces (kN) excluding γ_f are given for the lower two storeys only

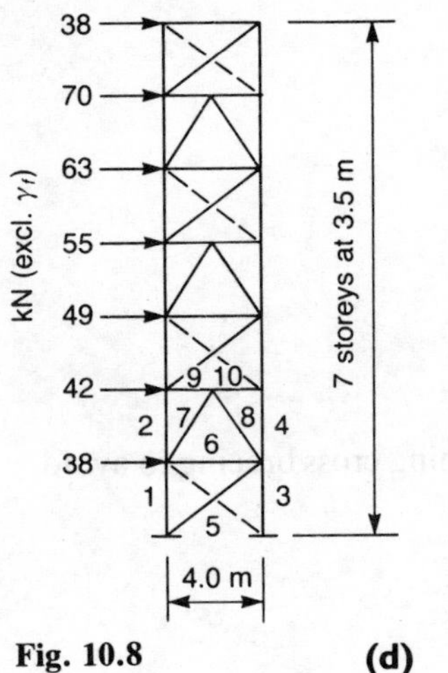

Fig. 10.8

1	−993
2	−716
3	1304
4	716
5	−472
6	197
7	−319
8	319
9	317
10	0

Note that wind loading can act in the reverse direction, which will generally reverse the force direction in each member. Member 10, however, will carry 317 kN in this wind reversal case.

(d) Cross bracing

Maximum tension = 1.4 × 472 = 661 kN
Use **203 × 133 × 30 UB** (grade 43)
section 10.6(e) Tension capacity = **888 kN**

(e) K-bracing

Maximum compression = 1.4 × 319 = 447 kN
Use **254 × 146 × 37 UB** (grade 43)
$\lambda = 0.85 \times 4030/34.7 = 116$
BS table 27b $p_c = 114\ \text{N/mm}^2$
Compression resistance $P_c = 47.5 \times 114/10 =$ **541 kN**

As in Section 10.6(g) the forces in columns 1, 2, 3 and 4 and beams 6, 9 and 10 must be taken into account in the overall design of these members which will include dead and imposed loading from floors, etc. The value of γ_f appropriate to each combination of loads must be used (Section 2.7).

STUDY REFERENCES

Topic	*Reference*
1. Frame stability	(1988) *Stability of Buildings.* (to be published) The Institution of Structural Engineers

PART

II

THE DESIGN OF STRUCTURAL STEEL FRAMEWORKS

The design of simple elements given in Part I is usually only part of the overall building concept. It is necessary to develop a spatial awareness of the structural framework in three dimensions. The action of the whole framework must be considered including its behaviour under lateral loading. In some cases frame action (continuous construction) may be preferred to connected element design (simple construction) from considerations of cost or appearance. Many of the design procedures used in Part II have been developed in Part I, and it is advisable for the student first to become familiar with element design.

11

DESIGN OF SINGLE-STOREY BUILDING – LATTICE GIRDER AND COLUMN CONSTRUCTION

11.1 INTRODUCTION

Over half of the total market share of the constructional steelwork fabricated in the United Kingdom is used in single-storey buildings. Therefore it is almost certain that an engineer will, at some time, have to design or check such a building. Whereas previous chapters have introduced the design of various simple elements, the next three chapters extend the design concept to the overall design procedure of whole structures, i.e. three-dimensional structural arrangements. Essentially, most members in frameworks are positioned so as to transfer load in space to other members and eventually down to the ground, by the simplest, economic 'structural' route. The next two chapters will be devoted specifically to the various design aspects of single-storey structures. Though these structures represent the simplest form of three-dimensional frameworks, they can illustrate most of the structural design criteria which an engineer might encounter during his design life. In particular, this chapter outlines the different considerations used in designing all the structural members for a complete building, based on the main frame being of lattice girder and column construction (see Fig. 11.1). In the next chapter the same building will be redesigned using portal frame construction. For brevity only the main supporting frame will be redesigned as a portal frame, as the design of the remaining structural members is common to both forms of construction.

In developing the structural arrangement for a single-storey building or even a multi-storey building it should be borne in mind that the shorter the span of a structural member the more economic it becomes. However, the client/owner of a single-storey building frequently stipulates, as in this design exercise, that the floor area should be free of internal columns in order to obtain the greatest flexibility of space which can readily accommodate any future modification to the usage of the floor area, without major structural alterations to the building. For

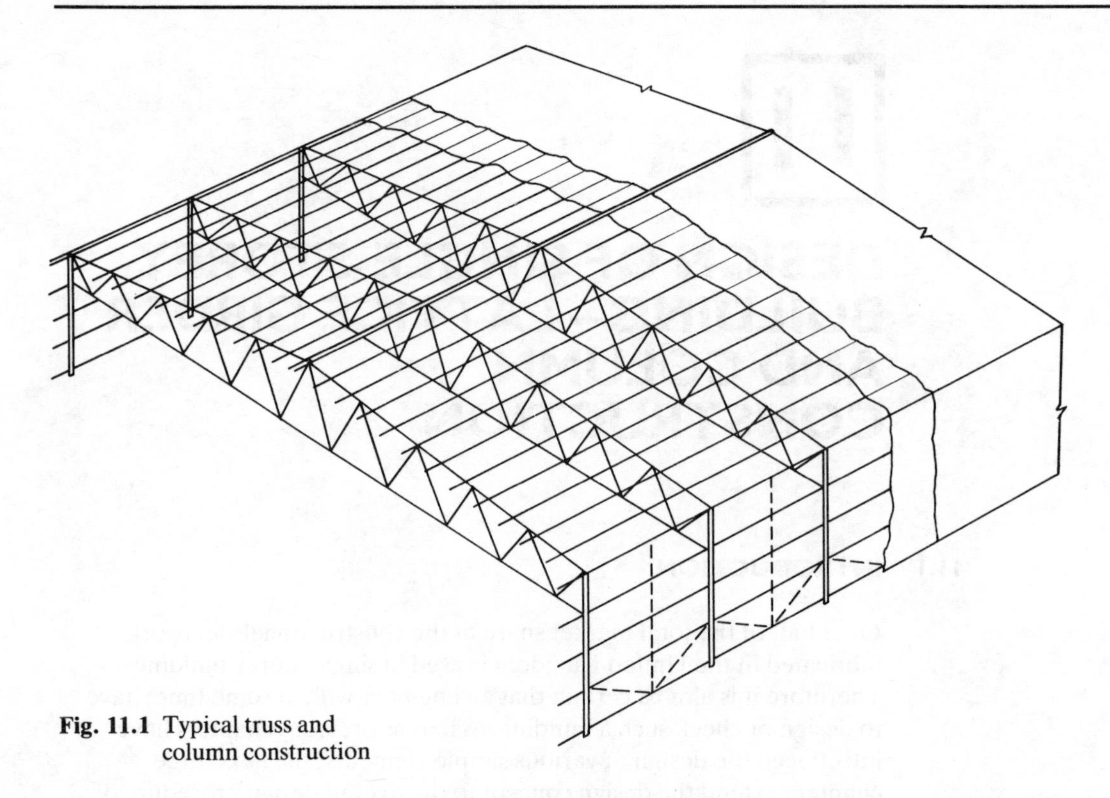

Fig. 11.1 Typical truss and column construction

economic reasons (such as having a flow-line production operation), most industrial buildings have a rectangular floor plan, therefore always arrange, where possible, to span the main frames across the shorter distance, thereby minimizing member sizes (see Fig. 11.2).

Though the client usually gives the designer a free choice of structural arrangement, he does nevertheless expect the best 'least cost' solution. In a paper[1] giving comparative costs of four different types of single-storey arrangements (roof truss, lattice girder, portal frame and space frame) the single-span portal frame seemingly did not produce the most economic answer for the **single-bay** frame, when assessed on initial cost. The study indicated that the lattice girder construction produces the most economic solution for the span being considered in the design example. However, when one takes into account the cost of maintaining the statutory minimum temperature within such buildings then the low roof construction of a portal frame could have a financial advantage over other forms of construction during the lifespan of a building (usually 50 years). This is probably why, together with its simple, clean lines, the most common form of single-storey building found on any modern industrial estate is that of portal frame construction, with over 90% of all single-storey buildings so constructed[1].

Despite the dominance of portal frame construction in the past, there is a growing demand from the hi-tech industries for higher quality and flexibility in the use of buildings. The structures of such buildings are frequently flat-roofed, utilizing solid/castellated beams or lattice girders. The advantage of the lattice girder or castellated form of construction is that they allow services to be accommodated within the depth of the roof construction, at the expense of deeper roof construction when compared with portal frames.

In order to understand the overall design procedure for a building, the following design exercise will deal with the complete design of a single-storey building based on lattice girder and column construction. However, in Chapter 12 an alternative approach, using portal frames as the main supporting structure will be considered; that is, simple lattice girders with universal sections as column members will be replaced by portal frames.

11.2 DESIGN BRIEF

A client requires a single-storey building, having a clear floor area, 90 m × 36.4 m, with a clear height to underside of the roof steelwork of 4.8 m, with possible extension to the building in the future. The slope of the roof member is to be at least 5°. It has been specified that the building is to be insulated and clad with BSC metal sheeting profile Long Rib 1000R (0.70 mm thick, necessary to prevent damage during maintenance access). A sub-strata survey of the site, located in a new development area on the outskirts of Hull, has shown that the ground conditions are able to sustain a foundation bearing pressure of 150 kN/m² at 0.8 m below existing ground level.

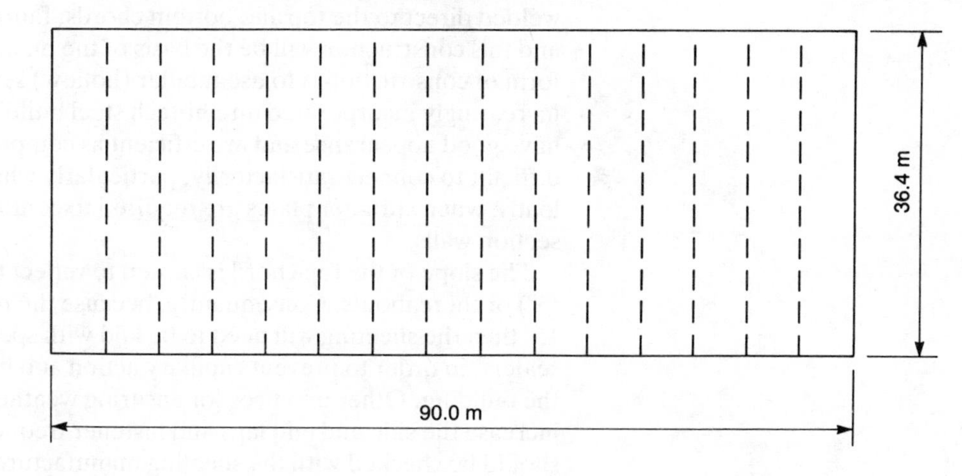

Fig. 11.2 Proposed main frame spacing

11.3 PRELIMINARY DESIGN DECISIONS

The two common arrangements for open web (lattice) girders are illustrated by the diagrams in Fig. 11.3, i.e. the Warren or Pratt truss girders. The difference between the two types is basically that the

Warren truss has pairs of diagonal members of approximately the same length, while the Pratt truss has short verticals and long diagonals. Under normal circumstances of gravity loading when there is no load reversal, then the Pratt truss is structurally more efficient because the short vertical members would be in compression and the long diagonals in tension. However, when there is load reversal in the diagonals then the Warren truss may prove to be the more economic arrangement. Also, the Warren truss may give larger access space for circular and square ducts and is considered to have a better appearance. Therefore, the Warren truss has been selected for this design exercise.

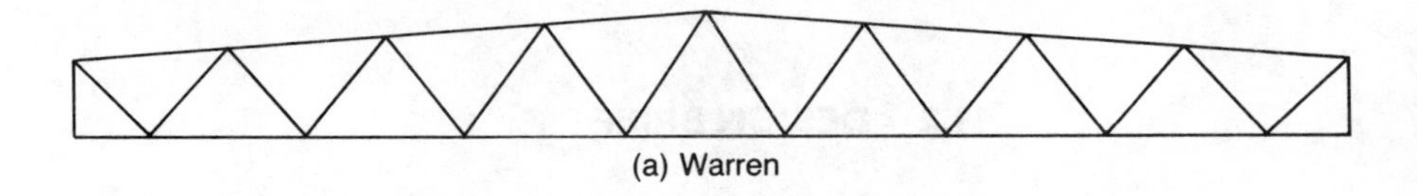

(a) Warren

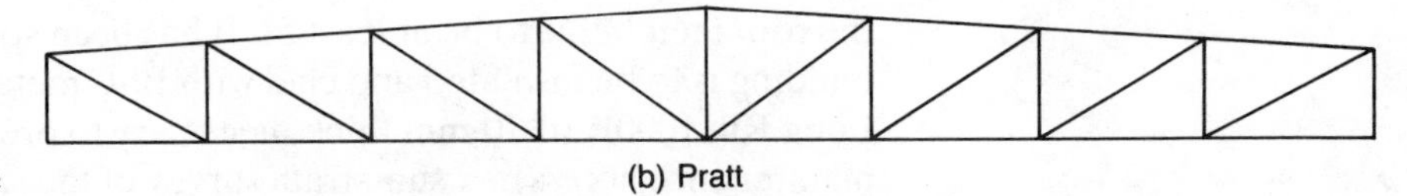

(b) Pratt

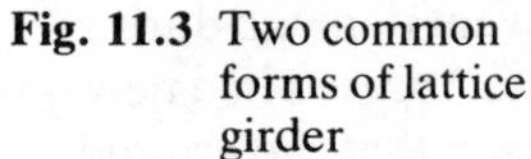

Fig. 11.3 Two common forms of lattice girder

One of the cheapest forms of construction for the lattice girders would be if the angle sections, used for the diagonals (web members), were welded direct to the top and bottom chords, fabricated from T sections, and this construction will be the basis of the main design. An alternative form of construction is to use tubular (hollow) sections, which are being increasingly incorporated into hi-tech steel buildings. Tubular sections have good appearance and are efficient as compression members but are difficult to connect satisfactorily, particularly when subject to high loads, when stiffener plates are required to control bending of the section walls.

The slope of the top chord is chosen to reflect the minimum specified (5°) or thereabouts. Consequently, because the roof slope is less than 15° then the sheeting will need to be laid with special strip mastic lap sealers, in order to prevent capillary action and hence rain leaking into the building. Other practices for ensuring weather-tightness are to increase the side and end laps and fastener frequency. Such details should be checked with the sheeting manufacturer's catalogue. The alternative is to use standing seam type sheeting with concealed fixings. As regards limiting deflection, there are no mandatory requirements, but commonly accepted criteria for a typical insulated building are $L/200$ for roofs and $L/150$ for vertical walls.

The required slope can be achieved by making the depth of the lattice girder at the eaves equal to 1/20 of the span and the depth at the centre of the span 1/10. Assuming the column depth to be 0.6 m then, with reference to Fig. 11.4:

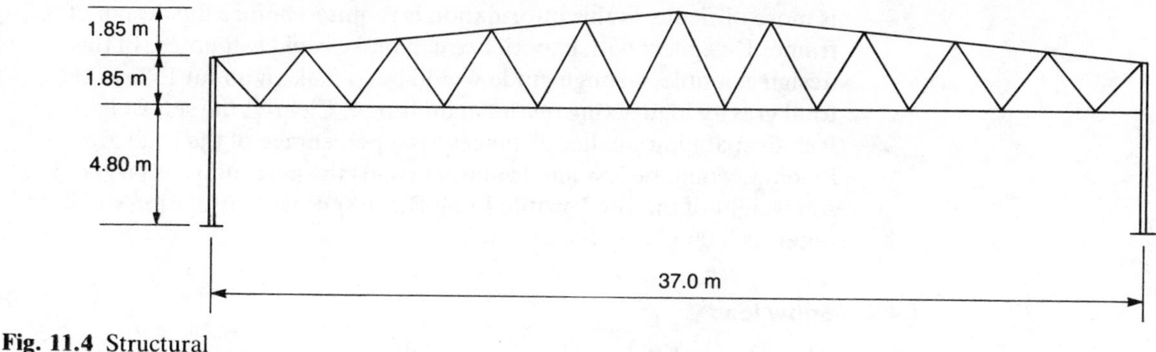

Fig. 11.4 Structural arrangement of main frame members

Span between column centres ($L = 36.4 + 0.6$)	= 37.0 m
Spacing of lattice frames	= 6.0 m
Height to underside of girder	= 4.8 m
Depth of girder at eaves	= 1.85 m
Depth of girder at ridge (apex)	= 3.70 m
Actual slope of rafter [$\theta = \tan^{-1}(1.85/18.5)$]	= 5.71°

11.4 LOADING

Before any design calculations can be undertaken, the loads that can occur on or in a building have to be assessed as accurately as possible. The loads which normally govern the design of a single-storey building are dead loads[(2)], snow load[(2)] and wind loading[(3,4)]. In addition, the designer should give thought to the possibility of unusual loadings, such as drifting of snow[(6)], and overloading of a gutter if the downpipes become blocked or cannot cope with a large volume of rainwater during a deluge. Modern buildings may also be required to accommodate services, such as ducting or sprinkler systems. The weight of these items can be significant and it is advisable that advice from the suppliers is sought. (Note that the snow load information referred to in references (2) and (6) is to be incorporated in BS 6399: Part 3 – *Code of Practice for Snow Loads*.) Also, depending on the function of a building, dynamic loading from crane operations can be an extra design consideration.

11.4.1 Dead load

The dead loads affecting the design of the building result from the self weight of the sheeting (including insulation), the secondary members and main frames and will be included in the design calculations, as and when they occur. Estimating the self weight of sheeting and secondary members is relatively easy as this information is contained in the manufacturers' catalogues. Assessing the self weight of the main frame is more difficult, as this information is required before the design of the frame. Designers with experience can make rapid estimates. In this design example, a rough guide would be to make it about 15% of the total gravity load acting on the main frame. Clearly, the self weight of frames spanning smaller distances, as a percentage of the total gravity loading, would be less and for larger spans the percentage is larger. The self weight of the BSC profile Long Rib 1000R with insulation will be taken as 0.09 kN/m^2.

11.4.2 Snow load

The relevant information regarding the snow loading is at present contained in BS 6399: Part 1. For this[2] design example, it is assumed that the snow load is 0.75 kN/m^2 (acting on plan), though it is anticipated that in Part 3 of BS 6399 there will be regional variations, as is already permitted in the Farm Buildings Code BS 5502. The equivalent snow load acting along the inclined roof member is $0.75 \cos \theta = 0.75 \times 0.995 = 0.75$ kN/m^2. The use of an equivalent load makes due allowance for the purlin spacing being given as a slope distance. However, at this slope it is seen that the difference in load on slope and on plan is negligible.

11.4.3 Wind load

From CP3 Chapter V Part 2 (also to be incorporated into BS 6399), the wind loading on the building being designed can be established. Also, the reader is directed to reference (4) which deals more fully with wind loading on buildings and contains the background information on which reference (3) is based. As the site is located in Hull, then from reference (3) the basic wind speed is estimated as being 45 m/s (Fig. 1 of (3); the factors S_1 and S_3 are both 1.0. From the information supplied, the ground roughness is assumed to be 3, and because the building is longer than 50 m, then the 'building size' is designated as class C. From Table 3[3], knowing the building height is in the 5–10 m region, then the factor S_2 is found to be 0.69, based on a height of 10 m. (A slightly lower value might be obtained by interpolation, as actual height is 8.5 m.) Hence the dynamic pressure (q) is:

$$q = 0.613\,(1.0 \times 0.69 \times 1.0 \times 45)^2/1000$$
$$= 0.59 \text{ kN/m}^2$$

The **external pressure coefficients** for the building (with a roof slope of 5.7°, a building height ratio $h/w = 8.5/37.0 = 0.23$ and a building plan ratio $l/w = 90.0/37.0 = 2.43$) are obtained from Table 7 (walls) and Table 8 (pitch roofs)[3]. Though the dimensions used in this example are based on the centre-lines of members, it is the usual practice to use the overall dimensions of a building. However, the latter would not materially affect the calculated values. The **internal pressure coefficients** are assessed from Appendix E[3], assuming that the two long faces of the building are equally permeable while the gable faces are not; that is, +0.2 when wind is normal to permeable face and −0.3 when normal to impermeable face. However, it could be argued that one should allow for the occurrence of dominant openings, particularly if details of openings have not be finalized at the design stage; that is, the designer makes the appropriate decision, depending on the information available regarding openings in the buildings. Therefore, use the clause in reference (3) which allows the designer to take the more onerous of +0.2 and −0.3. This covers the possibility of a dominant opening, provided it is closed during a severe storm. The resulting wind loading conditions for the building (cf Fig. 2.9 for frame only), are given in Fig. 11.5. The value of 0.95 is determined by interpolation. If there is no possibility of a dominant opening, then only wind cases B and C apply (see Fig. 11.5).

The diagrams show that the maximum pressure that can act on the sides and gables of the building is 1.0×0.59 kN/m^2. However, the maximum **local** pressure, for which the sheeting has to be designed, is generally significantly higher. In assessing the local pressure on the **roof** sheeting, a revised value of S_2 has to be obtained from Table 3[3], noting that cladding is defined as class A, hence $S_2 = 0.78$ and the appropriate value of $q = 0.613(1.0 \times 0.78 \times 1.0 \times 45)^2/1000 = 0.76$ kN/m^2. Attention is drawn to the larger values of C_{pe} which occur at the edge zones of roofs, as indicated in Table 8[3]. Therefore the maximum design pressure (in this case, suction) acting on the roof sheeting is

$$(C_{pe} + C_{pi})q = -(1.4 + 0.2)0.76 = -1.22 \text{ kN/m}^2.$$

Similarly for the **side** cladding where $S_2 = 0.70$ based on a height of 5 m (actual 6.65 m) and the local C_p value is 1.0[3], then the maximum pressure sustained by the side sheeting is

$$(1.0 + 0.2)0.61 = 0.73 \text{ kN/m}^2.$$

These wind loads, based on **pressure coefficients**, are used when determining the loads acting on a particular surface or part of the surface of a building, i.e. applicable in the design of the lattice girder or columns. However, in estimating the wind loads acting on the whole of the building, the **force coefficients** have to be used; that is, the total wind load on a building is calculated from:

$$F' = C_f q A_e$$

where C_f is obtained from Table 10[3] and A_e is the effective frontal area. This means that when the wind is blowing perpendicular to the longitudinal axis of the building, then $S_2 = 0.60$ and $q = 0.45$ kN/m^2, hence:

$$F' = 1.0 \times 0.45 \times (6.0 \times 6.65) = 18.0 \text{ kN per main frame}$$

For the case of the wind blowing on the gables, then:

$$F' = 0.7 \times 0.59 \times [37.0(6.65 + 1.85/2)] = 116 \text{ kN}$$

This force of 116 kN has to be distributed between the bracing systems (see Section 11.8).

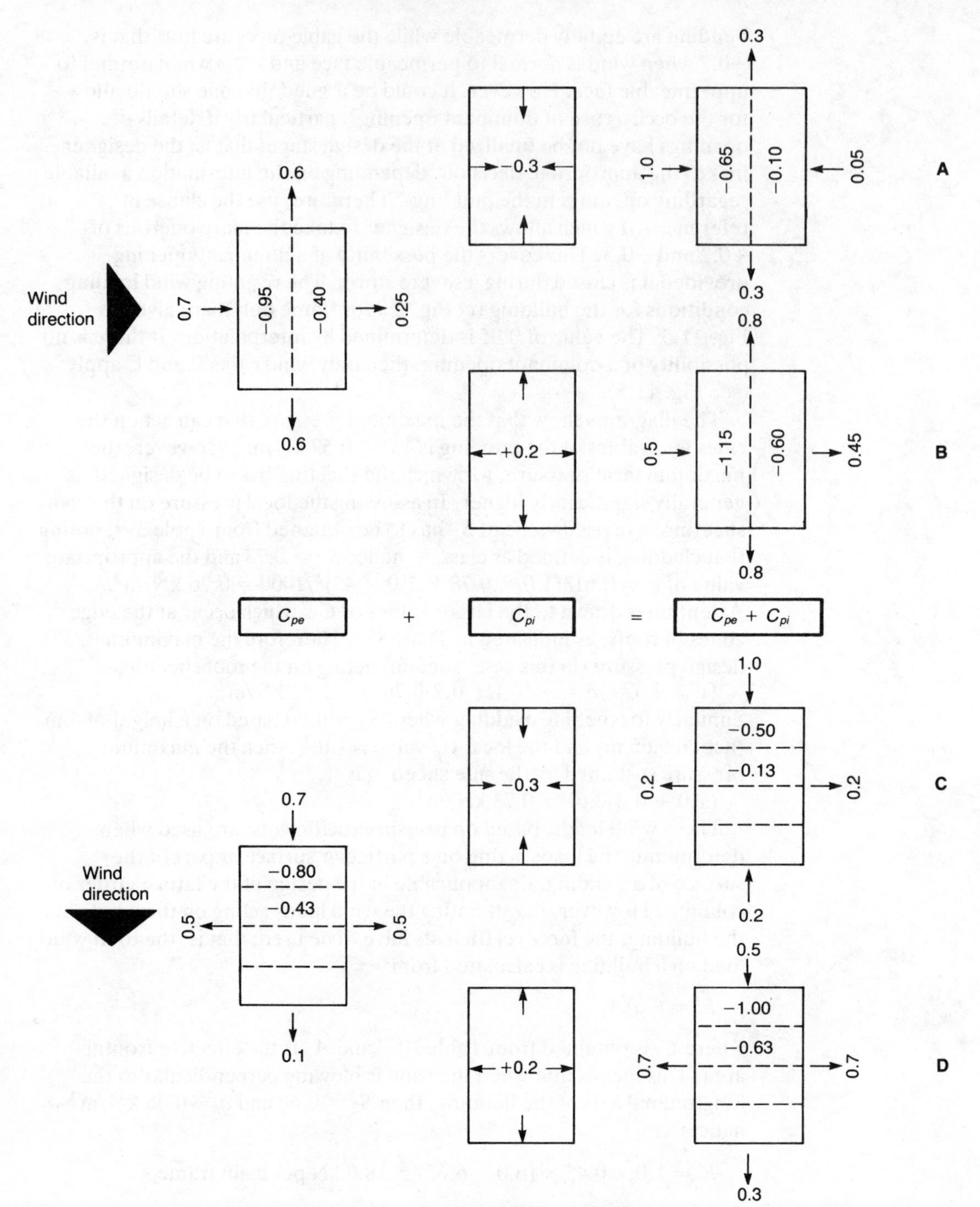

Fig. 11.5 External and internal wind coefficients for building

In addition to direct wind pressure, there can be frictional drag forces. For rectangular clad buildings, these drag forces need to be taken into account only where the ratio of the dimension in the wind direction (d) compared with the building height (h) or the ratio of dimension d compared with the dimension normal to the wind direction (b) is greater than 4. The drag force can be determined from[3]:

$$F' = C_{f'}\,[(\text{roof surface})q_1 + (\text{wall surface})q_2]$$

Consider the wind blowing parallel to the longitudinal axis of the building, then $d/b = 2.43$ and $d/h = 10.7$, i.e. drag force must be taken into account. Under this wind condition, the selected sheeting has ribs running across the wind and therefore $C_{f'} = 0.04$[3] and for the condition $h \leqslant b$, then the code of practice states that[3]:

$$\begin{aligned} F' &= C_{f'}[b(d - 4h)q_1 + 2h(d - 4h)q_2] \\ &= 0.04[37.0(90.0 - 4 \times 8.5)0.59 + 2 \times 6.65(90.0 - 4 \times 6.65)0.45] \\ &= 49.0\ (\text{roof}) + 15.2\ (\text{walls}) = 64.2\ \text{kN} \end{aligned}$$

which has to be resisted by the braced bay(s) (see Section 11.8).

When the wind is deemed to blow in the lateral direction, then $d/b = 0.41$ and $d/h = 4.40$. As the ribs of the sheeting do not run across the wind direction, then $C_{f'} = 0.01$, and the drag force is determined from[3]:

$$\begin{aligned} F' &= C_{f'}[b(d - 4h)q_1 + 2h(d - 4h)q_2] \\ &= 0.01[90.0(37.0 - 4 \times 8.5)0.59 + 2 \times 7.58(37.0 - 4 \times 7.58)0.45] \\ &= 1.6 + 0.5 = 2.1\ \text{kN}\ (= 0.2\ \text{kN per main frame}) \end{aligned}$$

That is, when the wind is blowing in the longitudinal direction of a single-bay building the drag force can be significant, while the drag force per frame in the lateral direction is comparatively small and is usually ignored.

11.5 DESIGN OF PURLINS AND SHEETING RAILS

One of the initial decisions that the engineer has to make is the spacing or centres of the main frames. Though the paper on costs[1] indicates that 7.5 m spacing would be more economic, it has been decided to use 6.0 m centres due to the practical consideration of door openings. Also, if large brick panel walls are used in the side elevations instead of sheeting, it is advisable in any case to limit the frame centres to about 6.0 m or less to avoid having to use thicker than standard cavity wall construction.

As indicated in Section 1.5, the imposed loading acting on a single-storey structure is due to snow and wind, which is carried initially by the cladding and transferred into the secondary members, purlins (roof) and side rails (vertical walls). These members, which are usually designed as double-span members, transfer the imposed loads plus their own self weight by flexural action on to the main frames as a series of point loads. Therefore, another decision to be made is the actual spacing of the purlins, which is dependent on the snow load and the profile and thickness of the metal sheeting selected for the cladding.

It has been shown[1] that the spacing of purlins has little effect on the total cost of purlins, though increased spacing would lead to an increase in cost of sheeting. Metal sheeting is generally used in long multi-span lengths to minimize the number of transverse joints. Such long lengths can be easily handled on the roof (as purlins give support), but not as side cladding because it is difficult to support vertically during erection even with scaffolding. In this example, the client has specified the sheeting. Nevertheless, the maximum length over which the sheeting can span has to be established. The BSC sheeting profile Long Rib 1000R when used as roof cladding has to support a snow load of 0.75 kN/m^2 and/or a local wind suction of 1.22 kN/m^2. *Note that the high local wind pressures/suctions apply only to the design of the cladding*. From the BSC catalogue and assuming that the length of sheet runs over at least two spans, it can be shown that the selected profile can sustain 1.65 kN/m^2 over a span of about 2.0 m, while complying with the deflection limitation for roof sheeting of $L/200$.

Today, the design of the secondary members is dominated by cold formed sections. Though there is a British Standard covering the design of cold formed members (BS 5950: Part 5)[5], the manufacturers tend to develop new profiles, based on the results of extensive testing. There are a number of manufacturers of purlins and sheeting rails and therefore in making a choice, one needs to consult the various manufacturers' catalogues.

The 'design' of cold formed members consists of looking up the relevant table for the chosen range of sections. The choice of a particular manufacturer's products is dependent on client's or designer's experiences and preferences. Table 11.1a illustrates a typical **purlin** load table from a manufacturer's catalogue (Ward Multibeam[7]) for the double-span condition. The loads shown in the table are based on lateral restraint being provided to the top flange of the purlin by the sheeting. Also, it should be noted that the loads quoted in Table 11.1a are for working load condition, i.e. **unfactored**, and that the self weight of the purlin has already been allowed for in the limiting values of load given in the table.

Assuming the overall distance between the outer faces of the column members is 37.6 m, which if divided into 24 equal portions would give purlin centres about 1.570 m (on the slope). The gravity loading (dead plus snow) to be supported by the purlins is $0.75 + 0.09 = 0.84$ kN/m^2, while the maximum uplift on the purlins is $-1.15 \times 0.59 + 0.09 = -0.59$ kN/m^2. From Table 11.1a, knowing the purlin length of 6.0 m, purlin spacing of 1.570 m and the gravity load to be supported by the purlin (0.84 kN/m^2), the A140/165 section seems adequate. (Usually purlin spacing tends to be cost-effective in the range 1.8–2.0 m.)

Table 11.1 Double span loads (kn/m²) – including single span and bay members with sleeved connection (reproduced by permission of Ward Building Systems Ltd, Sherburn)

Span	Section	UDL kN	Purlin Centres (mm)									
			1000	1200	1375	1500	1675	1800	2000	2200	2400	2600
4.5	B120/150	7.20	1.60	1.33	1.16	1.07	0.96	0.89	0.80	0.73	0.67	0.62
	A140/155	9.43	2.10	1.75	1.52	1.40	1.25	1.16	1.05	0.95	[illegible]	[illegible]
	A140/165	10.507	2.33	1.95	1.70	[illegible]	[illegible]	[illegible]	[illegible]	[illegible]	[illegible]	0.84
	*A140/180	[illegible]	[illegible]	[illegible]	[illegible]	[illegible]	1.35	1.26	1.13	1.03	0.94	0.87
[illegible]	[illegible]	7.82	1.42	1.18	1.03	0.95	0.85	0.79	0.71	0.65	0.59	0.55
	A140/165	8.69	1.58	1.32	1.15	1.05	0.94	0.88	0.79	0.72	0.66	0.61
	*A140/180	10.00	1.82	1.52	1.32	1.21	1.09	1.01	0.91	0.83	0.76	0.70
	A170/160	10.38	1.89	1.57	1.37	1.26	1.13	1.05	0.94	0.86	0.79	0.73
	A170/170	11.61	2.11	1.76	1.54	1.41	1.26	1.17	1.06	0.96	0.88	0.81
	A170/180	12.77	2.32	1.94	1.69	1.55	1.39	1.29	1.16	1.06	0.97	0.89
6.0	A140/155	7.19	1.20	1.00	0.87	0.80	0.72	0.67	0.60	0.54	0.50	0.46
	A140/165	7.99	1.33	1.11	0.97	0.89	0.80	0.74	0.67	0.61	0.56	0.51
	*A140/180	9.18	1.53	1.28	1.11	1.02	0.91	0.85	0.77	0.70	0.64	0.59
	A170/160	9.56	1.59	1.33	1.16	1.06	0.95	0.89	0.80	0.72	0.66	0.61
	A170/170	10.69	1.78	1.49	1.30	1.19	1.06	0.99	0.89	0.81	0.74	0.69
	*A170/180	11.75	1.96	1.63	1.43	1.31	1.17	1.09	0.98	0.89	0.82	0.75
	A200/160	12.11	2.02	1.68	1.47	1.35	1.21	1.12	1.01	0.92	0.84	0.78
	A200/180	13.49	2.25	1.87	1.64	1.50	1.34	1.25	1.12	1.02	0.94	0.87
6.5	A170/160	8.86	1.36	1.14	0.99	0.91	0.81	0.76	0.68	0.62	0.57	0.52
	A170/170	9.9	1.52	1.27	1.11	1.02	0.91	0.85	0.76	0.69	0.63	0.59
	A170/180	10.88	1.67	1.40	1.22	1.12	1.00	0.93	0.84	0.76	0.70	0.64
	A200/160	11.23	1.73	1.44	1.26	1.15	1.03	0.96	0.86	0.79	0.72	0.66
	A200/180	12.51	1.92	1.60	1.40	1.28	1.15	1.07	0.96	0.87	0.80	0.74
	*A200/200	14.59	2.25	1.87	1.63	1.50	1.34	1.25	1.12	1.02	0.94	0.86
	A230/180	15.02	2.31	1.93	1.68	1.54	1.38	1.28	1.16	1.05	0.96	0.89
7.0	A170/170	9.21	1.32	1.10	0.96	0.88	0.79	0.73	0.66	0.60	0.55	[illegible]
	A170/180	10.12	1.45	1.20	1.05	0.96	0.86	[illegible]	[illegible]	[illegible]	[illegible]	0.67
	A200/160	10.45	1.49	[illegible]	[illegible]	[illegible]	[illegible]	[illegible]	1.02	0.93	0.85	0.78
	[illegible]	[illegible]	[illegible]	[illegible]	1.88	1.73	1.55	1.44	1.29	1.18	1.08	1.00
[illegible]	A200/200	11.89	1.49	1.24	1.08	0.99	0.89	0.83	0.74	0.68	0.62	0.57
	A230/180	12.31	1.54	1.28	1.12	1.03	0.92	0.86	0.77	0.70	0.64	0.59
	A230/200	14.34	1.79	1.49	1.30	1.20	1.07	1.00	0.90	0.82	0.75	0.69
	*A230/240	18.21	2.28	1.90	1.66	1.52	1.36	1.26	1.14	1.03	0.95	0.88

(a) Purlin load table

Span	Section	UDL kN	Rail Centres (mm)							
			1400	1550	1700	1850	2000	2150	2300	2500
4.5	B120/150	9.00	1.43	1.29	1.16	1.08	1.00	0.93	0.86	0.80
	B140/150	11.30	1.79	1.61	1.47	1.35	1.25	1.16	1.09	1.00
5.0	B120/150	8.12	1.16	1.04	0.95	0.87	0.81	0.75	0.70	0.64
	B140/150	9.80	1.40	1.26	1.15	1.06	0.98	0.91	0.85	0.78
	B140/165	11.2	1.60	1.44	1.31	1.21	1.12	1.04	0.97	0.89
5.5	B140/150	8.93	1.15	1.04	0.95	0.87	0.81	0.75	0.70	0.64
	B140/165	10.20	1.32	1.19	1.08	1.00	0.92	0.86	0.80	0.74
6.0	B140/165	9.36	1.11	1.00	0.91	0.84	0.78	0.72	0.67	0.62
	*B140/180	9.73	1.16	1.04	0.95	0.87	0.81	0.75	0.70	0.64
	B170/155	11.02	1.31	1.18	1.07	0.98	0.91	0.85	0.79	0.73
	B170/165	12.02	1.43	1.29	1.17	1.08	1.00	0.93	0.86	0.80
6.5	B170/155	10.20	1.12	1.00	0.91	0.84	0.78	0.72	0.67	0.62
	B170/165	11.12	1.22	1.10	1.00	0.92	0.85	0.79	0.74	0.68
	*B170/180	12.57	1.38	1.24	1.13	1.04	0.96	0.89	0.83	0.77
	B200/160	12.60	1.38	1.24	1.13	1.05	0.96	0.89	0.83	0.77
7.0	B170/165	10.32	1.05	0.95	0.86	0.79	0.73	0.68	0.63	[illegible]
	B200/160	11.71	1.19	1.07	[illegible]	[illegible]	[illegible]	[illegible]	[illegible]	[illegible]
	[illegible]	[illegible]	[illegible]	[illegible]	1.06	0.97	0.90	0.84	0.78	0.72
8.0	B200/170	11.13	0.99	0.89	0.81	0.75	0.69	0.64	0.60	0.55
	B200/180	11.62	1.03	0.93	0.85	0.78	0.72	0.67	0.63	0.58
	A200/180	12.76	1.13	1.04	0.93	0.85	0.79	0.73	0.69	0.63
	*A200/200	14.86	1.32	1.19	1.08	1.00	0.92	0.86	0.80	0.74

(b) Cladding rail load table

If the design load is limited to 0.84 kN/m^2 (unfactored), then the maximum spacing for this particular profile would be:

$$L_s = \frac{\text{u.d.l.}}{\text{span} \times \text{max. applied load}}$$

where u.d.l. - see third column of Table 11.1a (7.99 kN)
span - purlin length, i.e. 6 m
applied load - 0.84 kN/m^2

$$L_s = \frac{7.99}{6.0 \times 0.84} = 1.585 \text{ m}$$

that is, the design spacing of 1.570 m is just within the capacity of the purlin section A140/165. However, Table 11.1a shows that this section is near its deflection limit, i.e. not included for 6.5 m span. Therefore it might be advisable to select a deeper section, i.e. use A170/160 profile. If a purlin 'safe load' table does not state the deflection limit, check with the manufacturer that the limit of $L/200$ is not exceeded.

For the majority of design cases, the design of the purlin section would now be complete. However, under some wind conditions the resultant uplift on the roof can produce a stress reversal in the purlin, thereby inducing compression in the outstand flange, e.g. in this exercise the uplift is 0.59 kN/m^2. As this flange is not laterally restrained by the cladding, then some form of restraint to the flange may be necessary; check with the manufacturer regarding any special restraint requirement for wind uplift. Indeed, under high wind loading, the wind uplift-no snow condition could result in a more severe loading for the purlin than that due to gravity loading. As the metal cladding is normally fixed by self-tapping screws (designer's choice) then the same load-span table (Table 11.1a) can be used for suction conditions, provided the anti-sag tie arrangements are adhered to (see next paragraph). Note that if the selected cladding had been metal or asbestos sheeting fixed with hookbolts, then a mid-span restraint would have been necessary[7], as the wind uplift condition exceeds this particular manufacturer's limit of 50% of the permissible gravity loading, i.e. 0.59/0.84 = 70%. Such limitations are dependent on individual manufacturer's recommendations.

Anti-sag ties at mid-span of the purlins spanning more than 6.1 m are recommended by the manufacturer. Such ties are required to prevent distortions and misalignment of purlins during the fixing of sheeting or where extreme axial loads exist. Under normal conditions, it would appear that sag bars are not required in this example, as the purlin span is less than 6.1 m. However, if any purlin forms part of the roof bracing system, then sag bars may become necessary.

Coming now to the design of the side rails for the vertical walls, as snow loading is not a design problem, then the section is usually chosen independent of the purlin. The wind conditions for the sides/gables (Fig. 11.5) indicate that a pressure of 1.0 × 0.59 kN/m^2 and a suction of −0.8 × 0.59 kN/m^2 are the appropriate design loading, which acts perpendicular to the sheeting (allowed implicitly by clause 4.12.4.4b);

that is, it is assumed that the vertical panel of sheeting (connected to the side rails) behaves as a deep girder, thereby imposing negligible flexural action (due to self weight) on the side rails in the vertical plane. (Try bending a flat sheet of paper in the plane of the paper.) However, care must be taken during erection to reduce any distortion that can occur in side rails before the cladding is attached. The reduction of such distortion is discussed in the next paragraph. To maximize the strength of the side rails, they are placed normal to the sheeting and column members. Wind load permitting, then the side rails can be spaced further apart. In this example, one could use the same purlin size (A170/160). However, the manufacturer of the Multibeam system produces special sections for cladding rails and reference to Table 11.1b would indicate the section size B170/155 is suitable for the same reasons given in selecting the purlin profile. There is also the possibility of having to restrain laterally the column member, which might cause the 'maximum' spacing of some rails to be reduced.

The maximum, unfactored wind suction ($-0.8 \times 0.59 = -0.41$ kN/m^2) acting on the sheeting would cause compression in the outstand flange, therefore mid-span restraints may become necessary. In using the cladding section (B170/155), the manufacturer limits suction loading to 80% of the allowable wind pressure load. By coincidence, the suction coefficient is 0.8 and therefore the same section selected to withstand the wind pressure can be used.

It is essential during erection that any distortion, which can occur in side rails before the cladding is attached, is minimized. This can be achieved by employing the 'single strut system' (for use up to 6.1 m frame centres), as recommended by the manufacturers[(7)]; that is, any distortion and levelling is controlled by adjusting the diagonal ties before the placement of the sheeting (see Fig. 11.1). The inherent benefit of the single strut system is the mid-span restraint it provides.

In practice, the joints of the double-spanning purlins/rails are staggered across each frame, thereby ensuring that each intermediate main frame receives approximately the same total purlin loading; that is, the larger 'central' reactions arising from the continuity are applied to alternative frames. The self weight of selected purlin section (A170/160) is 0.045 kN/m.

The purlins/sheeting rails are attached to the primary structural members by means of cleats, bolted or welded to the main members. As an integral part of the Multibeam system, the manufacturer supplies special cleats. If a manufacturer does not supply cleats, then they have to be designed (see Ch. 4). Nevertheless, it is essential that any standard hole arrangements stipulated by the manufacturer are complied with. Otherwise, extra cost could be incurred for a non-standard arrangement.

11.5.1 Summary of secondary member design

Purlin size: Ward Multibeam, **A170/160**
actual spacing of purlins (on slope) = 1.570 m

actual spacing of purlins (on plan) = 1.562 m

Side rail size: Ward Multibeam, **B170/155**

actual spacing of side rails: see Section 11.7

11.6 DESIGN OF LATTICE GIRDER

Taking the overall dimensions for the main frame as defined in section 11.3 then a good structural arrangement for the 'web' members is to make the inclination of these diagonals in the region of 45°–60° to the horizontal. By dividing the top chord member, over half the span, into five equal panel widths, then the diagonals in the end panels are at 45°, while those at the mid-span are at 63° (see Fig. 11.6). This means that the purlin positions along the top chord do not coincide with the member intersecting points (connections) of the frame panels. Therefore, in addition to the primary axial forces, the top chord member has to be designed to resist the bending action induced by the purlin loads.

In spite of the fact that the top and bottom chords will in practice be continuous members, a safe assumption is to analyse the lattice girder initially as a pin-jointed frame, thereby allowing the primary axial forces in the various members to be evaluated readily; that is, the flexural action in the top chord caused by the purlin loads can be ignored for the purpose of calculating the primary axial forces. Indeed, clause 4.10 permits such a procedure.

Consequently, any purlin load needs to be redistributed so that it is applied only at the panel points. This is simply done by dividing the total load acting on the girder by the number of panels in the top chord, i.e. as there are 10 panels, then the panel load is the total load/10. Note that the two outermost panel nodes carry just over half load, as they support only half a panel width of roof, plus any sheeting overlap. It can be shown that this apparent redistribution of load does not materially affect the magnitudes of the axial forces in the members. There is an implied assumption that the self weight of the girder is uniformly distributed throughout the frame. This approximation would have a negligible effect on the outcome of the design of a girder of this size.

Having decided the geometry of the girder and the different patterns of loading required, the next stage is to calculate the unfactored loads acting on the girder; see Table 11.2. The noted wind loads (w_w) are based on a wind coefficient of −1.0; wind loads for other wind coefficients are obtained by multiplying the noted values with the appropriate coefficient. Total self weight of girder is estimated, based on previously suggested figure of 15% of the total dead load; that is, the dead load, excluding self weight, is 20.1 + 7.0 + 166.5 = 193.6 kN, then the estimated self weight is 193.6 × 15/(100 − 15) = 34.0 kN, i.e. 0.92 kN/m.

Consideration of the various load combinations (Section 2.7(f)) indicates that there are only two load conditions for which the girder

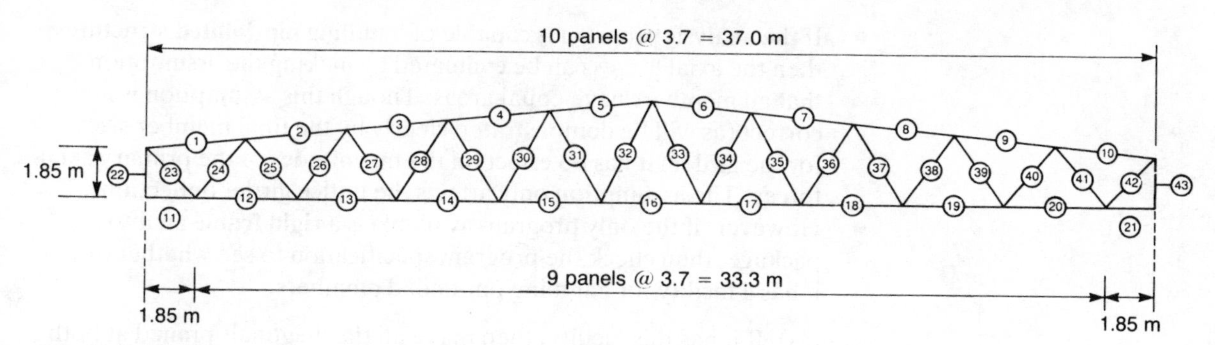

Fig. 11.6 Member numbering for lattice girder

Table 11.2 Unfactored loads (kN) on lattice girder

			Total load	Panel load	Purlin load
Dead load (w_d):					
sheeting and insulation	0.09 × 6.0 × 37.18	=	20.1	2.01	0.848 (on plan)
purlins	26 × 0.045 × 6.0	=	7.0	0.70	0.296 (on plan)
self weight	0.92 (est) × 37.0	=	34.0	3.40	1.435 (on plan)
Snow load (w_i):	0.75 × 6.0 × 37.0	=	166.5	16.65	6.966 (on plan)
Wind load (w_w):	0.59 × 6.0 × 37.18	=	131.6	13.16	5.557 (on slope)

needs to be designed, i.e:

- $1.4w_d + 1.6w_i$ – maximum gravity loading
- $1.0w_d + 1.4w_w$ – maximum uplift loading

As wind loading on the girder always produces an uplift condition for this example, then the load combination $1.2w_d + 1.2w_i + 1.2w_w$ will produce conditions between the combinations A and B (see Fig. 11.5) and therefore need not be considered for the design of the girder.

11.6.1 Forces in members

Knowing the 'panel loads', an elastic analysis of the girder can now easily be undertaken manually, by resolving forces at a joint or by other well-known methods[8]. Note that **gravity loading** (dead plus snow) acts in the **vertical** direction, while the **wind loading** acts **perpendicular** to the inclined roof members.

Alternatively, the primary axial forces in the members can be determined by a computer analysis. However, the kind of analysis program to which the designer may have access, can vary and the following observations might prove helpful:

- Always minimize the maximum difference between **adjacent** node numbers, e.g. in Fig. 11.7 the maximum difference is 2. By this simple rule, computing costs are kept to a minimum.

- If the analysis program is capable of handling **pin-jointed** structures, then the axial forces can be evaluated by making the assumption that all members have equal areas. Though this assumption is not correct (as will be demonstrated clearly by the final member sizes for the girder) it has no effect on the magnitudes of the primary axial forces. The assumption only affects the lattice girder deflections.
- However, if the only program available is a **rigid frame** analysis package, then check the program specification to see whether or not it has a facility for handling pin-ended members:

  (a) If it has this facility, then make all the diagonals pinned at both ends; make top and bottom chords continuous, except for the ends (adjacent to the verticals) which should be pinned (see Fig. 11.7). Do not also make the end verticals pin-ended, as this produces numerical instability. Make the chords relatively stiff by putting the second moment of area (inertia) for the chords equal to, say, 100 cm^4. This will ensure that any moments generated in these members are nominal. Such nominal moments are induced due to the small relative movements of the panel nodes. (A similar effect would occur with slight settlement of the supports for a continuous beam.) Again make all members have equal area.

  (b) If the rigid frame analysis program does not have this facility, then make the second moment of area of the web members very small, say 0.01 cm^4 and make all members have equal area. An analysis will result in the same numerical values obtained from other analyses. This simple 'device' of using virtually zero inertia, in fact, prevents the web members from attracting moment, thereby producing effectively a pin-ended condition for the members so designated; that is, although the top chord is continuous, the loads are being applied only at the nodes (panel points) and because the connected web members are made to act as pin-ended, then only nominal moments can be induced into the top chord. The chord behaves essentially as a series of pin-ended members between panel points.

Due to the symmetry of the components of vertical loading (dead and snow) on the girder, the girder need only be analysed for panel loads equivalent to the condition $1.0w_d(= 2.01 + 0.70 + 3.40 = 6.11\text{ kN})$. The resulting member forces can then be proportioned to give the appropriate forces due to $1.4w_d(8.554\text{ kN})$ and $1.6w_i(26.64\text{ kN})$. However, separate analyses are required for wind loading due to the non-symmetrical nature of this type of loading.

The results from the different elastic analyses are shown in Figs 11.8a (dead load), 11.8b (wind on side) and 11.8c (wind on end). The axial forces in each member, duly factored, are summarized in Table 11.3. Though the analysis for wind on the sides is executed for wind blowing from left to right, it should be borne in mind that the wind can blow in the reverse direction, i.e. right to left. Therefore, if wind affects the

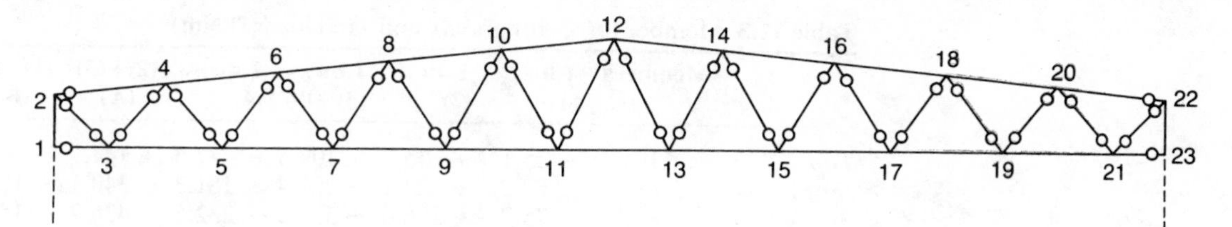

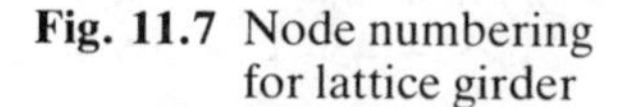

Fig. 11.7 Node numbering for lattice girder

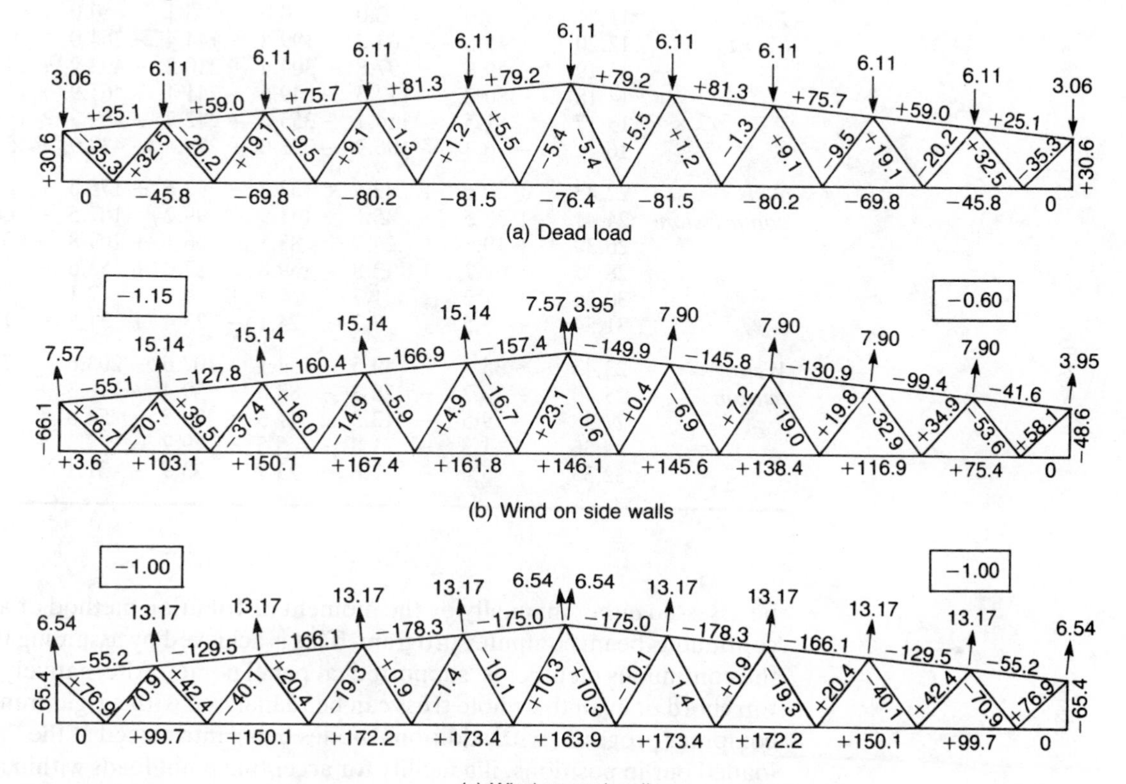

Fig. 11.8 Lattice girder – unfactored member loads (kN)

design (as in this case), members must be designed for the worst condition, irrespective of wind case, i.e. only the **worse load** from either Fig. 11.8b or Fig. 11.8c is recorded in Table 11.3.

The various members making up the Warren truss have been grouped, so that a common member size for any individual group of members can readily be determined.

Having established the individual factored forces, then the design loads for each member (resulting from the two load combinations being considered) can be obtained; see columns (A) and (B) in Table 11.3.

An assessment must now be made of the flexural action in the top chord caused by the purlin loading being applied between panel joints. By taking account of the fact that the top chord in the half-span will be fabricated continuous, then the bending moments in the top chord can

Table 11.3 Member forces for gravity and wind loads (kNm)

	Members	$1.0w_d$ (1)	$1.4w_d$ (2)	$1.6w_i$ (3)	$1.4w_i$ (4)	(2)+(3) (A)	(1)+(4) (B)
Top chord	1,10	+ 25.1	+ 35.2	+ 109.5	− 77.3	+ 144.7	− 52.2
	2,9	+ 59.0	+ 82.7	+ 257.4	− 181.3	+ 340.1	− 122.3
	3,8	+ 75.7	+ 106.0	+ 330.2	− 232.5	+ 436.2	− 156.8
	4,7	+ 81.3	+ 113.8	+ 354.3	− 249.6	+ 468.1	− 168.3
	5,6	+ 79.2	+ 110.9	+ 345.2	− 245.5	+ 456.1	− 166.3
Bottom chord	11,21	0.0	0.0	0.0	+ 5.1	0.0	+ 5.1
	12,20	− 45.8	− 64.2	− 199.8	+ 144.4	− 264.0	+ 98.6
	13,19	− 69.8	− 97.8	− 304.4	+ 210.2	− 402.2	+ 140.4
	14,18	− 80.2	− 112.3	− 349.6	+ 241.1	− 461.9	+ 160.9
	15,17	− 81.5	− 114.1	− 355.2	+ 242.7	− 469.3	+ 161.2
	16	− 76.4	− 106.9	− 333.0	+ 229.4	− 439.9	+ 153.0
Web compression	22,43	+ 30.6	+ 42.8	+ 133.2	− 92.6	+ 176.5	− 62.0
	24,41	+ 32.5	+ 45.6	+ 141.9	− 99.2	+ 187.5	− 66.5
	26,39	+ 19.1	+ 26.7	+ 83.1	− 56.1	+ 109.8	− 37.0
	28,37	+ 9.1	+ 12.8	+ 39.8	− 27.0	+ 52.6	− 17.9
	30,35	+ 1.2	+ 1.7	+ 5.4	+ 6.9	+ 7.1	+ 8.1
	31,34	+ 5.5	+ 7.7	+ 24.1	− 23.4	+ 31.8	− 17.9
Web tension	23,42	− 35.3	− 49.5	− 154.1	+ 107.6	− 203.6	+ 72.3
	25,40	− 20.2	− 28.3	− 88.0	+ 59.4	− 116.3	+ 39.2
	27,38	− 9.5	− 13.3	− 41.5	+ 28.5	− 54.8	+ 19.0
	29,36	− 1.3	− 1.8	− 5.5	− 9.7	− 7.3	− 11.0
	32,33	− 5.4	− 7.6	− 23.5	+ 23.9	− 31.1	+ 18.5

be assessed either manually by the moment distribution method or a continuous beam computer program. This is achieved by assuming that the continuous member is 'supported' at panel points. Alternatively, the top chord or even the whole truss can be reanalysed with a rigid frame computer program, with additional nodes being introduced at the loaded purlin positions, if a facility for accepting point loads within a member length is not available. See Fig. 11.9 for positions of purlins relative to panel joints along the top half-chord member.

The self weight of the top chord can be ignored in the determination of the bending moments along the chord as it will have minimal effect on the moments for this size of frame. Also, for the analysis it is assumed that the ends of the half-chord are pin-ended, which again is a safe assumption, as it could be argued that though the 'apex' end is continuous with the other half-chord, there remains the possibility of a site connection at the apex. Therefore the end might not achieve full continuity, depending on fabrication details.

Alternatively, the two separate computer operations (for the primary axial forces and for the moments in the top chord) could have combined to run as one loading condition, i.e. purlin loads being applied at correct positions, with the top and bottom chords made continuous and all web members pin-jointed.

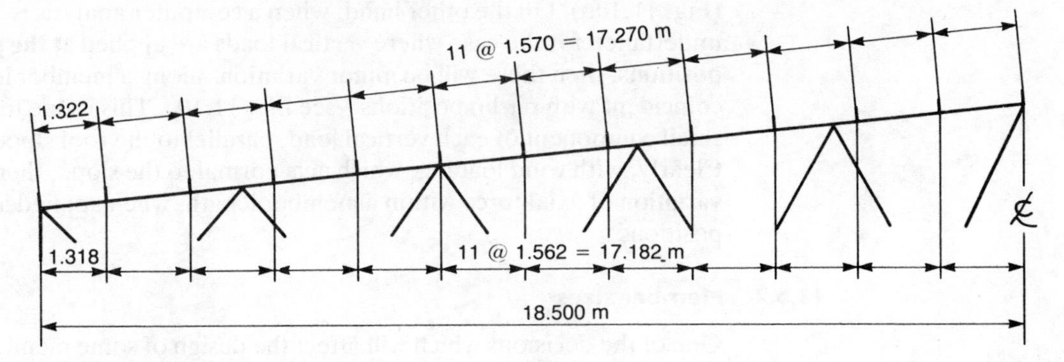

Fig. 11.9 Positioning of purlins along top chord member

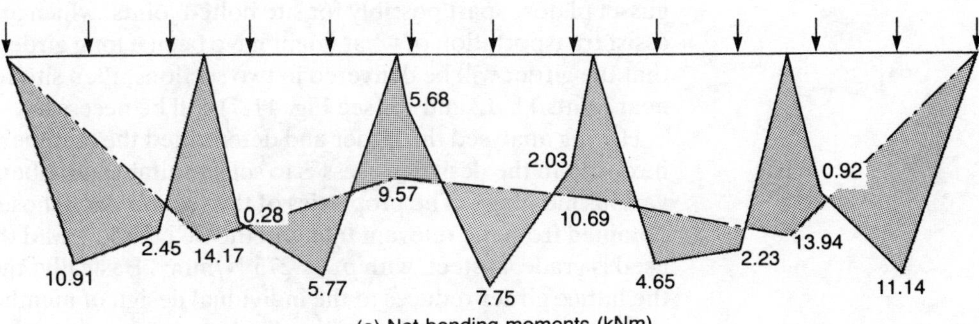

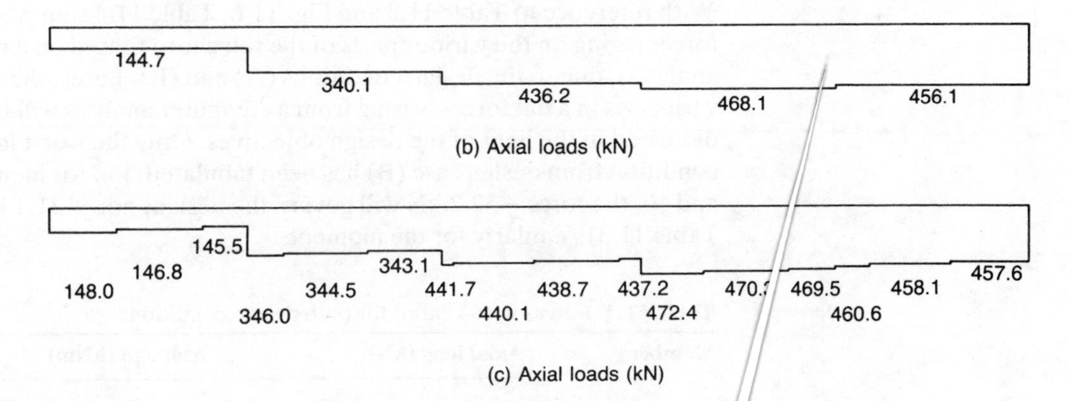

Fig. 11.10 Axial load and bending moments in top chord member

Where the exact positions of the purlin loads relative to panel points are not known, clause 4.10 allows the local bending moment to be taken as equal to $WL/6$. Fig. 11.10a shows the bending moment diagram for the half-chord as a result of the analysis for gravity loading, together with a diagram indicating the variation of primary axial load (based on panel load analysis) in each member (panel) length of the top chord

(Fig. 11.10b). On the other hand, when a computer analysis is undertaken for the case where **vertical** loads are applied **at the purlin** positions, then there will be minor variations along a member length, coincident with purlin positions – see Fig. 11.10c. This is due to the small component of each vertical load, parallel to the roof slope. Clearly, with wind loading, which acts normal to the slope, there is no variation of axial force within a member length, when applied at purlin positions.

11.6.2 Member sizes

One of the decisions which will affect the design of some members is the actual construction of the girder. It has already been noted (Section 11.3) that an economic form for lattice girders is to weld the diagonal members (angles) to the chord members (T sections). This eliminates gusset plates, apart possibly for site bolted joints, which are needed to assist transportation of what might have been a long girder. Assuming that the girder will be delivered in two sections, then site connections near joints 11, 12 and 13 (see Fig. 11.7) will be necessary.

Having analysed the girder and determined the member forces, the next step in the design process is to select suitable member sizes for the various members. The properties of the section sizes chosen are obtained from the relevant tables in the SCI guide[9] and the steel to be used is grade 43 steel, with $p_y = 275\ \text{N/mm}^2$. Basically, the design of the lattice girder reduces to the individual design of member elements, and follows the principles outlined in Part 1.

11.6.2.1 TOP CHORD MEMBER

With reference to Table 11.3 and Fig. 11.6, Table 11.4 summarizes the forces acting on the various parts of the top chord (based on a manual analysis), due to the design conditions (A) and (B). Later, the slight variations in axial forces arising from a computer analysis will be discussed in the light of the design objectives. Only the worst load condition from design case (B) has been tabulated, i.e. for members 1 and 10, the force −52.2 kN will govern the design, not − 31.1 kN (see Table 11.3), similarly for the moments.

Table 11.4 Forces in top chord for both design conditions

Members	Axial load (kN)		Moment (kNm)	
	(A)	(B)	(A)	(B)
1,10	+ 144.7	− 52.2	− 14.17	+ 6.04
2, 9	+ 340.1	− 122.3	− 14.17	+ 6.14
3, 8	+ 436.2	− 156.8	− 10.69	+ 4.63
4, 7	+ 468.1	− 168.3	− 13.94	+ 6.04
5, 6	+ 456.1	− 166.3	− 13.94	+ 6.04

As the top chord is to be fabricated in one length, i.e. continuous, then the most critical portion of that length is 'member' 4 or 7 which has an axial compression of 468.1 kN and a bending moment of 13.94 kNm for the design case (A). The member is primarily a **strut** and the design will be based on that fact, i.e. design for load case (A) and check for case (B). The design procedure is essentially a trial and error method, i.e. a section is selected and then its adequacy checked against various design criteria. If the section size is inadequate, then a different size is chosen.

Try **191 × 229 × 41 Tee** (cut from 457 × 191 × 82 UB).

(a) Classification This check has to be done from first principles as T sections are not classified in the SCI guide[9]. The worst design conditions for the top chord occur at a panel point, i.e. the stem of the T section is in compression for the design case (A). This means that the special limitations, noted in BS table 7, for stems of T sections, apply.

$$\varepsilon = \sqrt{275/275} = 1.0$$

BS table 7 $$\frac{b}{t} = \frac{191.3}{2} \times \frac{1}{16.0} = 6.0 \leq 8.0\varepsilon \quad \textbf{plastic}$$

BS table 7 $$\frac{d}{t} = \frac{230.1}{9.9} = 23.2 \leq 19\varepsilon \quad \textbf{slender}$$

i.e. the T section is **slender** and is governed by clause 3.6. Referring to clause 3.6.4 in particular, then the design strength p_y has to be modified by the stress reduction factor given in BS table 8 for stems of T sections:

$$\text{factor} = \frac{14}{\dfrac{d}{t\varepsilon} - 5} = \frac{14}{23.2 - 5} = 0.77$$

Therefore the reduced design strength for the section is

$$0.77 \times 275 = 212 \text{ N/mm}^2.$$

This reduced value of $p_y (p'_y)$ has to be used for the T section whenever the **stem** is in compression. However, the code deems that such a reduction is not necessary when designing connections associated with the stem.

(b) Local capacity check The design criterion to be satisfied is:

clause 4.8.3.2a $$\frac{F_c}{A_g p'_y} + \frac{M_x}{M_{cx}} \leq 1.0$$

$$A_g = 52.3 \text{ cm}^2$$

clause 4.2.5 $$M_{cx} = p'_y Z \text{ (slender section)}$$
$$= 0.212 \times 142 = 30.10 \text{ kNm}$$

$$\frac{468.1 \times 10}{52.3 \times 212} + \frac{13.94}{30.10} = 0.422 + 0.463 = 0.885 < 1.0$$

Section OK

(c) Member buckling check

clause 4.8.3.3.1

$$\frac{F_c}{A_g p_c} + \frac{mM_x}{M_b} \leq 1.0$$

This is the simplified approach; one really has no choice as a more exact solution cannot be considered due to the reduced plastic moment of a T section not being readily available.

In order to derive p_c and M_b the slenderness of the chord member about the major and minor axes has to be evaluated. In assessing the effective lengths in these directions, i.e. in-plane (x–x axis) and out-of-plane (y–y axis), it should be understood that the top chord will attempt to buckle in-plane between the panel connections and out-of-plane between purlin positions, as shown in Fig. 11.11. This point is reinforced by clause 4.10. This clause further states that for the purpose of calculating the effective lengths of members, the fixity of connections and the rigidy of adjacent members may be taken into account.

For in-plane buckling it can be seen from Fig. 11.11a that the web members effectively hold each panel length of the top chord in position at the connections and supply substantial end restraint to these lengths. Therefore the effective length in the x–x direction can conservatively be assumed to be 0.85 × panel length (on slope) (BS table 24). Out-of-plane, the buckling mode (Fig. 11.11b) is such that the member behaves as though it was pin-ended between purlin positions which hold the chord effectively at those points, i.e. effective length is 1.0 × distance between purlins (on slope).

$$\frac{L_{Ex}}{r_x} = \frac{0.85 \times 3718}{68.9} = 46$$

$$\frac{L_{Ey}}{r_y} = \frac{1.0 \times 1570}{42.3} = 37$$

These values of slenderness comply with clause 4.7.3.2a, which states that for members resisting loads other than wind loads, then their slenderness should not exceed 180.

According to clause 4.7.6 the compressive strength p_c depends on the larger of the two slenderness values, i.e. 46, the reduced design strength of 212 N/mm^2 and the relevant strut table. In the strut selection table (BS table 25) it states that for T sections, table 27(c) has to be used, irrespective of the axis about which buckling occurs. Referring to table 27(c) reveals that the lowest design strength tabulated is 225 N/mm^2. One can either extrapolate down to a value of 212 N/mm^2 in order to obtain p_c, or alternatively p_c may be calculated from the formulae given in appendix C:

by extrapolation $p_c = 180$ N/mm^2
by appendix C $p_c = 179.8$ N/mm^2

Note that the value of E used in the formulae has the units N/mm^2, not kN/mm^2 as defined in section 3.1.2 of the code. Next p_b has to be evaluated, in order to define M_b.

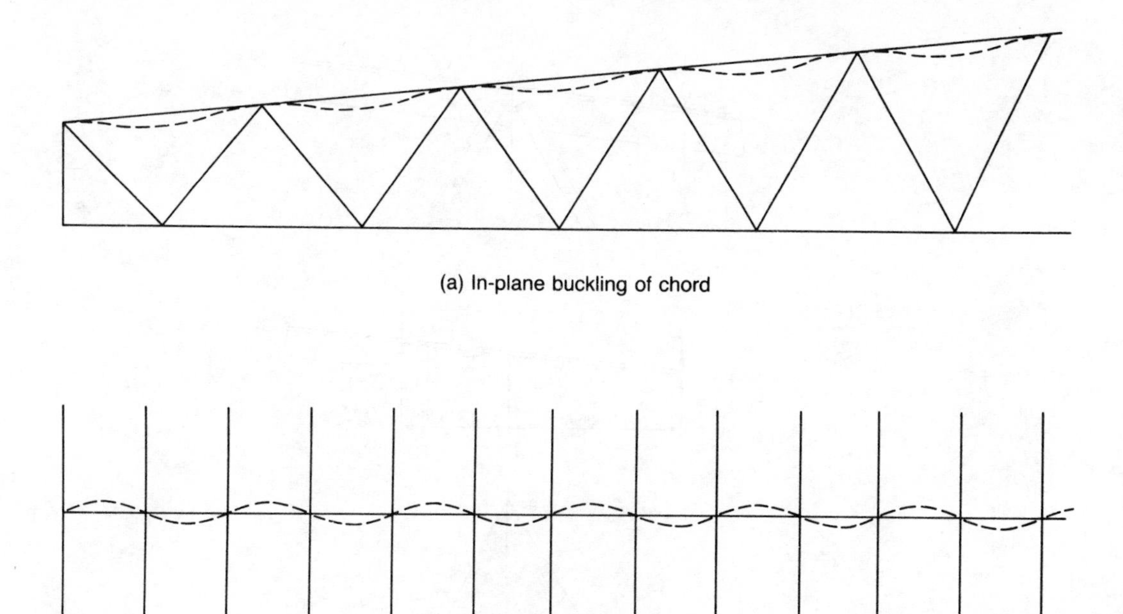

Fig. 11.11 Buckling modes of top chord member

clause 4.3.7.5 Now $\lambda_{LT} = nuv\lambda$

In evaluating n, it must be recognized that the member is 'loaded' within its unrestrained length between adjacent restraints (purlins) as indicated in Fig. 11.12. Also, the load does not have a destabilizing effect on the member. BS table 13 states that under these conditions the value of n can be derived from BS table 15 or 16. It can be shown by reference to Fig. 11.12 that the load does not lie within the middle fifth of the member and therefore table 15 does not apply. The problem with table 16 is that the diagram associated with it indicates a moment distribution, representative of a uniformly distributed load, whereas the case being considered is 'peaky'. Guidance might be obtained from BS table 20, though there is no direct reference to it in clause 4.3.7.5. The design case lies approximately half-way between that for a load at mid-span and that at the quarter point. Therefore, taking a mean of 0.86 and 0.94 (see table 20) as the value of n, then:

BS table 13

$$n = 0.90$$

$$m = 1.00$$

From the SCI guide[9] the following values have been obtained:

$$u = 0.584$$

$$x = 15.5$$

$$\lambda/x = 37/15.5 = 2.39$$

Note that λ is defined in clause 4.3.7.5 as the **minor axis** slenderness,

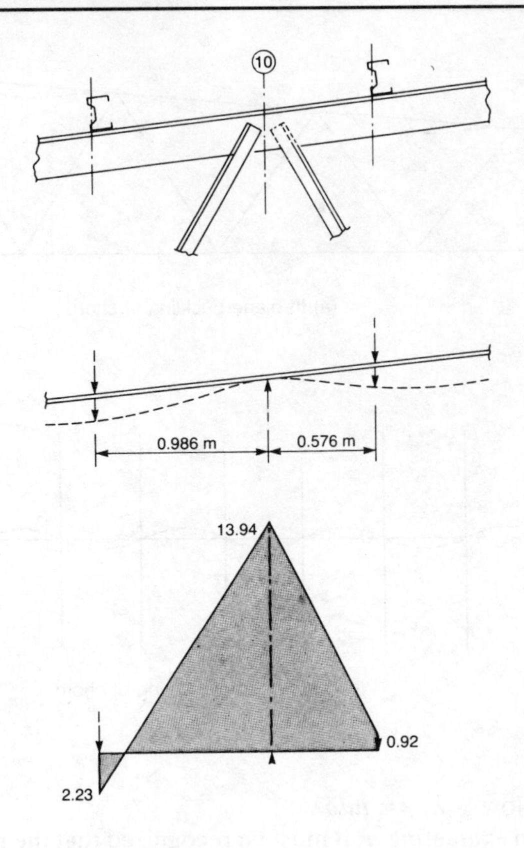

Fig. 11.12 Local bending moment distribution near node 10

i.e. 37. With reference to BS table 14 it can be seen that, because the stem of T section is in compression for the critical part of the length of member being considered, then $N = 0.0$; hence by interpolation:

$$v = 2.76$$

Hence $\quad \lambda_{LT} = 0.90 \times 0.584 \times 2.76 \times 37.18 = 54$

From BS table 11 by extrapolation down to $p_y = 212\ \text{N/mm}^2$ (or by using BS appendix B):

$$p_b = 185\ \text{N/mm}^2$$

$$M_b = p_b S_x$$

$$= 0.185 \times 251 = 46.44\ \text{kNm}$$

$$\frac{468.1 \times 10}{52.3 \times 180} + \frac{1.0 \times 13.94}{46.44} = 0.497 + 0.300 = 0.797 < 1.0$$

Section is OK for lateral-torsional buckling

Now, if the more accurate value of axial load (472.4 kN), obtained from a computer analysis (see Fig. 11.10c), is used then the axial load ratio would marginally increase from 0.497 to 0.502; that is, the small differences in axial forces between different analyses are not significant and can be ignored.

(d) Reverse load condition As a result of the wind suction on the girder, the design case (B) causes the top chord to act as a tie, i.e. 168.3 kN tension and a moment of −6.04 kNm which produces compression in the table of T section.

clause 4.8.3.2a

$$\frac{F_t}{A_e p_y} + \frac{M_x}{M_{cx}} \leqslant 1$$

As the connections are welded at the panel point at which worst design forces occur, then $A_e = A_g$. Also, because the stem of the T section is not in compression, then the design strength is 275 kN/mm^2, hence:

$$M_{cx} = p_y Z = 0.275 \times 142 = 39.05 \text{ kNm}$$

$$\frac{168.3 \times 10}{52.3 \times 275} + \frac{6.04}{39.05} = 0.117 + 0.155 = 0.272 < 1.0$$

This means that the member is more than adequate to cope with the tension arising from reversal of load conditions.

Use 191 × 229 × 41 Tee.

Clearly, when the primary mode of a member is as a strut, then those loading conditions that produced the compression govern the design.

11.6.2.2 BOTTOM CHORD MEMBER

Coming to the design of the bottom chord member, it can be seen that its primary function is a **tie**, with reversal of stress producing compression. One finds that in spite of the primary tension being significantly larger than the compression, the latter will probably govern. Even more so with a member like the bottom chord, where seemingly there are no secondary members to hold the chord at intervals along its length (37 m), which could result in an extremely large slenderness. However, clause 4.7.3.2c states that any member normally acting as a tie, but subject to reversal of stress resulting from the action of wind, is allowed to have a slenderness up to 350. Nevertheless, in order to comply with this requirement, the bottom chord would need a substantial member size and therefore would prove uneconomic.

This apparent problem can be overcome by the use of longitudinal ties, which run the full length of the building and hold the bottom chord at selected positions. In this exercise, assume longitudinal ties occur at connections 9 and 15, i.e. length divided roughly into thirds (see Fig. 11.7). This means that the unrestrained lengths in the y–y direction are 12.95 m for the outer lengths (nodes 1–9 and 15–23) and 11.10 m for the middle length (nodes 9–15). In the x–x direction, the member will buckle in-plane between panel points (3.70 m), similar to the top chord. The longitudinal ties are designed in Section 11.8.3.4.

The bottom chord sustains only axial load under either design case (A) or (B). From Table 11.3, the maximum design forces in the outer lengths are 461.9 kN (tension) and 160.9 kN (compression) and for the middle length 469.3 kN and 161.2 kN respectively. Though the outer lengths have slightly smaller loads than that of the middle length, their length is longer, giving a larger slenderness and hence a lower compressive strength. Therefore, design for compression in the outer lengths and if satisfactory, check the design size against the conditions that prevail for the middle length.

Outer length of bottom chord In choosing the section for the bottom chord at this stage, it should be borne in mind that the end lattice girders form part of the braced bays (Section 11.8.3.2) and therefore attract additional loading.

Try **254 × 127 × 37 Tee** (cut from 254 × 254 × 73 UB).

(a) Classification

$$\frac{b}{T} = \frac{254.0}{2} \times \frac{1}{14.2} = 9.0 \leqslant 9.5\varepsilon \qquad \textbf{compact}$$

$$\frac{d}{t} = \frac{127.0}{8.6} = 14.8 \leqslant 19.5\varepsilon \qquad \textbf{semi-compact}$$

The T section is **semi-compact** and therefore the design strength $p_y = 275\ \text{N/mm}^2$.

(b) Check compression resistance Assume the connections to the columns at nodes 1 and 23 give some restraint in the y–y direction to one of the ends of each outer length, i.e. make $L_{Ey} = 0.95L$, hence:

$$\frac{L_{Ey}}{r_y} = \frac{0.95 \times 12\,950}{64.6} = 190 < 350$$

$$\frac{L_{Ex}}{r_x} = \frac{0.85 \times 3700}{30.0} = 105$$

As the section selected is a T section, use BS table 27(c), from which:

$$p_c = 46\ \text{N/mm}^2$$

$$A_g = 46.4\ \text{cm}^2 \text{ (reference (9))}$$

$$P_c = A_g p_c$$

$$= 46.4 \times 46/10 = 213.4\ \text{kN} > F_c\ (160.9\ \text{kN})$$

(c) Check tension capacity Now check the selected member section for the primary function as a tie carrying 461.9 kN. As there are no site splices in this length, then:

$$A_e = A_g$$

$$P_t = A_g p_y$$

$$= 46.4 \times 275/10 = 1276 \text{ kN} > F_t \text{ (461.9 kN)}$$

Section OK

Clearly, the smaller compression force dominates the design of the bottom chord, the primary function of which is as a tie.

Middle length of bottom chord Use the same section size for the middle length of the bottom chord as for the outer lengths, i.e. 254 × 127 × 37 Tee. The design loads have been noted as 469.3 kN (tension) and 161.2 kN (compression).

(a) Check compression resistance From the previous check on the compression resistance it is clear that only the slenderness about y–y axis needs to be examined, as the value of the x–x slenderness will be lower.

$$\frac{L_{Ey}}{r_y} = \frac{1.0 \times 11\,100}{64.6} = 172$$

BS table 27c

$$p_c = 54 \text{ N/mm}^2$$

$$P_c = 46.4 \times 54/10 = 250.6 \text{ kN} > F_c \text{ (161.2 kN)}$$

(b) Check tension capacity As site splices occur in this portion of the bottom chord, it is anticipated that the T section is connected through both its table (flange) and stem (see Fig. 11.27(b)). BS clause 4.6.3.3 indicates the derivation of the **net** area of a T section connected in this manner is governed by clause 3.3.2, i.e:

clause 3.3.2

$$A_{net} = 46.4 - (2 \times 24 \times 8.6 + 2 \times 24 \times 14.2)10^{-2} = 35.5 \text{ cm}^2$$

However, clause 3.3.3 states the **effective** area A_e of each element at a connection where fastener holes occur, may be taken as K_e times its **net** area, but no more than its gross area. K_e for grade 43 steel is 1.2, therefore:

$$A_e = 1.2 A_{net}$$

$$= 1.2 \times 35.5 < A_g \text{ (46.4)}$$

$$= 42.6 \text{ cm}^2$$

$$P_t = 42.6 \times 275/10 = 1172 \text{ KN} > F_t \text{ (469.3 KN)}$$

Section OK

Use 254 × 127 × 37 Tee

If the T section had been bolted only through its flange, then clause 4.6.3 would apply. As that particular clause defines the effective area of a T section, then the modification allowed by clause 3.3.3 cannot be used.

An alternative approach for the bottom chord is to use a lighter section, but this would require additional longitudinal ties, which would produce a less pleasing appearance and probably cost more overall.

11.6.2.3 WEB MEMBERS

The diagonal members are to be fabricated from angle sections. From practical considerations the **minimum size** of angle used should be **50 × 50 × 6**. These diagonals will be welded direct to the stem of the T sections forming the top and bottom chords (see Fig. 11.21). Though the nominal lengths of members range from 2.89 m to 4.14 m, there is a greater variation in the axial forces (Table 11.5) and therefore the diagonals will designed according to the magnitude of the member load. Table 11.5 summarizes the various axial forces in the diagonal members. The members carrying the heaviest compressive force will be designed first, followed by the other members in descending order of loading until the minimum practical size is reached.

Table 11.5 Axial forces in the diagonal members (kN)

Members	Length	Load case		Members	Length	Load case	
	(m)	(A)	(B)		(m)	(A)	(B)
23,42	2.62	− 203.6	+ 72.3	24,41	2.89	+ 187.5	− 66.5
25,40	2.89	− 116.3	+ 39.2	26,39	3.18	+ 109.8	− 37.0
27,38	3.18	− 54.8	+ 19.0	28,37	3.49	+ 52.6	− 17.9
29,36	3.49	− 7.3	− 11.0	30,35	3.81	+ 7.1	+ 8.1
32,33	4.14	− 31.1	+ 18.5	31,34	3.81	+ 31.8	− 17.9

DIAGONAL MEMBER 24, 41

These particular members have to sustain an axial compression of 187.5 kN or a tension of 66.5 kN. Because the members are discontinuous, then an equal angle is preferred as it is structurally more efficient (weight for weight) than an unequal angle and therefore more economic. Using the SCI guide[9], refer to p.101, which gives a table (for members welded at ends) listing the compression resistance of equal angle struts for different nominal lengths. Given the nominal length is 2.89 m, the table indicates that a 100 × 100 × 12 angle can support 192 kN over a nominal length of 3.0 m. This section size is selected, because it provides sufficient compression resistance for least weight, i.e. compare areas of other sections. Although a check on the adequacy of this section is not necessary, one is now undertaken to illustrate the validity of the values given in the SCI guide.

Check **100 × 100 × 12 Angle.**

(a) Check compression resistance Since the strut is connected directly to another member by welding, then for single angle members (clause

4.7.10.2) the slenderness λ should not be less than:

$$0.85L/r_{vv} \quad \text{or} \quad 0.7L/r_{aa} + 30$$

where r_{vv} and r_{aa} are defined in BS table 28 for discontinuous angle struts (see also Fig. 6.5).

$$\lambda = \frac{0.85 \times 2890}{19.4} \quad \text{or} \quad \frac{0.70 \times 2890}{30.2} + 30 = 127 \quad \text{or} \quad 97$$

As with the T sections, BS table 27c must be used for rolled angle sections, irrespective of buckling axis. Therefore, knowing that $\lambda = 127$ and $p_y = 275$ N/mm^2, p_c is evaluated as 89 N/mm^2, hence:

$$P_c = 22.7 \times 89/10 = 202.0 \text{ kN} > F_c \text{ (187.5 kN)}$$

However, the strut tables for equal angles, given in the SCI guide[9], can be used direct and by interpolation of the tabulated values a very good estimate of compression resistance can be obtained. Thus, referring to p.101[9]:

$$P_c \text{ (est)} = 252 - \frac{(2.89 - 2.50)(252 - 192)}{(3.00 - 2.50)} = 205.2 \text{ kN}$$

Such interpolation gives a slight overestimate of the value of P_c. Therefore, if the interpolated value of P_c within 2–3% of F_c, one might have to resort to first principles, using BS 5950.

(b) Check tension capacity Now check the tension capacity of the selected member. With reference to clause 4.6.3.1, the effective area of a single angle is given by:

$$A_e = a_1 + \frac{3a_1 a_2}{3a_1 + a_2}$$

where a_1 = net area of connected leg

a_2 = gross area of unconnected leg

As one leg of the angle section is connected by welds to the chords, then there are no holes to be deducted in the connected leg, i.e:

$$a_1 = \left(b - \frac{t}{2}\right)t = (100 - 6)12 = 11.28 \text{ cm}^2$$

$$a_2 = a_1 = 11.28 \text{ cm}^2$$

which means for this special case for equal angles then:

$$A_e = 1.75a_1 = 1.75 \times 11.28 = 19.74 \text{ cm}^2$$

Hence $P_t = 19.74 \times 275/10 = 543 \text{ kN} > F_t$ (66.5 kN)

Section OK

This is the identical value for P_t for a 100 × 100 × 12 Angle noted in the SCI guide; see table for tension capacity of equal angle ties connected with one leg given on p.111[9].

Use 100 × 100 × 12 Angle.

The **rest of the diagonals** have been proportioned direct from the guide[(9)] and the relevant design details and member sizes are summarized in Table 11.6. Note that the **end vertical members** are in fact the upper portions of the column members, as can be seen from Fig. 11.13.

Table 11.6 Design details of diagonal members

Members	Length (m)	Design loads compres-sion (kN)	Design loads tension (kN)	max. λ	Angle size	Allowable compres-sion (kN)	Allowable tension (kN)	Actual λ
24,41	2.89	187.5	66.5	180	**100 × 100 × 12**	205.2	543	127
26,39	3.18	109.8	37.0	180	**100 × 100 × 8**	118.3	370	138
23,42	2.62	72.3	203.6	350	**90 × 90 × 6**	85.0	251	125
28,37	3.49	52.6	17.9	180	**90 × 90 × 6**	55.3	251	167
25,40	2.89	39.2	116.3	350	**90 × 90 × 6**[2]	73.6	251	138
31,34	3.81	31.8	17.9	180	**90 × 90 × 6**	53.0	251	182[1]
32,33	4.14	18.5	31.1	350	**90 × 90 × 6**[3]	41.8	251	198
27,38	3.18	19.0	54.8	180	**90 × 90 × 6**	64.0	251	152
30,35	3.81	7.1[4]	—	180	**90 × 90 × 6**	53.0	251	182[1]
29,36	3.49	—	11.0	—	**90 × 90 × 6**	—	251	—

[1] The slenderness of 182 just exceeds the limitation of 180. It can be argued due to practical considerations that the nominal length (distance between intersections) would be reduced by at least 10 mm at each end of the member, which would result in $\lambda < 180$.

[2] A 70 × 70 × 8 Angle section would suffice for this member, but a 90 × 90 × 6 section has the same area. The latter has been selected in an attempt to standardize on section sizes wherever possible as this leads to economy. The benefit is that the larger angle has improved properties.

[3] Though an 80 × 80 × 6 could have been used, a 90 × 90 × 6 has been preferred in order to standardize.

[4] The 7.1 kN load as noted is for the design case (A) and is less than the 8.1 kN load for case (B); see Table 11.3. However, the slenderness limitation of 180 for case (A) as opposed to 350 for (B) will control the design of the member.

11.6.3 Additional design checks

When a lattice girder forms part of a braced bay (see Section 11.8), then it has to sustain additional 'bracing' forces. After the relevant forces have been determined (Section 11.8.2), the top and bottom chords have to be checked to see whether or not the chosen member sizes require modification (see Sections 11.8.3.1 and 11.8.3.2). Also, dependent on the gable framing chosen (see comments in Section 11.8.1.1), one or two lines of sheeting rails may have to be supported by some of the diagonal members in the end girders. The transference of loads from the rail(s) would induce additional forces into these diagonals, in which case the appropriate members need to be checked (see Section 11.10).

11.6.4 Self weight of girder

Having designed the various members of the lattice girder, the designer is now in a position to check the estimated total self weight of the girder, i.e. 34.0 kN, used in the determination of the dead load forces. All that is needed is a rapid assessment, i.e:

chords:	$(41 + 37) \times 37.0$	= 2886
diagonals:	$2 \times 9.6 \times 2.6$	= 50
	$2 \times (17.8 + 8.3) \times 2.9$	= 151
	$2 \times (12.8 + 8.3) \times 3.2$	= 135
	$2 \times (8.3 + 8.3) \times 3.5$	= 116
	$2 \times (8.3 + 8.3) \times 3.8$	= 126
	$2 \times 8.3 \times 4.1$	= 68
	total	= 3532 kg = 34.6 kN

There has been a small underestimate of the self weight, i.e. 0.6 kN, which represents an error of 0.25% in a total dead load of 227.6 kN. A quick scan of the girder design indicates that member sizes would not be affected. Though there is no need to revise the calculations, 0.3 kN should be added to both vertical reactions.

11.7 DESIGN OF COLUMN MEMBERS

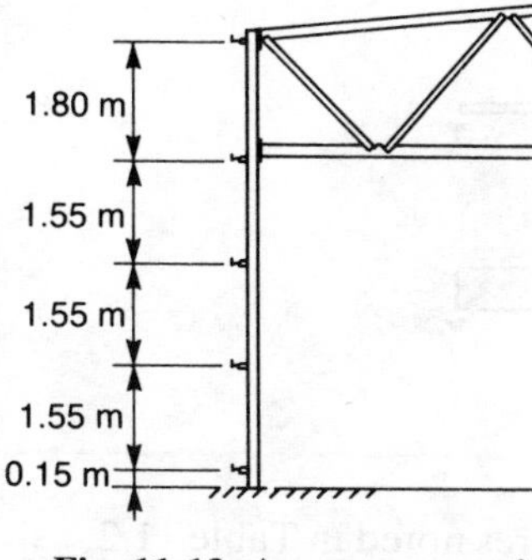

Fig. 11.13 Arrangement of sheeting rails

The columns, in supporting the lattice girders in space, transfer the load from the ends of the lattice girder down to the ground and into the foundation. In addition to the roof loads and the weight of the wall cladding, the column has to be designed to resist the side wind loading (see Fig. 11.5). Assuming that the same sheeting rail (B170/155) is used for both side and end walls and with the probability of the gable post spacing being in the 6.0–6.5 m range (see Section 11.8.1 for exact positioning), the maximum rail spacing is limited to about 2.0 m[7]. As a sheeting rail is required at the bottom chord level to provide restraint to the girder, then Fig. 11.13 shows a possible arrangement of the rails along the side elevations, whilst Fig. 11.2 indicates the rail spacing for the gable walls.

11.7.1 Forces in column members

The axial load acting on either column is the cumulative total of:

- the end reactions from lattice girder, depending upon which design case is being considered – see next paragraph;
- weight of vertical cladding: $0.13 \times 6.0 \times 6.8$ = 5.3 kN (including insulation + liner panel)
- weight of side rails: $5 \times 0.045 \times 6.0$ = 1.4 kN
- self weight of column: $50 \times 6.8 \times 9.81/1000$ = 3.3 kN
- weight of gutter[10,11]: 0.15×6.0 = 0.9 kN

Total = 10.9 kN

The end reactions from the lattice girder and the wind induced moment in the column are interrelated, being dependent on the design case, i.e:

(A) $1.4w_d + 1.6w_i$
(B) $1.0w_d + 1.4w_w$
(C) $1.2w_d + 1.2w_i + 1.2w_w$

and the wind pressure distribution (see Fig. 11.5). In determining the side wind load acting on an individual column, the combined wind load acting on both columns is shared in proportion to their stiffnesses (I/L), i.e. in this case equally. The end reactions of the lattice girder can be evaluated readily by assuming the girder is a simply supported beam (see Fig. 11.14 for different roof loading cases). Note that the loads shown for the wind cases represent the vertical components of the applied wind loads.

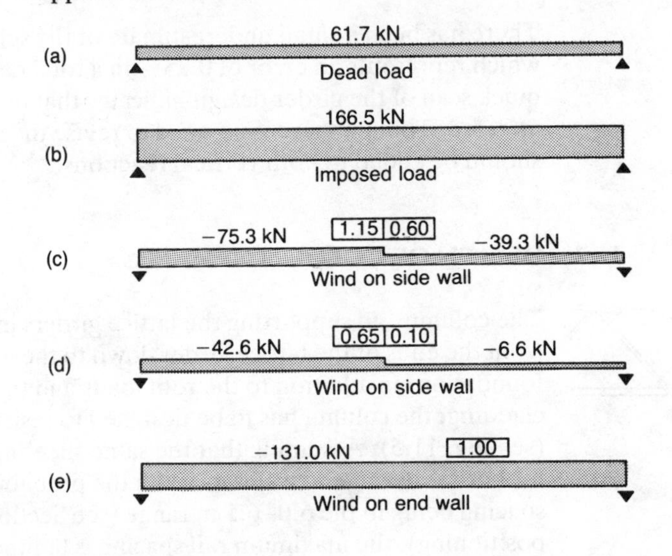

Fig. 11.14 Unfactored vertical loads acting on roof (kN)

11.7.1.1 DESIGN CASE (A)

The vertical load is based on the total load figures noted in Table 11.2, with 0.6 kN added to the dead load value to compensate for underestimation of the girder's own weights (see Fig. 11.14a and b). Therefore, the cumulative axial load in the column member is:

$$F_c = 1.4(10.9 + 61.7 \times 0.5) + 1.6(166.5 \times 0.5) = 191.7 \text{ kN}$$

As there is no wind loading in the design case, there is no associated moment, i.e:

$$M_x = 0.0$$

11.7.1.2 DESIGN CASE (B)

For this design case, select the wind condition that maximizes the uplift force on the roof, whilst maximizing the wind moment in the column member, i.e. $1.0w_d + 1.4w_w$. Loads due to wind on side walls are shown in Fig. 11.14c and the loads due to wind on end walls in Fig. 11.14e.

(i) For wind blowing on side walls With reference to Fig. 11.14a and c:

$$F_t = 1.0(10.9 + 61.7 \times 0.5) - 1.4(75.3 \times 0.75 + 39.3 \times 0.25)$$
$$= -51.1 \text{ kN (i.e. uplift)}$$

The **combined** wind pressure acting on the main building elevations is:

$$(0.5 + 0.45)q \quad \text{or} \quad (1.00 - 0.05)q = 0.95q \text{ (Fig. 11.5), therefore:}$$

$$F = 1.4(0.95 \times 0.59 \times 6.0 \times 6.8) = 32.0 \text{ kN}$$

giving a base shear of 32.0/2 = 16.0 kN per column.

Though it is assumed that the base is 'pinned', the main frame will be subject to 'portal action'. Advantage can therefore be taken of clause 5.1.2.4b which allows a nominal base moment of 10% of maximum column moment. Hence the maximum wind moment, acting at bottom chord level (see Fig. 11.15) is:

$$M_x = 90\% \text{ (base shear} \times \text{moment arm)}$$
$$= 0.90(16.0 \times 4.8) = 69.1 \text{ kNm}$$

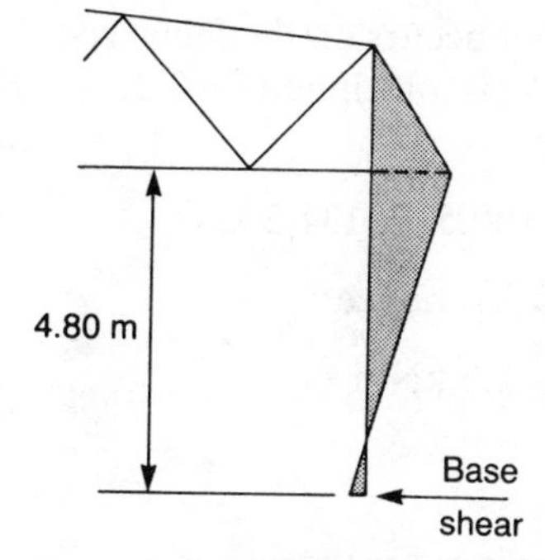

Fig. 11.15 Bending moment in column member (leeward side)

(ii) Wind blowing on end walls (gables) With reference to Fig. 11.14a and e:

$$F_t = 1.0(10.9 + 61.7 \times 0.5) - 1.4(131.0 \times 0.5) = -50.0 \text{ kN}$$

Any lateral wind load acting on the end column is taken by vertical bracing in the side walls (see Section 11.8.1.3). However, wind on the gables causes a suction of $0.2q$ or $0.7q$ acting simultaneously on both side walls. This leads to a zero combined wind load at bottom chord level. Nevertheless, the columns are subject to bending due to the suction. Assume that the column acts as a simply supported member between the base and bottom chord level, with the maximum moment occurring approximately mid-span. The effect of the 10% base moment is to reduce the mid-span moment by 5%, i.e:

$$M_x = 0.95[1.4(0.7 \times 0.59 \times 6.0 \times 4.8^2/8)] = 9.5 \text{ kNm}$$

11.7.1.3 DESIGN CASE (C)

For this design case, choose the wind condition that minimizes the uplift forces on the roof, whilst maximizing, where possible, the wind moment on the column, i.e. $1.2(w_d + w_i + w_w)$. See Fig. 11.14d and e for the appropriate roof loads.

(i) Wind blowing on side walls With reference to Fig. 11.14a, b and d: maximum axial compression occurs in leeward column, as the effect of the wind suction is smaller on the leeward side of the roof (see Fig. 11.14d):

$$F_c = 1.2[10.9 + (61.7 + 166.5)0.5 - (42.6 \times 0.25 + 6.6 \times 0.75)$$
$$= 131.3 \text{ kN}$$
$$F = 1.2(0.95 \times 0.59 \times 6.0 \times 6.8) = 27.4 \text{ kN}$$
$$M_x = 0.90(27.4/2) \times 4.8 = 59.2 \text{ kNm}$$

(ii) Wind blowing on end walls With reference to Fig. 11.14a, b and e:

$$F_c = 1.2[10.9 + (61.7 + 166.5 - 131.0)0.5] = 71.4 \text{ kN}$$

$$M_x = 0.95[1.2(0.7 \times 0.59 \times 6.0 \times 4.8^2/8)] = 8.2 \text{ kNm}$$

This case is generally not critical.

(iii) Wind blowing on end walls With reference to Fig. 11.14a, b and e, but considering the wind condition of $-0.2q$ that occurs on the leeward part of the roof, i.e. the load given in Fig. 11.14e is multiplied by 0.2, then:

$$F_c = 1.2[10.9 + (61.7 + 166.5 - 0.2 \times 131.0)0.5] = 134.3 \text{ kN}$$

However, the corresponding side suction is $-0.2q$, hence:

$$M_x = 0.95[1.2(0.2 \times 0.59 \times 6.0 \times 4.8^2/8) = 2.3 \text{ kNm}$$

11.7.2 Column member section

Table 11.7 summarizes the different design loadings for the column member and the design checks for a **254 × 146 × 37 UB**, which is a **plastic** section[9]. The design checks are based on the following:

LOCAL CAPACITY CHECK

$$\frac{F_c}{P_z} + \frac{M_x}{M_{cx}} \leqslant 1.0 \quad or \quad \frac{F_t}{P_t} + \frac{M_x}{M_{cx}} \leqslant 1.0$$

MEMBER BUCKLING CHECK

(a) For the combination of **axial compression** and **moment**, the following condition must be satisfied:

$$\frac{F_c}{P_c} + \frac{mM_x}{M_b} \leqslant 1.0$$

(b) For the combination of **axial tension** and **moment** clause 4.8.1 implies that the member buckling check is based solely on the following condition:

$$\frac{mM_x}{M_b} \leqslant 1.0$$

The noted values of $P_t(= P_z$ as there are no deductions for holes) and M_{cx} are listed in the left-hand column under the appropriate section size on p.171 of the SCI guide[9]. Also, from the tables on p.85 and p.134, the respective values of P_c and M_b are extrapolated, knowing the effective length about y–y axis, i.e:

$$L_{Ex} = 1.5 \times 4800 = 7.2 \text{ m}$$

$$L_{Ey} = 1.0 \times 4800 = 4.8 \text{ m}$$

It is assumed that the side rails do not restrain the inside flange of the column, except at bottom chord level (see Section 11.8.3.5). (If the designer needs to account for the restraint offered by the rails to the tension side of the column, then conditions outlined in BS appendix G govern.) The value of 0.85 indicated in BS appendix D1, for calculating the effective length about y–y axis, is not used as the base is 'pinned', i.e. conditions at column base do not comply with D1.1c. However, this condition is covered by D1.2d.

Thus, L_{Ey} is 4.8 m and by interpolation of the appropriate values listed in the table on p.134 of the SCI guide[(9)], M_b is evaluated to be 63 kNm. The value of m is defined by BS table 18, based on the ratio of the end moments (β) being −0.10, i.e. 0.54.

The values in Table 11.7 depicted in bold indicate those allowable values used in the member buckling check. Clearly, the wind on the side walls governs the design. The **254 × 146 × 37 UB** section appears to be more than adequate and is adopted. (Note that a 254 × 146 × 31 UB fails to meet the buckling check for the worst design condition, i.e. 1.085.) However, when the wind blows on the end walls, forces are produced and induced into the 'bracing system' (see next section) which are transmitted down to the foundations. As the columns in the end bays form an integral part of the bracing system, the selected section has to be checked for the additional forces when they are calculated.

Table 11.7 Details of column member design

	Axial load (kN)	Wind moment (kNm)	Allowable P_z (kN)	P_c (kN)	M_{cx} (kNm)	M_b (kNm)	Local capacity check	Member buckling check
Design case (A)	191.7	0.0	1310	**408**	—	—	0.146	0.470
Design case (B):								
wind on side	− 51.1	69.1	**1310**	—	133	**63**	0.559	0.592[1,2]
wind on end	− 50.0	9.5	**1310**	—	133	**63**	0.116	0.081[1,2]
Design case (C):								
wind on side	131.3	59.2	1310	**408**	133	**63**	0.545	0.829[2]
wind on end	71.4	8.2	1310	**408**	133	**63**	0.116	0.245[2]
wind on end	134.3	2.3	1310	**408**	133	**63**	0.120	0.349[2]

[1] For this design condition of axial tension plus moment, clause 4.8.1 implies that the member buckling check is based solely on the moment component, i.e. mM_x/M_b. The combined effect of tension plus moment is covered by the local capacity check.

[2] Note that in the member buckling check the moments are multiplied by m (= 0.54).

11.8 OVERALL STABILITY OF BUILDING

The designer must always ensure the structural stability of the building. At this stage, the main frames have been designed to cater for in-plane stability, particularly with respect to side wind loading. However, in order to give stability to the building in the longitudinal direction, all frames need to be connected back to a braced bay. Generally, the **end bay(s)** of the building are braced, so that the wind loads acting on the gables can be transferred to the foundations as soon as possible and thereby the rest of the structure is not affected. Another function of a braced bay is that it ensures the squareness and verticality of the structural framework, both during and after erection.

The bracing system for a single-storey building usually takes the form of rafter bracing in the plane of the roof (Fig. 11.16a). Occasionally another wind girder becomes necessary at bottom chord level, when the roof steelwork is deep, as with this design (Fig. 11.16b). Vertical bracing (located in the side walls) conveys the rafter bracing/wind girder reactions from eaves/wind girder level down to the foundations (Fig. 11.16c). It is assumed that each end of the building is braced.

This means that the top chords of the appropriate lattice girders form part of the rafter bracing, and the bottom chords are part of the wind girder. Hence, the only members that have to be designed are the diagonals. However, the top and bottom chords of the lattice girders must be checked against any additional effect from the 'bracing' forces. Also, the main column members asociated with the vertical bracing system must be checked for the effect of bracing forces.

In this example, the client has indicated that the building might be extended (design brief – Section 11.2), hence normal main frames are to be used at the gables. Consequently, there is no need for vertical bracing in the plane of the gable walls (cf. 'gable framing' of portal frame building, Section 12.10). The inherent in-plane stiffness of the main frame supplies the necessary stability to the walls, apart from any stressed skin action that might exist.

11.8.1 Arrangement of gable posts and bracing systems

11.8.1.1 GABLE POSTS

The loading on, and the structural arrangement of, the different bracing systems are dependent on the positioning of the gable posts and the relationship between the posts and the lattice girders. The gable posts, in addition to supporting the side rails and cladding, resist the wind loads acting on the gables. There are two basic choices when dealing with gable posts:

- either the posts run past the lattice girder and are connected sideways to the top and bottom chord members of the girder; or
- the posts run up and connect to the underside of the bottom chord member. This means that those lines of sheeting rails which occur

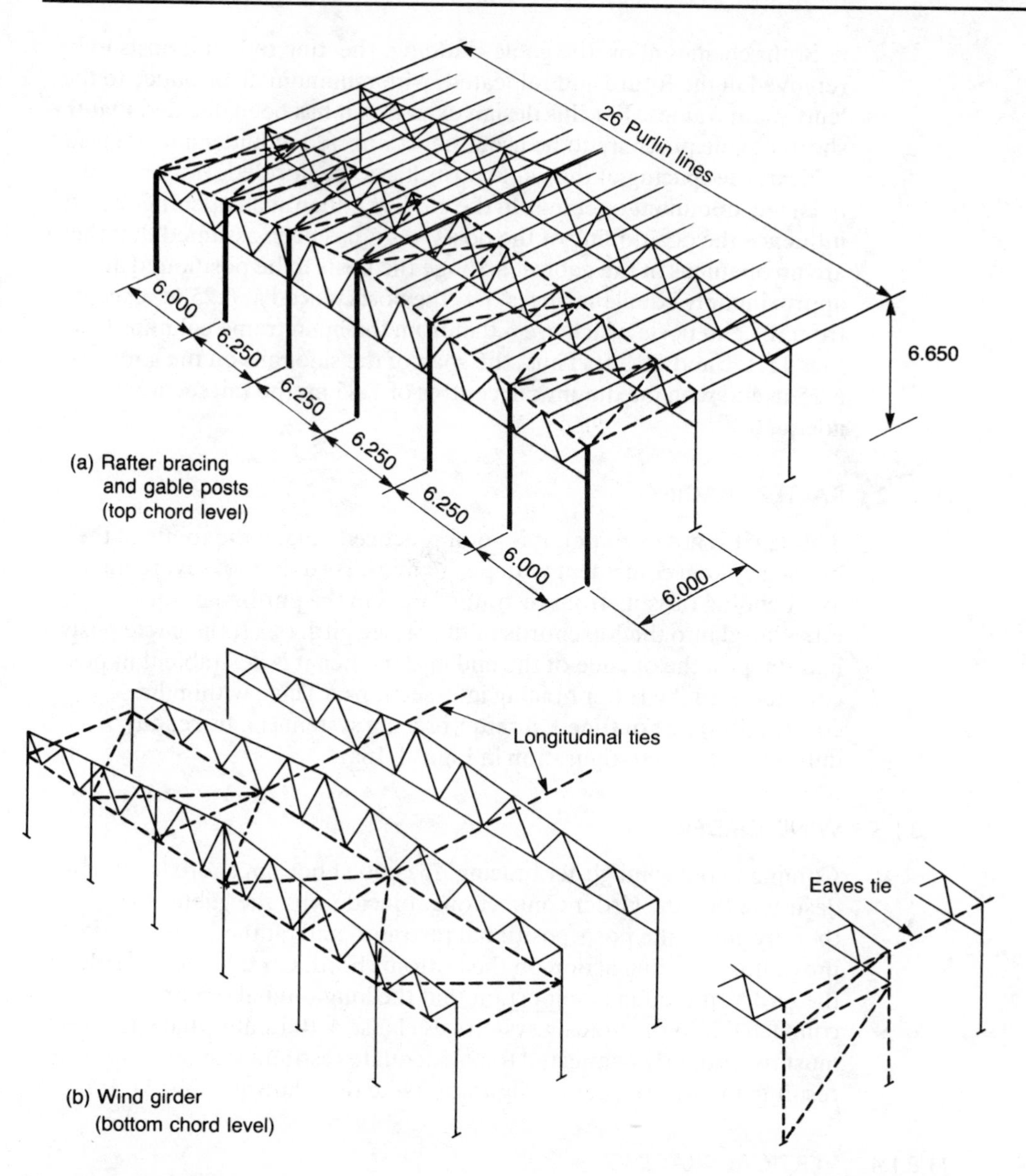

Fig. 11.16 Details of bracing systems

within the depth of the girder have to be supported (connected) by some of the diagonals of the girder, thereby inducing flexural action in those members due to both the eccentricity of the vertical loading, relative to the centroidal axes of the diagonals, and the horizontal wind loading. However, it has to be remembered that the end lattice girders carry only half load, compared with any intermediate girder, and therefore it might be possible to accommodate these additional moments without changing any of the member sizes of the end girders.

Both schemes allow the gable cladding, sheeting rails and posts to be removed in the future and relocated, with minimum disturbance to the 'end' main frames. For this design example, it has been decided that the shorter gable posts are to be used, i.e. connected to underside of girder.

Next, the spacing of the gable posts has to be decided. If there had been any dominant openings in the gable(s), then such openings could influence the positioning of the posts. As it has been assumed that there are no openings in the gables, arrange the posts to be positioned at approximately equal distances, i.e. the posts placed at 6.25 m, except from the end posts which are 6.0 m from the main frame columns (see Figs 11.17c and 11.20). Thus, the span of the side rails on the gable is 6.25 m and with maximum rail centres of 1.85 m, the rail section is adequate[7].

11.8.1.2 RAFTER BRACING

Though it is not essential, it is good practice to make the joints of the bracing system coincident with purlin lines. By doing so, severe minor axis bending (arising from restraint forces in the purlins) is not introduced into the top chords of the lattice girders. (If the gable posts had run past the outside of the end girders then it is desirable that posts coincide with the rafter bracing intersections.) Thus, within the constraint(s) just outlined, a rafter bracing system can be readily defined; see the configuration in Fig. 11.16a.

11.8.1.3 WIND GIRDER

Coming to the wind girder bracing, located at bottom chord level, it is desirable that the girder connections coincide with the gable posts, thereby giving the posts positional restraint, and, at the same time, preventing bending action on the bottom chord. On the internal side of the girder, it is equally important that the longitudinal ties are connected into the bracing system, as clause 4.10d states that such ties must be 'properly connected to an adequate restraint system'. This can result in the wind girder configuration like that shown in Fig. 11.16b.

11.8.1.4 VERTICAL BRACING

There are a number of ways of providing vertical bracing in the side elevations, as discussed in Chapter 10. Current practice for single-storey buildings is to use single member struts, instead of cross bracing. Figure 11.16c shows how the diagonal members have been arranged, so that wind pressure on gable causes the members to go into tension. This means that, under the smaller wind suction condition the members have to resist proportionally smaller compression forces; that is, bracing is arranged to minimize the compression that diagonal members have to sustain, leading to economic sizes.

If the client designates that openings are required in the end bay, then the vertical bracing arrangement may have to be modified if any bracing

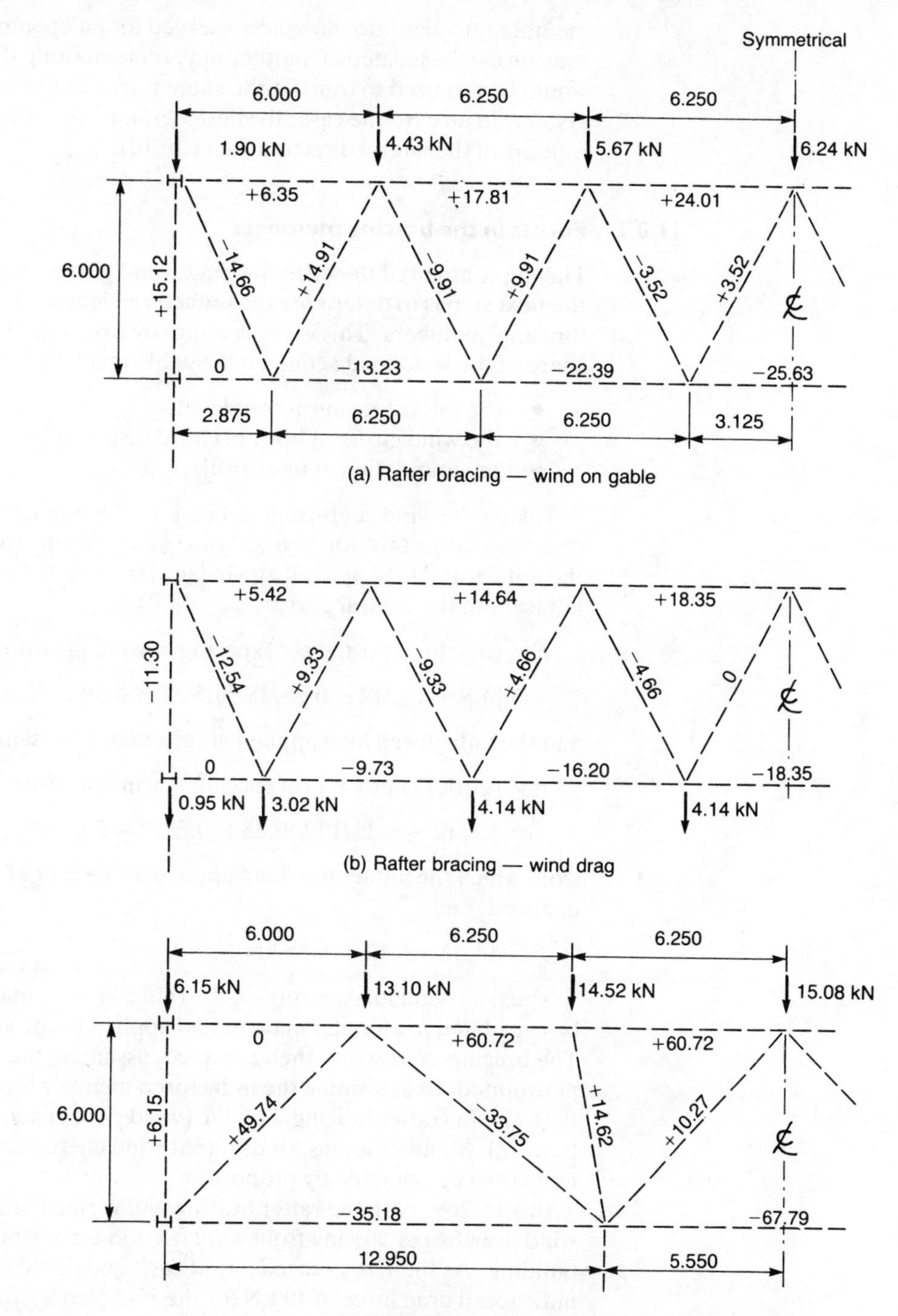

Fig. 11.17 Rafter bracing and wind girder – unfactored member loads (kN)

member intrudes into the space reserved for an opening. The vertical bracing can be located in another bay, remembering that an eaves strut would be required to transfer the gable forces to the vertical bracing system. In an extreme case, the designer may need to provide 'portal' framing in the lateral direction (see Ch. 10).

11.8.2 Forces in the bracing members

Having configured the rafter bracing, wind girder and vertical bracing, the next step is to determine the **unfactored** loads acting in the various 'bracing' members. This is easily done, by assessing the proportional share of the wind load acting on the gable taken by:

- the rafter bracing at roof level;
- the wind girder at bottom chord level; and
- the foundation at base level.

Taking the wind coefficient as being 1.0 (assuming no dominant openings subject to storm conditions) and referring to Fig. 11.16a, then the unfactored load applied at wind girder level, for, say, the post adjacent to the central post:

$$F = (\text{height of post} \times \text{post spacing} \times \text{wind pressure})/2$$
$$= [4.8 + 1.85(2 - 6.25/18.5)]6.25 \times 0.59/2 = 14.52 \text{ kN}$$

and the unfactored load applied at rafter level for same post:

$$F = (\text{girder depth} \times \text{post spacing} \times \text{wind pressure})/2$$
$$= 1.85(2 - 6.25/18.5)6.25 \times 0.59/2 = 5.67 \text{ kN}$$

from which the unfactored load applied at the foot of the column can be deduced, i.e:

$$F = 14.52 - 5.67 = 8.85 \text{ kN}$$

Thus, all wind loads acting on the rafter bracing and wind girder can be calculated in a similar manner and applied to the appropriate joints. The bracing systems are then analysed, assuming the diagonals are pin-jointed, to determine the unfactored member forces; see Figs 11.17a (rafter bracing), 11.17c (wind girder) and 11.18a (vertical bracing). Member forces for different wind coefficients (other than unity) can be obtained by proportion.

In addition, both the rafter bracing and vertical bracing have to resist **wind drag forces**, arising from wind friction across the surfaces of the building. As there is a braced bay at each end of the building, the unfactored drag force of 49 kN for the **roof** (see Section 11.4) is divided equally between the two braced bays. The rafter bracing is analysed for the wind drag force, assuming that the force is uniformly distributed across the rafter bracing system (see Fig. 11.17b). Next, analyse the vertical bracing for both the roof drag forces from the rafter bracing and the forces due to frictional drag on the side walls (= 15.2/4 = 3.8 kN

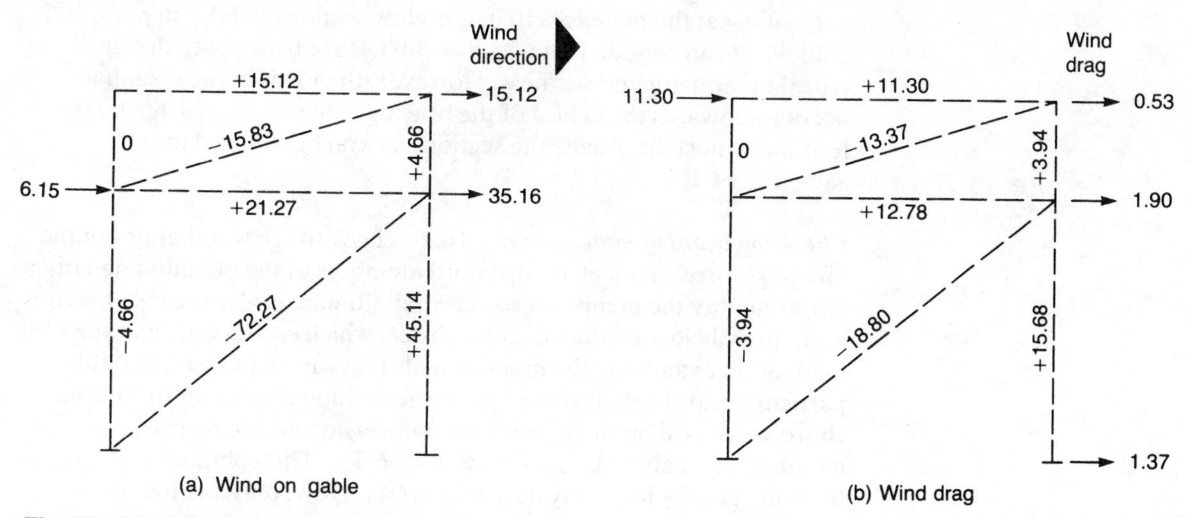

Fig. 11.18 Vertical bracing – unfactored member loads (kN)

per bracing; see Section 11.4.3). The unfactored member forces (due to wind drag) for the vertical bracing are given in Fig. 11.18b.

11.8.3 Design of bracing members and ties

Structural hollow sections usually prove to be the economic solution for bracing members, due to their structural efficiency in both the x–x and y–y directions. An alternative profile is an angle section, which is much less efficient but easier to connect. For members resisting very small loads in tension only (no stress reversal), then solid round bars could be used.

The slenderness for members resisting self weight and wind loads only, has to be less than 250 (clause 4.7.3.2b). This is particularly relevant when designing the diagonal members in the bracing systems.

11.8.3.1 RAFTER BRACING

First, the diagonal members are considered separately from the main booms, which are in fact the top chords of the end two lattice girders and therefore subject to multiple loading. Where possible, standardize on a common size for groups of members, so that an economic solution is produced. Discussion with a fabricator may establish preferred sizes.

The maximum forces in the diagonals occur when the wind loading on the gables is a maximum, i.e. when wind coefficient for gable is +1.0, together with the effect of wind drag. Therefore, design the diagonals based on the member with the largest combined compression force. From the unfactored loads indicated in Figs. 11.17a and b, the design load is calculated as:

$$F_c = 1.4(14.66 + 12.54) = 38.1 \text{ kN}$$

which is sustained by a member with a 'discontinuous' length of 6.66 m.

Looking at the table for circular hollow sections (CHS) on p.89, SCI guide[9], it can be seen that a **88.9 × 4.0 CHS** satisfies both design criteria – strength and stiffness. However, the loads in the diagonals decrease towards the centre of the bracing. Therefore, apart from the four outermost diagonals, the section size can be reduced to **88.9 × 3.2 CHS.**

Check top chord of lattice girder Next, check to see whether or not the 'bracing' forces induced in top chord members of the penultimate lattice girder modify the member size. The penultimate girder is chosen, as it is more heavily loaded than the end girder, which carries only half the roof loading. In examining the bracing loads that can be induced into this particular top chord, it should be borne in mind that the design of the chord was based on the maximum compression occurring in that member; see Table 11.4 and Section 11.6.2.1. This compression arose from the gravity load only design case (A). Now, only suction on the gable ($-0.8q$) appears to induce further compression into the top chord of the penultimate girder. However, this suction only exists as a result of the wind uplift on the roof, which in this example represents a maximum uplift condition. The resulting tension in the chord (from the uplift) more than nullifies the compression from the bracing system; that is, in this particular design example the bracing forces do not create a more severe design situation than those already considered.

The other condition which could be examined is the cumulative tension in the top chord which arises from design case (B) (Table 11.4), plus tension bracing forces. The latter occur in the top chord of the penultimate girder when there is wind pressure on the gable. However, the wind loads involved are lower than the 'compression' condition just examined. Therefore, there is no need for further calculations.

11.8.3.2 WIND GIRDER

Again, the diagonals are treated separately from the boom members of the wind girder, which are the bottom chords of the lattice girders. From Fig. 11.17c, it can be seen that the maximum compression load for a diagonal is 49.74 kN (unfactored), hence:

$$F_c = 1.4 \times 49.74 = 69.9 \text{ kN}$$

for a member having a discontinuous length of 8.49 m. Again, with reference to p.89[9], the selection is a **114.3 × 6.3 CHS** for the two outermost diagonals. A **114.3 × 5.0 CHS** is suitable for the other diagonals, even for the reversed load conditions of $-0.8q$. For example, the member carrying 33.75 kN (unfactored) tension for wind pressure on gable (see Fig. 11.17), has to sustain 27.0 kN (unfactored) compression for wind suction.

Check bottom chord of lattice girder This time, it is the maximum uplift conditions for the roof which produce compression in the bottom chord of the lattice girder. Consequently, the effect of the additional compression in the bottom chord of the penultimate lattice girder (resulting from suction on the gable) must be checked. The forces noted on Fig. 11.17c need to be multiplied by 1.4(−0.8) to obtain the required bracing forces in the bottom chord; see column (4) in Table 11.8.

Table 11.8 Bracing forces in bottom chord

	Axial forces (kN)				
Members	**Roof loading**		**Design load**	**Gable wind**	**Design load**
	1.0$_d$ **(1)**	**1.4**w_w **(2)**	**(1)+(2) (3)**	**1.4**w_w **(4)**	**(3)+(4) (5)**
11,21	0.0	+ 5.1	+ 5.1	+ 39.4	+ 44.5
12,20	− 45.8	+ 144.4	+ 98.6	+ 39.4	+ 138.0
13,19	− 69.8	+ 210.2	+ 140.4	+ 39.4	+ 179.8
14,18	− 80.2	+ 241.1	+ 160.9	+ 39.4	+ 200.3
15,17	− 81.5	+ 242.7	+ 161.2	+ 75.9	+ 237.1
16	− 76.4	+ 229.4	+ 153.0	+ 75.9	+ 228.9

On checking the design calculations for the bottom chord in Section 11.6.2.2, it can be seen that the compression resistance of both the outer and middle lengths is just adequate, i.e. 200.3/213.4 = 0.939 and 237.1/250.6 = 0.946 respectively. Therefore, adopt the **245 × 127 × 37 Tee** for **all** bottom chords of lattice girders.

11.8.3.3 VERTICAL BRACING

Figure 11.18a shows the unfactored forces in the members due solely to the gable wind with a coefficient of +1.0, while Fig. 11.18b indicates the unfactored member forces arising from wind drag on both the roof and side walls. These wind drag forces in the same direction as the wind. Figure 11.19 gives the factored forces (with $\gamma_f = 1.4$) in the members for those wind conditions which might affect the design of the members carrying bracing forces. The load factor of 1.4 is used as the loads in the non-vertical members are entirely due to wind. Also, these members must have a limiting slenderness of 250. The following information summarizes the design of these non-vertical members, based on the maximum loads that can arise from wind on side, with insignificant drag forces (Fig. 11.19a) or the combined effect of wind on gable plus drag (Fig. 11.19b, c and d).

Horizontals 6.0 m long; 47.7 kN compression; 26.8 kN tension
Use 88.9 × 4.0 CHS[(9)]

Upper diagonal 6.3 m long; 25.4 kN compression; 40.9 kN tension
Use 88.9 × 3.2 CHS[(9)]

Lower diagonal 7.7 m long; 80.9 kN compression; 127.5 kN tension
Use 114.3 × 5.0 CHS[(9)]

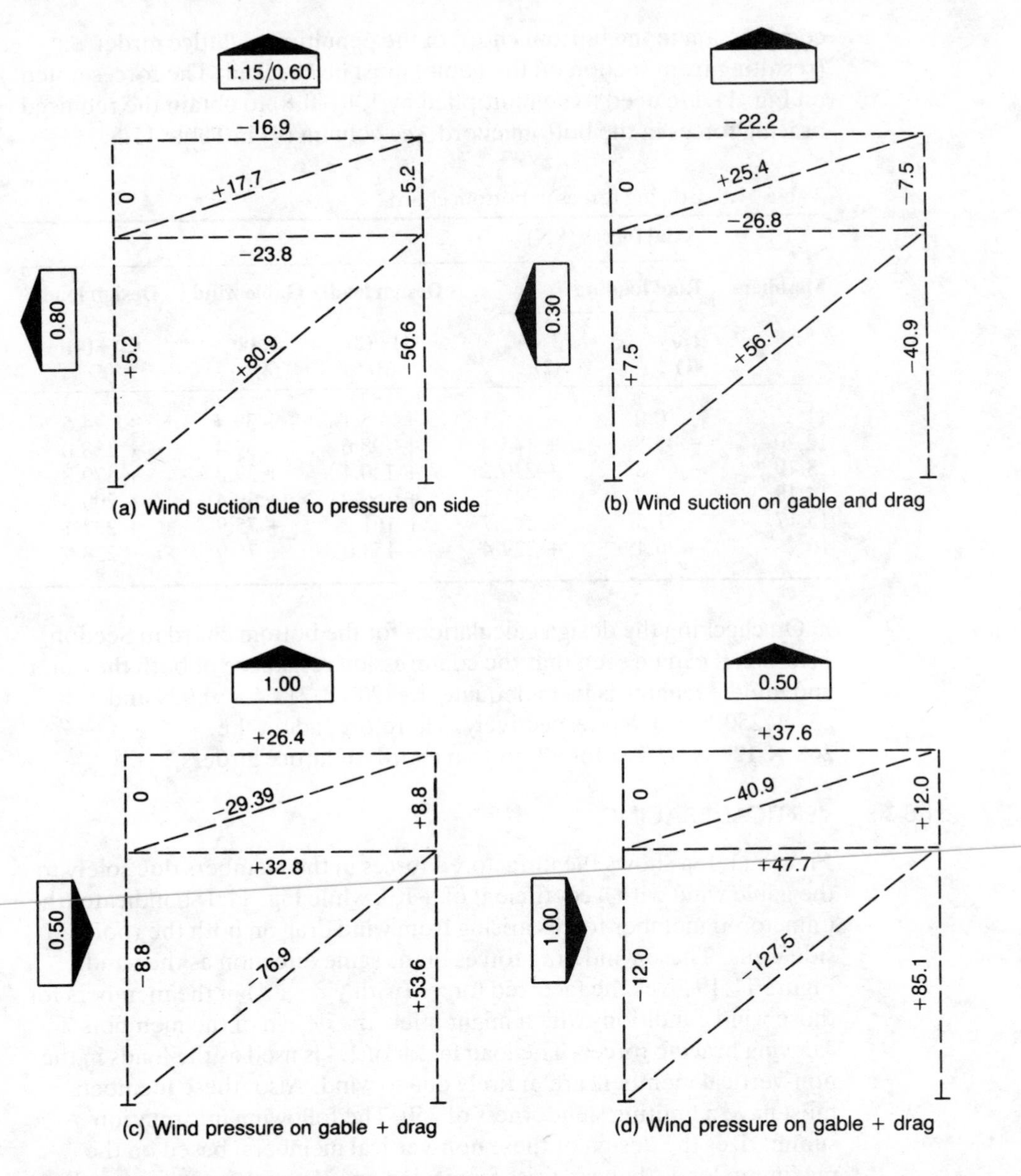

Fig. 11.19 Vertical bracing – factored member loads (kN)

Check column member Now the penultimate column member has to be checked for the maximum axial compression load from the bracing forces, i.e. 45.14 + 15.68 = 60.8 kN (unfactored). This load arises from a wind condition which causes a pressure of $+1.0q$ on the gable and $-0.2q$ uplift on the roof. In fact, this is the design case (C) (iii) noted in Section 11.7.1.3 and Table 11.7. Therefore, total axial compression is:

$$F_c = 1.2[10.9 + (61.7 + 166.5 - 0.2 \times 131.0)0.5 + 60.8] = 207.2 \text{ kN}$$

$$M_x = 2.3 \text{ kN (Section 11.7.1.3 (iii))}$$

giving $$\frac{207.2}{408} + \frac{0.54 \times 2.3}{63} = 0.508 + 0.020 = 0.528 < 1.0$$

Though wind suction on the gable and the associated wind drag can cause additional tension for design case (B), noted in Section 11.7.1.3, the effect is marginal in terms of the local check. As the buckling check does not apply for the 'tension' case, the adopted section remains satisfactory.

11.8.3.4 LONGITUDINAL TIES

As stated, the purpose of the longitudinal ties is to restrain the bottom chord member of the lattice girders (see Fig. 11.16b). Assume that the 'ties' need to provide a restraining force equal to at least 2% of the maximum load in the chord, i.e. $0.02 \times 469.3 = 9.4$ kN, which may cause compression in the 'ties'. Over a discontinuous length of 6 m and requiring a slenderness not exceeding 250, a **76.1 × 3.2 CHS** is suitable[9].

11.8.3.5 EAVES TIES

These members give positional restraint to the column heads and also provide restraint to the ends of the top chords of the lattice girder. Use same size as that for the longitudinal ties, i.e. **76.1 × 3.2 CHS**. Restraint at bottom chord level is to be provided by diagonal braces from the nearby sheeting rail to the inner flange of the column.

11.9 DESIGN OF GABLE POSTS

The design of the gable posts is straightforward. They are self-supporting and are only connected to the main frame at top and bottom chord level (see Fig. 11.16a or b). Having established the spacing of the posts, i.e. 6.25 m, the post are assumed (for simplicity) to be simply supported between the column base and the connection at wind girder level, i.e. 4.8 m span. Apart from the gravity load of the sheeting, plus insulation, plus liner, plus sheeting rails and the column's own weight (which produces a nominal axial load in the post), the main loading is a bending action, due to wind blowing on the gable.

Design the central post; see Fig. 11.20. The axial load includes that for cladding plus self weight of the side rails and post, i.e:

$$F_c = 1.4[\text{cladding} + \text{insulation}) + (\text{post} + \text{side rails})]$$

$$= 1.4[0.13 \times 5.72 \times 6.25 + 0.5\ (\text{est.}) \times 4.8] = 9.9\ \text{kN}$$

Wind load on a 4.8 m length of post is:

$$F = 1.4 \times 1.0 \times 0.59 \times 4.8 \times 6.25 = 24.8\ \text{kN}$$

Hence $M_x = 24.8 \times 4.8/8 = 14.9$ kNm

Select a **203 × 133 × 25 UB**. From the SCI guide[9], $P_z = 779$ kN and $M_b = 29$ kNm, from which a buckling check can be undertaken, i.e:

$$\frac{9.9}{779} + \frac{14.9}{29} = 0.013 + 0.514 = 0.527 < 1.0$$

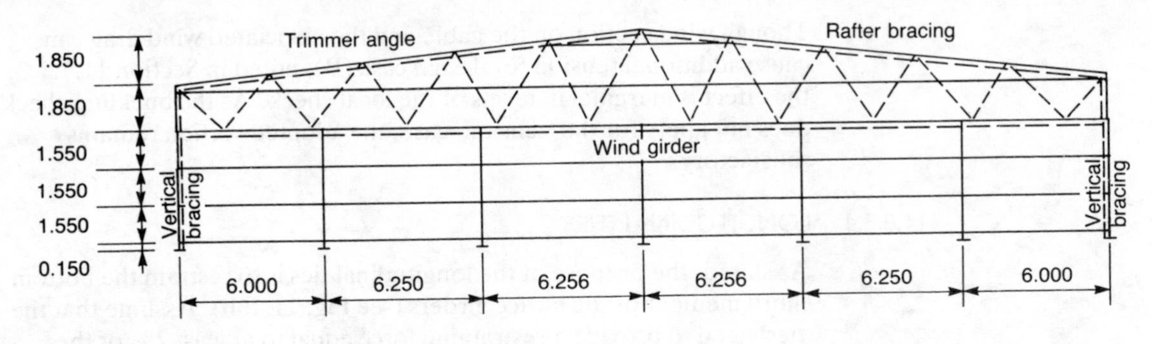

Fig. 11.20 Gable framing

This neglects (for simplicity) any lateral restraint from sheeting rails.

Use 203 × 133 × 25 UB.

11.10 DESIGN OF CONNECTIONS

Having designed the members that make up the lattice girder there remains the design of the connections to be completed. Basically, the function of a connection is to transfer loads efficiently from one member to another without undue distress, while minimizing the cost of fabrication and ensuring ease of erection.

When designing connections, it is always useful to draw connection details to scale. For bolted joints, such details can demonstrate readily whether or not the bolts can be inserted easily at the appropriate locations. In the case of welded joints, it is essential to check that the welder is able to deposit the weld metal without impedance.

Though the girder was analysed assuming that the diagonals were pin-ended, connecting the diagonals to the chords results in a degree of fixity. Independent of whether or not the connection is designed bolted (with at least two bolts) or welded, clause 4.10 implies that secondary stresses could be ignored, provided that the slenderness of the chord members in the plane of the girder is greater than 50 and that of the web members is greater than 100. In this example, the slenderness of the chord is just below 50 (46), while the web members have slenderness well above 100. Therefore, it could be deemed that such stresses are insignificant.

By arranging alternate diagonal members to be connected on the opposite side of the stem of the T section, then the number of ends cut on a skew is reduced and hence cost of fabrication. The other benefit is that each half girder can be manufactured in an identical manner (Fig. 11.21) and when spliced together, the alternate pattern for the diagonals continues across from one half to the other half; that is, all diagonals sloping one way are connected to one side of the T section, while the remaining diagonals sloping in the opposite direction are attached to the other side of the chord. This arrangement allows any sheeting rail(s) in the gable, located within the depth of the end girders, to be supported by the oustand legs of every other diagonal member.

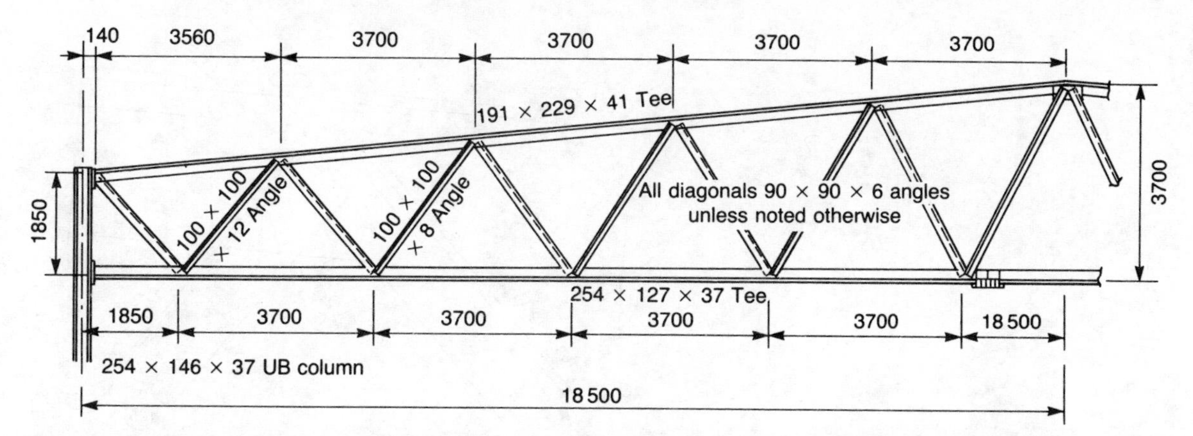

Fig. 11.21 General details of lattice girder

However, the disadvantage is that the additional loading imposed on these members might cause their sizes to be modified (see Section 11.10.5). The design of typical connections for the lattice girder is now examined in detail.

11.10.1 Typical diagonal to chord connections

The typical internal connection chosen to be designed is node 3, where member 23 (90 × 90 × 6 Angle) and member 24 (100 × 100 × 12 Angle) intersect with the bottom chord (254 × 127 × 37 T section); see Fig. 11.22b. When designing welds, it is immaterial whether the axial load is tension or compression (apart from fatigue or brittle fracture considerations, which do not apply to this example). The welds have to be designed to transfer the largest possible load occurring in the member under any loading conditions, i.e. member 23 carries a force of 204 kN and member 24 a force of 188 kN (see Table 11.5). Almost invariably with this form of construction (using angle sections), the centre of the weld group is eccentric to the centroidal axis of angle section, i.e. the weld group is not balanced. Therefore, in addition to the axial load, the weld group has to be designed to resist the in-plane torsional moment generated by this eccentricity, as will be illustrated by the ensuing calculations.

For example, take the connection between member 23 and the bottom chord: the maximum possible lengths along the toe and heel of the angle are 135 mm and 55 mm respectively (see Fig. 11.22b). Also, the weld across the end of the angle is 90 mm. As the stem of the T section is relatively short it is probably advisable to weld the angle to the edge of the stem, i.e. on the reverse side to the other welds, producing the weld group shown in Fig. 11.23.

First, determine the distance of the longitudinal axis of weld group from heel of angle by taking moments of the weld lengths about the heel. The inclined weld length (120 mm) is treated in the same manner as the other welds.

$$\bar{x} = (90^2/2 + 135 \times 90 + 120 \times 45)/(55 + 135 + 120 + 90)$$

$$= 54.0 \text{ mm}$$

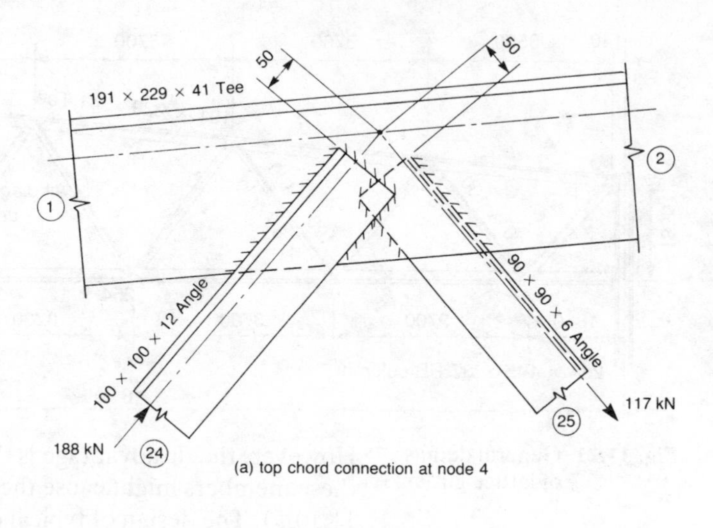

(a) top chord connection at node 4

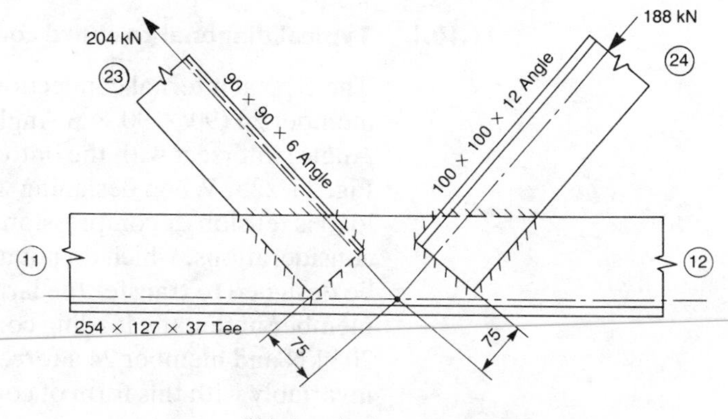

(b) bottom chord connection at node 3

Fig. 11.22 Details of typical internal connections

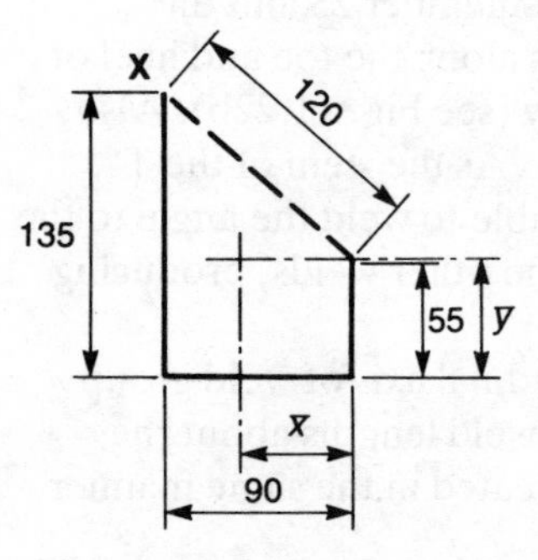

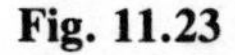
Fig. 11.23

The lateral axis of the weld group, relative to the end, is determined in a similar manner, i.e:

$$\bar{y} = (55^2/2 + 135^2/2 + 120 \times 95)/(55 + 135 + 120 + 90) = 55.1 \text{ mm}$$

From the SCI guide[9] the centroidal axis is 24.1 mm from heel of the angle. Hence, the eccentricity of the load is 54.0–24.1 = 29.9 m resulting in a moment of 204 × 29.9/1000 = 6.10 kNm. Thus, the weld group for member 23 has to be designed for an axial load of 204 kN and an in-plane moment of 6.10 kNm. Therefore the equivalent polar second moment of area (inertia) of the weld group needs to be evaluated.

The general practice is to ignore the local inertia about their own axes for welds parallel to the global axis being considered – a conservative assumption. Note the inertia about its local axis for the component of an inclined weld (of length d), perpendicular to the global axis, is significant and must be taken into account, i.e:

$$I'_{cg} = \frac{dh^2}{12}$$

where h is the projected length of the inclined weld, normal to the global axis about which the equivalent inertia is being determined (see Fig. 11.24). Therefore, the total inertia for an inclined weld about a global axis, say the x–x axis, is:

$$I'_x = \frac{dh^2}{12} + dy^2$$

where y is the distance from the centroid of the inclined weld to the global axis, normal to the axis being considered (see Fig. 11.24). The total inertia for an inclined weld about the other global axis can be obtained in a similar manner, as is demonstrated in the following calculations for the weld group shown in Fig. 11.23.

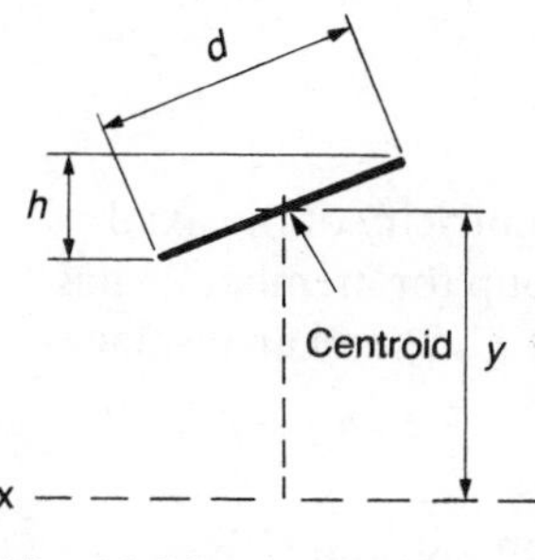

Fig. 11.24

$$I'_p = I'_x + I'_y$$

$$I'_x = 90 \times 55.1^2 + 2 \times 55.1^3/3 + (135 - 55.1)^3/3 + 120 \times 80^2/12 + 120(95 - 55.1)^2 = 810 \times 10^3 \text{ mm}^4$$

$$I'_y = 54.0^3/3 + (90 - 54.0)^3/3 + 55 \times 54.0^2 + 135(90 - 54.0)^2 + 120 \times 90^2/12 + 120(54.0 - 45)^2 = 494 \times 10^3 \text{ mm}^4$$

$$I'_p = (810 + 494) \times 10^3 = 1304 \times 10^3 \text{ mm}^4$$

The shear per unit length due to axial load (F_S) is:

$$F_S = 204/(55 + 135 + 120 + 90) = 0.510 \text{ kN/mm}$$

and the shear per unit length due to the eccentric moment (F_T) is obtained by considering that part of the weld furthest away from the centre of weld group, i.e. point X in Fig. 11.23:

$$F_T = MR_{max}/I'_p$$

where $R_{max} = \sqrt{[(90 - 54.0)^2 + (135 - 55.1)^2]} = 87.6$ mm
and $\cos\theta = (90 - 54.0)/87.6 = 0.411$

$$F_T = 6.10 \times 87.6/1304 = 0.410 \text{ kN/mm}$$

Now $F_r = \sqrt{[F_S^2 + F_T^2 + 2F_SF_T\cos\theta]}$
$= \sqrt{[0.510^2 + 0.410^2 + 2 \times 0.510 \times 0.410 \times 0.411]}$
$= 0.775 \text{ kN/mm} < 0.903 \text{ kN/mm}$ (p.205 reference 9)

Use 6 mm FW.

Now examine the connection at node 4, where member 24 (100 × 100 × 12 Angle) and member 25 (90 × 90 × 6 Angle) intersect with the top chord (191 × 229 × 41 T section); see Fig. 11.22a. Designing the connection between member 24 and the chord, it is noted that the stem of the top chord is deeper than that for the bottom chord, with the result that the longitudinal welds can have lengths up to 240 mm (heel) and 130 mm (toe), while the end weld is 100 mm long (Fig. 11.25). In this case, the weld group will be designed without the inclined weld length as it might not be necessary.

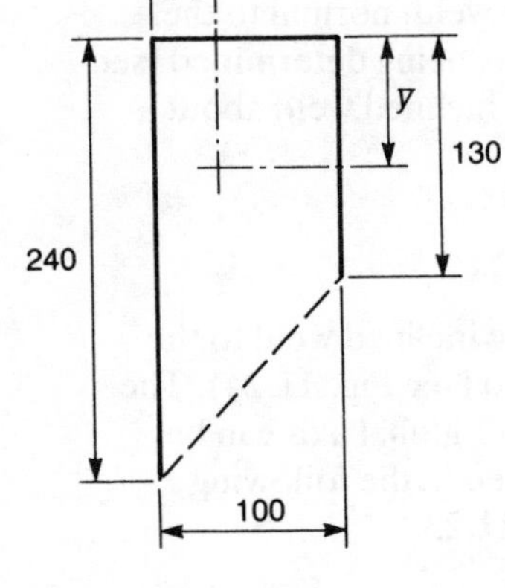

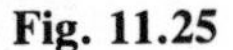
Fig. 11.25

The determination of the weld size is identical to previous calculations, except that there is no inclined weld. For brevity, only an results are given:

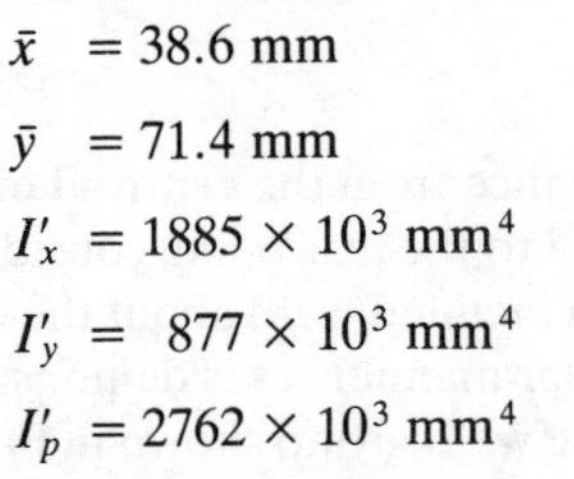

$$\bar{x} = 38.6 \text{ mm}$$

$$\bar{y} = 71.4 \text{ mm}$$

$$I'_x = 1885 \times 10^3 \text{ mm}^4$$

$$I'_y = 877 \times 10^3 \text{ mm}^4$$

$$I'_p = 2762 \times 10^3 \text{ mm}^4$$

Relative to the centre of the weld group, the eccentricity of the axial load is 38.6 − 29.0 = 9.6 mm. Thus the weld group for member 24 has to be designed for an axial load of 188 kN (Table 11.5) and an in-plane moment of 188 × 9.6/1000 = 1.80 kNm.

The shear per unit length due to axial load is:

$$F_S = 188/(120 + 100 + 220) = 0.427 \text{ kN/mm}$$

Now $R_{max} = 153.5$ mm
and $\cos\theta = 0.251$

Therefore the shear due to the torsional moment is:

$$F_T = 1.80 \times 153.5/2762 = 0.100 \text{ kN/mm}$$

$$F_r = \sqrt{[0.427^2 + 0.100^2 + 2 \times 0.427 \times 0.100 \times 0.251]}$$
$$= 0.462 \text{ kN/mm} < 0.602 \text{ kN/mm (p.205 reference 9)}$$

In fact, a 6 mm fillet weld (FW) is the minimum practical size for welds.

Use 6 mm FW.

Other internal connections can be designed in a similar way.

11.10.2 Lattice girder to column connections

The design of the welded connection between member 1 and member 23 at node 2 (see Fig. 11.26a), follows the same method already outlined in Section 11.10.1. There remains the transfer of the vertical shear 176 kN

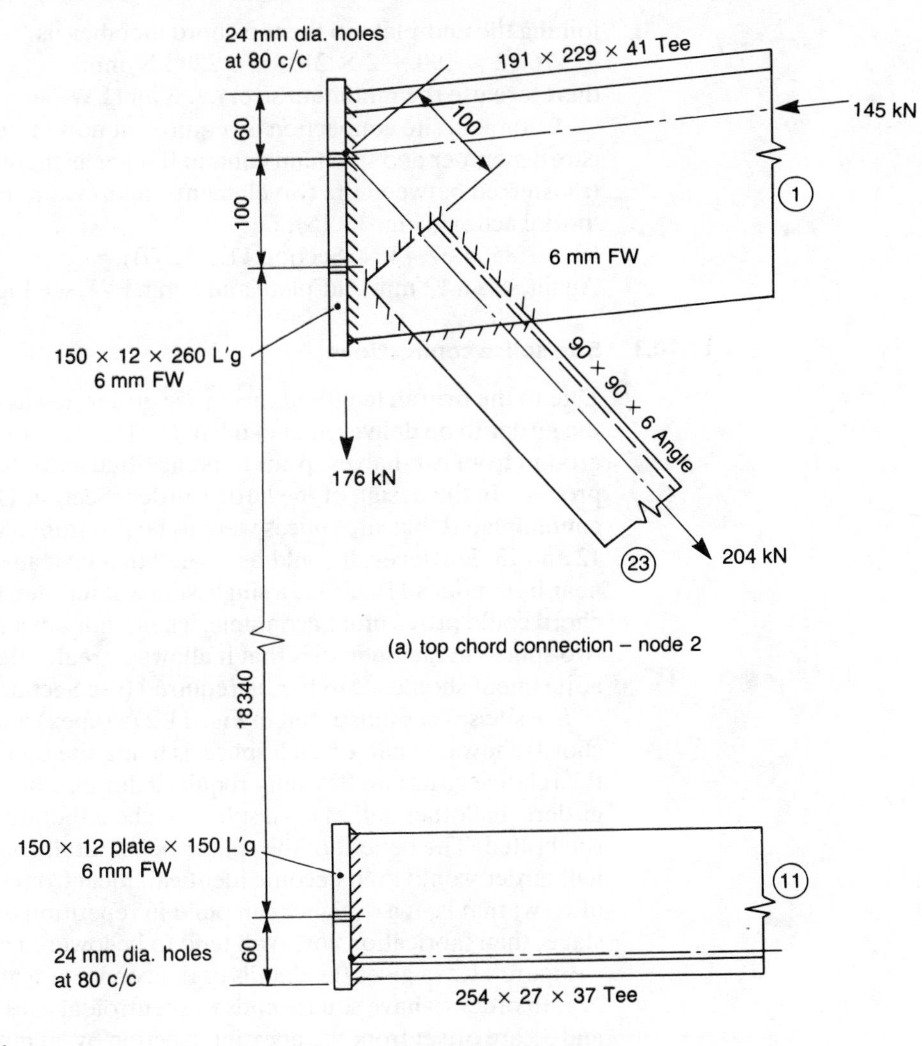

Fig. 11.26 Lattice girder to column connections

from the end-plate connection into the column flange. In fact, this is a site connection, which is generally bolted (more cost-effective), and anticipating other site connection details, 22 mm diameter bolts (grade 4.6) are used. As the single shear strength of this bolt size is 48.5 kN[(9)], then the number of bolts required is 176/48.5 = 3.6, four bolts (see Fig. 11.26a). Bearing strength requirements indicate a minimum plate thickness of 6 mm[(9)]. This is more than satisfied by using an end-plate thickness of 12 mm, which would restrict any excessive out-of-plane deformation of the plate. Also, the shear per unit length for the weld,

joining the end-plate to the top chord member, is
$176/(2 \times 190 + 2 \times 210) = 0.220$ kN/mm;
therefore use the minimum size, i.e. 6 mm FW.

A nominal site connection is required at node 1 between the bottom chord member and the main column flange, as there is little load being transferred between the two elements, apart from the axial load due to 'portal action' (Fig. 11.15), i.e.
$69.1/1.85 = 37.4$ kN (Section 11.7.1.2(i)).
Again, use a 12 mm end-plate and 6 mm FW; see Fig. 11.26b for details.

11.10.3 Site splice connections

Due to the overall length of the lattice girder, it was decided to arrange the girder to be delivered in two halves. The girder is assembled on the ground from two halves, prior to being lifted aloft during the erection process. In the design of the lattice girder (Section 11.6.2) it had been contemplated that site splices were to be positioned at or near nodes 11, 12 and 13. However, it could be argued that instead of making splices near both nodes 11 and 13, a single splice at mid-length of the bottom chord could prove more economic. The minor advantage of having the two splices in member 16 is that it allows a greater flexibility in adjustment should a camber be required (see Section 11.12).

The site splices illustrated in Fig. 11.27a (apex) and 11.27b (bottom chord) show one half of each splice as being site bolted. Depending on the relative costs and flexibilty required during assembly of the lattice girders, the other half of each splice can be either shop welded or also site bolted. The benefit of the splices being totally bolted is that each half girder would then become identical, ideal from the fabrication point of view; that is, if a designer can build in repetition during the design stage, then fabrication costs will tend to be lowered as a direct result.

Figure 11.27a gives the details of the connection at the apex (node 12). In order to have square ends the centroidal axes of the diagonals 32 and 33 are offset from the apex intersection by 50 mm. As the load in these members is relatively small, the slight eccentricity should not have a significant effect. Also, though one bolt would suffice to transmit the load of 31 kN from the diagonal to the chord, it is good practice to use a minimum of two bolts. Note that in the case of the lower bolt in the diagonal, its load is transferred via the gusset plates and pack into the stem of the chord.

The size and shape of the gusset plates has been arranged so that it can be square cut from a standard plate section. It makes good sense to avoid too many cuts in a gusset plate, otherwise costs will increase. Too many cuts indicates a lack of appreciation of fabrication processes and costs. The splice has to transmit 456 kN between the two halves of the chord. Using grade 4.6 bolts, the number required is $456/48.5 = 9.4$, say 10. This is accomplished by locating (in one half of the splice) 6 bolts in the table of the T section and the other 4 bolts in the stem, i.e. roughly proportional to the areas of the table and stem. Figure 11.27a shows a possible arrangement for the 10 bolts.

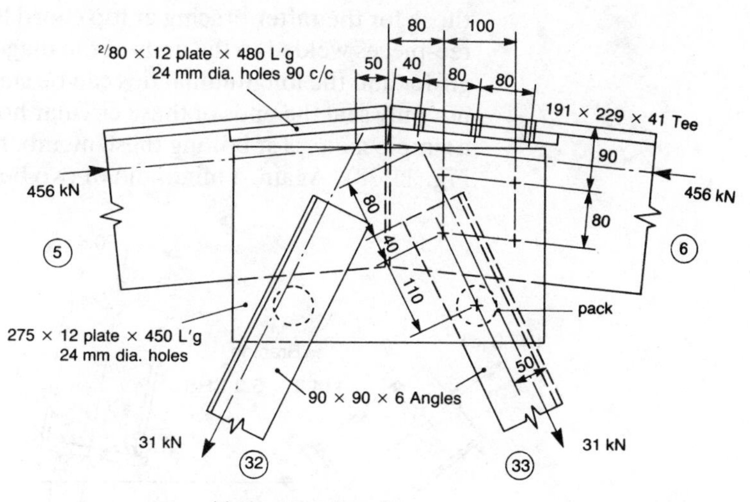

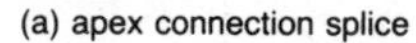
(a) apex connection splice

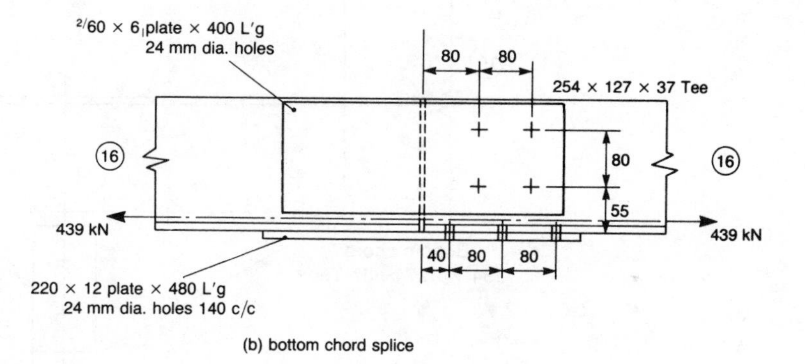

(b) bottom chord splice

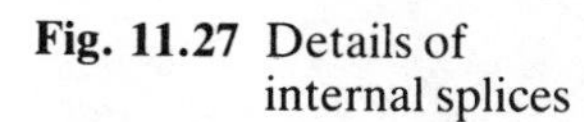
Fig. 11.27 Details of internal splices

Coming to the bottom chord splices, the design of them is not too dissimilar from that of the apex splice, i.e. 10 bolts are also required on each side of the splice in order to transfer 439 kN (see Fig. 11.27b for details of splice).

11.10.4 Typical bracing connections

The details of a connection between the diagonals of the wind girder and the bottom chord of the penultimate girders (given in Figs 11.28 and 11.29) illustrate the design basis for other bracing connections, including those for the rafter bracing at top chord level. By careful detailing, the tee-pieces welded to the ends of the diagonal members of the wind girder and the longitudinal ties can be standardized. These tee-pieces not only seal the ends of these circular hollow sections, but also provide a simple means for bolting these members to a gusset plate (see Fig. 11.28). Again, a minimum of two bolts per joint is used.

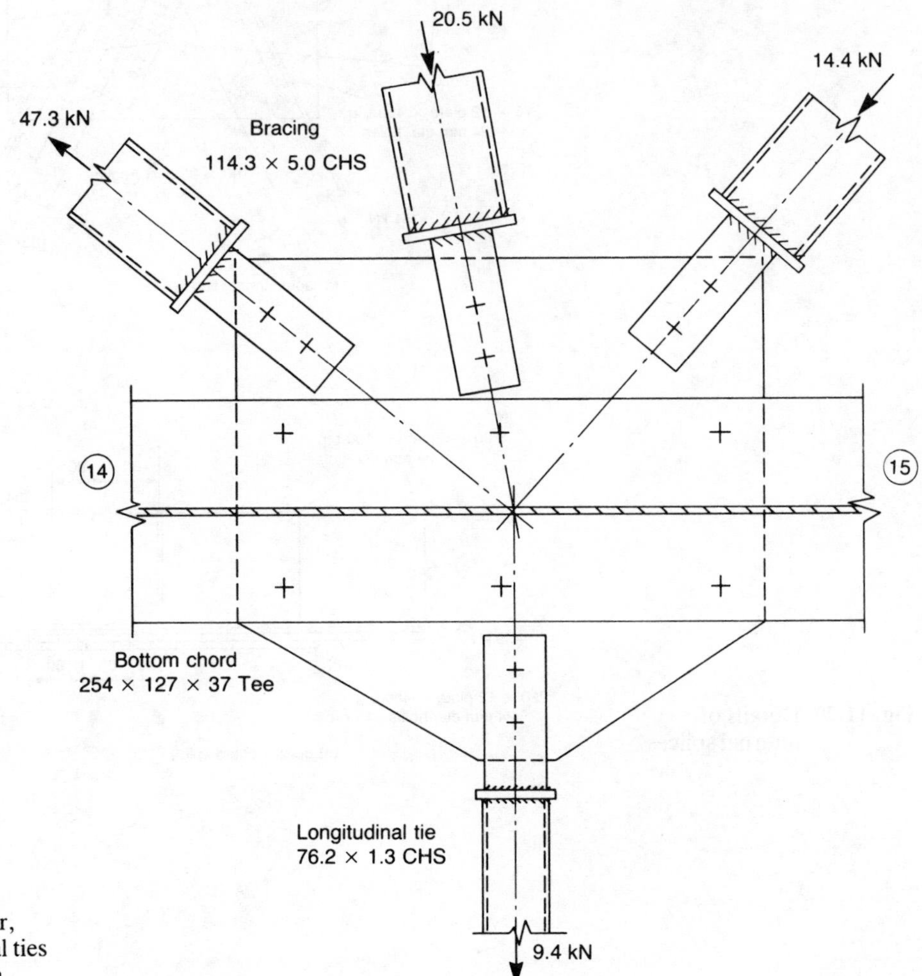

Fig. 11.28 Wind girder, longitudinal ties and bottom chord connection

These tee-pieces can either be made by welding two plate elements together or manufactured from a cutting from a suitable I section. The plate element (welded direct to the tubular members) must be stiff enough, otherwise dishing can occur, causing high stress concentrations at the junction between the two plate elements comprising the tee-piece. Such high stress can lead to material rupture in the form of tearing, particularly if the elements are welded together. It is recommended that, for the size of the hollow section selected, at least a 12 mm thick plate should be used for the table of the tee-pieces.

Figure 11.28 shows the gusset plate (12 mm thick) bolted to the table of the bottom chord, for ease of removal if and when the building is extended. Though no holes were deducted in the design of the outer lengths of the bottom chord (Section 11.6.2.2), an examination of those particular calculations indicates the reduced area is more than adequate to resist the design loads in tension. Alternatively, this plate could be welded into position, though this might cause some problems should removal be required.

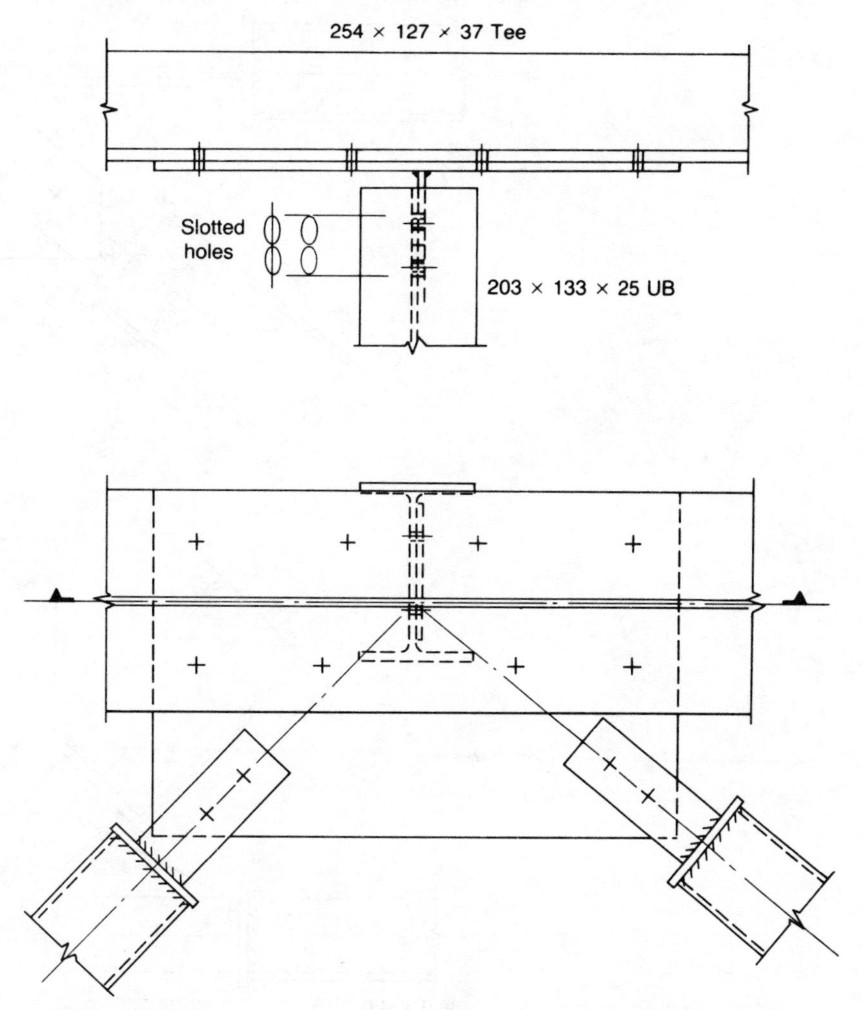

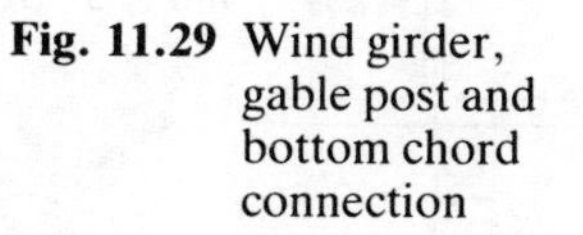
Fig. 11.29 Wind girder, gable post and bottom chord connection

The connection between the diagonals of the wind girder and the bottom chord of the end girders is similar to the connection just outlined, except that there is an additional factor. A previous design decision was to run the gable posts up to the underside of the end girders, coinciding with the nodal points of the wind girder (see Fig. 11.17c). Each gable post is to be connected into the wind girder system via a plate welded underneath and to the gusset plate (see Fig. 11.29). However, the end lattice girder (even though only carrying half the load) deflects vertically under imposed and/or wind loading and could cause problems in the gable posts. Therefore the plate to which the web of the post is bolted, has slotted holes which allow the girder to deflect the calculated movement due to imposed loading and/or wind uplift.

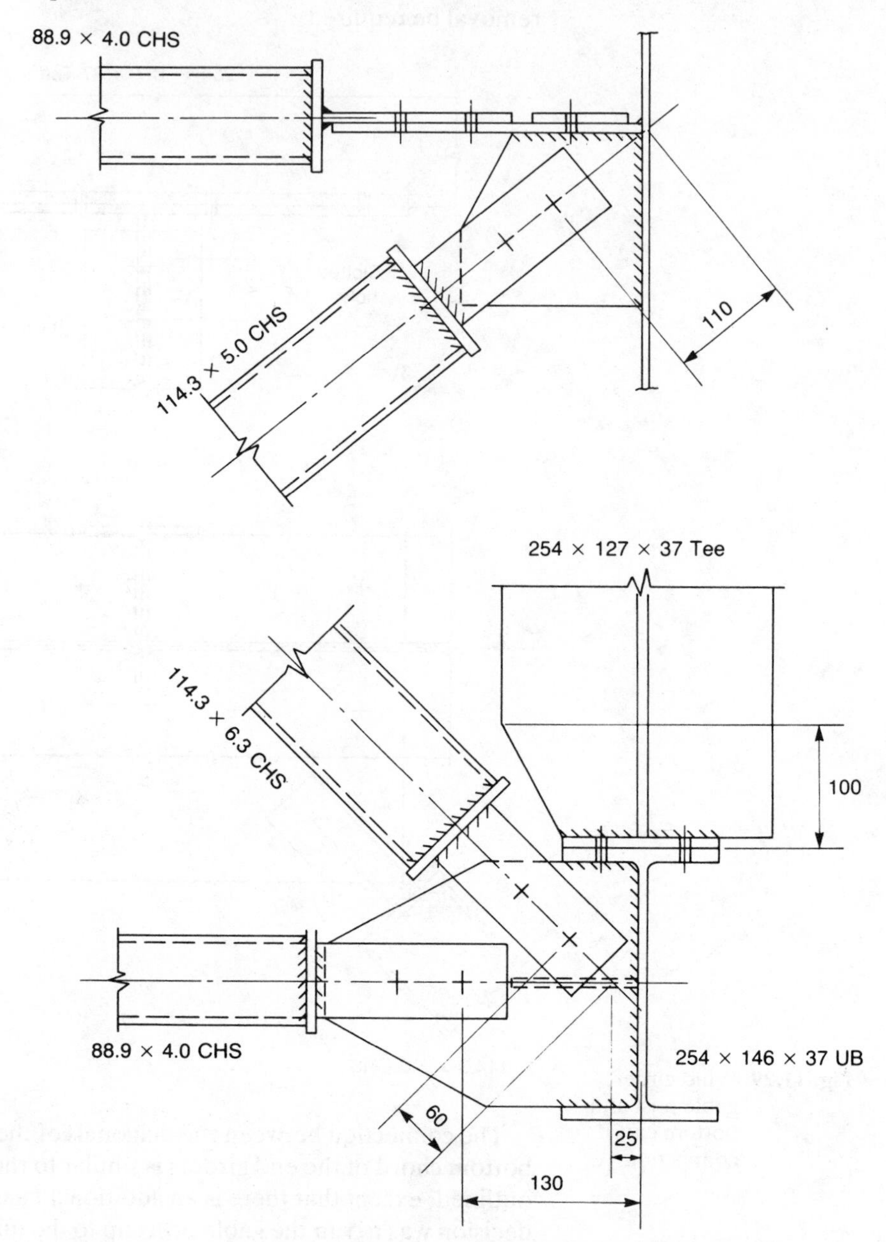

Fig. 11.30 Wind girder, vertical bracing and column connection

Finally, the connection detailed in Fig. 11.30 is typical of those connections in which two bracing systems acting in different planes intersect at a common point. In the example shown, the intersection of the wind girder and vertical bracing with the column member is considered. To achieve a correct transfer of forces, the various bracing members should intersect at a common point, otherwise moments are induced. However, practical considerations dictate that the horizontal diagonal could be offset by 25 mm.

11.10.5 Check diagonals of the end girders

As the gable posts run up to the underside of the end girders, this means that the sheeting rails located within the depth of an end lattice girder have to be supported via the alternate diagonal members of the girder. Due to the eccentricity of the rails relative to the vertical plane of the girder (Fig. 11.31), additional loading in the diagonals is induced. Therefore the sizes of these diagonals need to be checked for their adequacy to withstand this extra loading. It is noted that the end girder carries only half the load for which the intermediate girders were designed (Table 11.6).

Assuming that the rails are supported by every other diagonal, it can be seen from Fig. 11.21 that, if the vertical spacing of sheeting rails is continued at 1.55 m, then the span of these rails is about 3.7 m. (Doubling the span would result in making the section B170/155 unsuitable.) Examination of the central diagonal (member 33), shows that the additional loading arises from two sheeting rails. This load is mainly due to the wind suction on the gable ($-0.8q$) in the design case (B); that is, referring to Fig. 11.32:

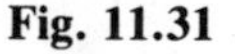

Fig. 11.31

Horizontal load at A due to wind

$$= 1.4 \times 0.8 \times 0.59 \times 1.55 \times 3.7 = 3.79 \text{ kN}$$

Vertical load at A due to cladding, insulation, liner and rails

$$= 1.4(0.13 \times 1.55 + 0.043)3.7 \quad = 1.28 \text{ kN}$$

The loads at B can be derived by proportion, i.e:

$$= [1.55 + 0.50(\text{av.})]/(2 \times 1.55) \ = 0.66$$

Hence the horizontal load at B

$$= 0.66 \times 3.79 = 2.50 \text{ kN}$$

and the vertical load at B

$$= 0.66 \times 1.28 = 0.84 \text{ kN}$$

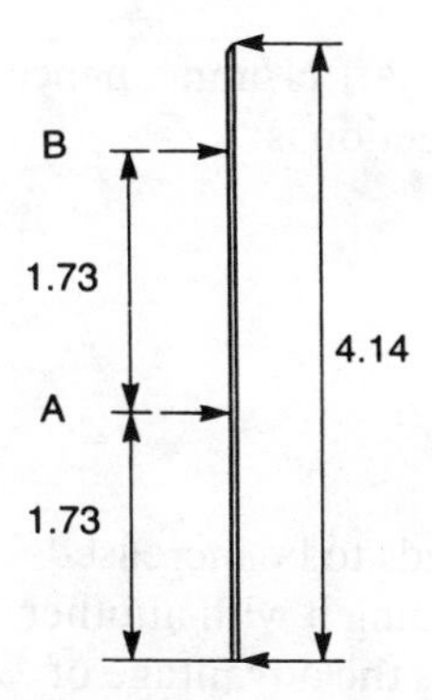

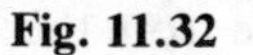

Fig. 11.32

The resulting moment in the diagonal due to the horizontal load is noted in Table 11.9, based on the dimensions indicated. The moment due to the vertical load is not significant (2%) and is ignored. The appropriate loads can be determined for all the other diagonal members supporting sheeting rails (see Table 11.9).

Table 11.9 Additional loading on diagonal members

Member	Size	Length (m)	Distance from bottom chord along diagonal (A)	Distance from bottom chord member (B)	Max. moment (kNm)	Axial load (kN)
23	90 × 90 × 6	2.62	2.20	—	0.83	+ 36.2
25	90 × 90 × 6	2.89	2.02	—	1.65	+ 19.6
27	90 × 90 × 6	3.18	1.90	—	2.42	+ 9.5
29	90 × 90 × 6	3.49	1.83	—	3.26	− 5.5
31	90 × 90 × 6	3.84	1.77	3.54	3.84	+ 9.0
33	90 × 90 × 6	4.14	1.73	3.46	4.54	+ 9.3
35	90 × 90 × 6	3.84	1.77	3.54	3.84	+ 4.1
37	90 × 90 × 6	3.49	1.83	—	3.26	− 9.0
39	100 × 100 × 8	3.18	1.90	—	2.42	− 18.5
41	100 × 100 × 12	2.89	2.02	—	1.65	− 33.3

In checking the adequacy of member 33, it is assumed that the combined out-of-plane stiffness of the cladding and sheeting rails constrains the diagonal member (via the rail attachments) to bend about its x–x axis, rather than y–y axis. The axial load is 9.3 + 1.28 + 0.84 = 11.4 kN and the maximum bending moment in the member is 4.54 kNm. The SCI guide[9] indicates that the member size (90 × 90 × 6 Angle) is **slender**, therefore the stress reduction factor (BS table 8) is the lesser of:

$$\frac{11}{\frac{d}{T\varepsilon} - 4} \quad \text{or} \quad \frac{19}{\frac{b+d}{T\varepsilon} - 4}$$

$$= 1.00 \quad \text{or} \quad 0.73$$

This gives a reduced design strength of 0.73 × 275 = 201 N/mm², hence the local capacity check for the 90 × 90 × 6 Angle section is:

$$\frac{F_c}{P_z} + \frac{M_x}{M_c} \leqslant 1.0$$

$$\frac{11.4 \times 10}{10.6 \times 201} + \frac{4.54}{0.201 \times 12.2} = 0.054 + 1.851 > 1.0$$

Therefore the angle is not adequate – the section needs to be increased substantially or its properties enhanced by compounding it with another angle. The latter solution is the one adopted, as it has the advantage of providing support to the sheeting rails at the required distance from the vertical plane of the girder, i.e. 135 mm. If an additional angle is bolted

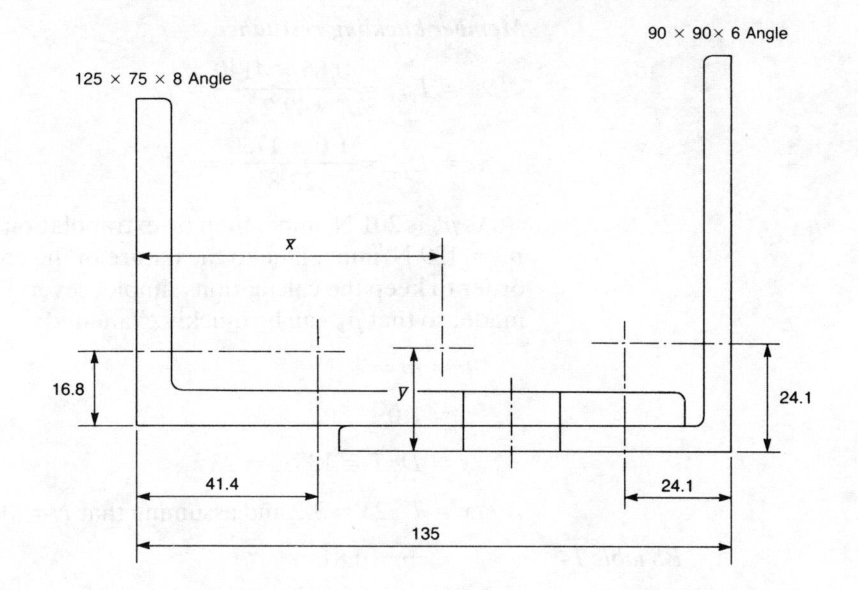

Fig. 11.33 Details of compound section for diagonal member

to the diagonal (see Fig. 11.33), then it can easily be removed in the future, at which time the original end girder in fact becomes an intermediate frame.

A 125 × 75 × 8 Angle is selected, resulting in the compound section shown in Fig. 11.33. The properties of the compound section now have to be calculated, i.e:

$$\bar{x} = \frac{15.5 \times 4.14 + 10.6(13.5 - 2.41)}{15.5 + 10.6} = 6.96 \text{ cm}$$

$$I_x = [247 + 15.5(6.96 - 4.14)^2] + [80.3 + 10.6(13.5 - 2.41 - 6.96)^2] = 631 \text{ cm}^4$$

$$Z_x = 631/6.54 = 96.5 \text{ cm}^3$$

$$r_x = \sqrt{\frac{I_x}{A}} = \sqrt{\frac{631}{26.1}} = 4.92 \text{ cm}$$

$$\bar{y} = \frac{15.5(1.68 + 0.6) + 10.6(2.41)}{26.1} = 2.33 \text{ cm}$$

$$I_y = [67.6 + 15.5(2.33 - 1.68 - 0.60)^2] + [80.3 + 10.6(2.41 - 2.33)^2] = 148 \text{ cm}^4$$

$$r_y = \sqrt{\frac{I_y}{A}} = \sqrt{\frac{148}{26.1}} = 2.38 \text{ cm}$$

Local capacity check

$$\frac{11.4 \times 10}{26.1 \times 201} + \frac{4.54}{0.201 \times 69.5} = 0.022 + 0.234 < 1.0$$

Member buckling resistance

$$\lambda_x = L_{Ex} = \frac{0.85 \times 4140}{49.2} = 72$$

$$\lambda_y = L_{Ey} = \frac{1.0 \times 1730}{23.8} = 73$$

As p'_y is 201 N/mm^2, then by extrapolation from BS table 27(c) $p_c = 139$ N/mm^2. Due to the nature of the compound section, and in order to keep the calculations simple, several safe assumptions are made, so that p_b can be quickly evaluated:

$$n = m = 1.0$$

$$u = 1.0$$

$$x \simeq D/T = 135/6 = 23$$

$\lambda/x = 72/23 = 3.2$ and assuming that $N = 0.7$

BS table 14 $\quad v = 0.81$

$$\lambda_{LT} = 1.0 \times 1.0 \times 0.81 \times 73 = 59$$

BS table 11 $\quad p_b = 168$ N/mm^2

$$\frac{11.4 \times 10}{26.1 \times 139} + \frac{4.54}{0.168 \times 69.5} = 0.031 + 0.389 < 1.0$$

Though wind pressure on the gable (1.0q) would generate a larger moment, it can be seen that the compound section has adequate reserves of strength and therefore no further check is necessary. It can be shown that the other diagonals when compounded with a 125 × 75 × 8 Angle are more than adequate.

The only other check that has to be made is to ensure that the net sectional area of each alternate diagonal (which would result in the future with the removal of the 125 × 75 × 8 Angles) can support the loads noted in Table 11.6. Such a check indicates that the members are satisfactory.

An alternative to this solution would be to increase the size of the diagonals in the end girders and then use short lengths of angles to connect the sheeting rails to the diagonals. Also, the reader is reminded that the gable posts could have been run up past the end girders, but this has the disadvantage of constraining the deflections of the end lattice girders relative to the other girders.

11.10.6 Corner columns

Unlike the other main column members, there is lateral loading on the corner columns from the gable sheeting rails. However, the end frames only carry half the load compared with that carried by the intermediate frames. A check would show that the total loading regime acting on the corner columns represents a less severe condition than that for which the columns were originally designed.

11.10.7 Design of column bases

The column base has to be designed for the factored loads arising from the three design cases:

(A) $F_h = 0.0$ kN; $F_v = 191.7$ kN; $M = 0.0$ kNm
(B) $F_h = 16.0$ kN; $F_v = -51.1$ kN; $M = 7.7$ kNm
(C) $F_h = 13.7$ kN; $F_v = 131.3$ kN; $M = 6.6$ kNm

The values of moment in cases (B) and (C) represent the nominal 10% base moment allowed by the code (see clause 5.1.2.4b).

11.10.7.1 DESIGN OF COLUMN BASE-PLATE

The column member size is a 254 × 146 × 37 UB. If the base had to sustain a substantial moment, then the base-plate size should have a minimum protection of 100 mm beyond the column section's overall dimensions of 256 mm × 146 mm to allow bolts to be positioned outside the flanges. However, the base carries only a nominal moment and the usual detail in these circumstances is either to place two holding bolts along the neutral axis of the column section, at right angles to the column web, or to position four bolts just inside the section profile (see Fig. 11.34). The latter detail is to be used as it affords a certain amount of moment resistance which could prove useful in case of fire, as well as helping the erectors in positioning the columns. Therefore, make the base-plate wide enough for the plate to be welded to the column member, i.e. 275 mm × 160 mm.

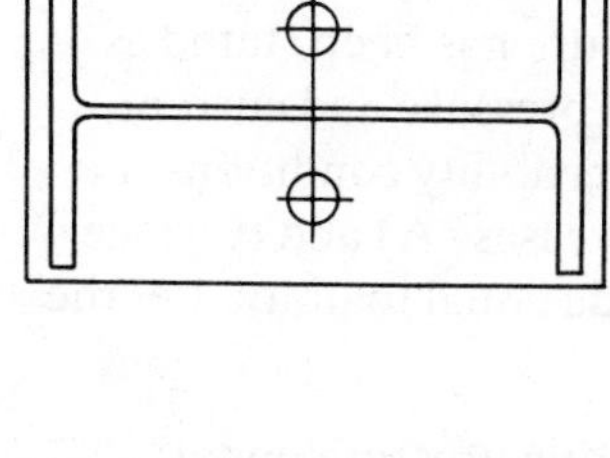

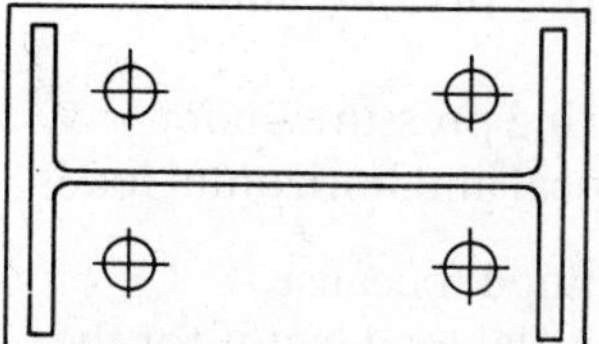

Fig. 11.34 Column base detail

Consider initially the loading from design case (A). Using a concrete mix for the foundation which has a cube strength of $f_{cu} = 30$ N/mm^2, then the bearing pressure should not exceed $0.4 \times 30 = 12$ N/mm^2 (clause 4.13.1):

$$\text{Bearing pressure} = \frac{191.7 \times 10^3}{275 \times 160} = 4.4 \text{ N/mm}^2$$

The plate thickness, t, may be determined from the formula given in clause 4.13.2.2, i.e.

$$t = \sqrt{\left[\frac{2.5w}{p_{yp}}(a^2 - 0.3b^2)\right]}$$

As the projections of the base-plate beyond the profile of the column section are minimal, the theoretical value of t would be small. In these circumstances, it is recommended that the base-plate thickness > column flange thickness, say 15 mm thick plate (grade 43 steel). The welds connecting the column member to the base-plate need to transfer 191.7 kN. Assuming that the column is fillet welded all the way round its profile then the weld length would be about 1 m, hence the required design strength of weld is 192/1000 = 0.19 kN/mm, i.e. nominal size required – use 6 mm FW.

Use 160 mm × 15 mm plate × 275 mm long
6 mm FW.

11.10.8 SIZING OF HOLDING DOWN BOLTS

Where axial load is transmitted by the base-plate (without moment) then nominal holding down bolts are required for location purposes (Section 8.2 and reference (12)). In this example the bolts have to resist an uplift in 51.1 kN, coupled with a moment of 7.7 kNm and a nominal horizontal shear of 16.0 kN. Assume **four** 24 mm diameter bolts, grade 4.6 steel; a smaller diameter is more prone to damage.

Use four 24 mm diameter bolts (grade 4.6 steel)

11.11 DESIGN OF FOUNDATION BLOCK

Generally, the design of the foundations for any structure is dependent on the ground conditions that exist on site, the maximum load conditions that can arise from any combination of loads and the strength of concrete used. Therefore it is important that the engineer has data regarding soil conditions or some reasonable basis for estimating the soil capacity. In this example, the soil bearing pressure has been stated as 150 kN/m^2 (Section 11.2). This is a **permissible** pressure and current practice for foundation design is based on serviceability conditions, i.e. working load level. Therefore, for the two load cases (A) and (C) (see Section 11.8.1), the partial load factors are made equal to unity, i.e. the loads for these cases become:

(A) $1.0w_d + 1.0w_i$ – maximum bearing pressure under vertical load

(C) $1.0w_d + 1.0w_i + 1.0w_w$ – maximum bearing pressure under combined vertical and horizontal loads

However, the third load case (B) must be examined, because it produces the maximum uplift condition. The partial load factor for the wind load must be the largest possible, so that a foundation block of sufficient weight can be selected to counterbalance any uplift force, i.e:

(B) $1.0w_d + 1.4w_w$ – maximum uplift condition

Clause 2.4.2.4, BS 5950 states that the design of foundations should be in accordance with CP2004 and be able to accommodate all forces imposed on them. Usually, foundation design is governed by gravity loading, but in this example, the uplift force in the main columns is significant. Clause 2.4.2.2, BS 5950 indicates that factored loads should not cause the structure (including the foundations) to overturn or lift off its seating; that is, the weight of the foundation block must be sufficient to counter-balance any uplift force. Therefore, it might be prudent to proportion a **mass concrete** foundation block, based on the worst uplift condition, i.e design case (B) (i) (wind blowing on the side walls) (see Section 11.7.1.2) and then check the resulting foundation block size against the other critical design cases. The loads for design case (B) are:

$$F_v = 1.0(10.9 + 61.7 \times 0.5) - 1.4(75.3 \times 0.75 + 39.3 \times 0.25)$$
$$= -51.1 \text{ kN (uplift)}$$

$$F_h = 16.0 \text{ kN}$$

Taking the specific weight of concrete as 23.7 kN/m^3, the minimum volume of mass concrete necessary to prevent uplift of the column member is 51.1/23.7 = 2.16 m^3. A base of 1.7 m × 1.7 m × 0.9 m thick, which has a volume of 2.60 m^3 and weighs 61.6 kN, provides sufficient mass. Assuming a spread of the vertical load of 45° through the concrete from the edges of the base-plate to the sub-strata, then the block provides 1.7 × 1.7 = 2.89 m^2 of bearing area. The 45° spread line should cut the vertical sides of the block, otherwise the depth of the block has to be increased. In this example the block appears to be adequate, i.e. 0.85 − 0.14 = 0.61 m < 0.90 m (see Fig. 11.35a).

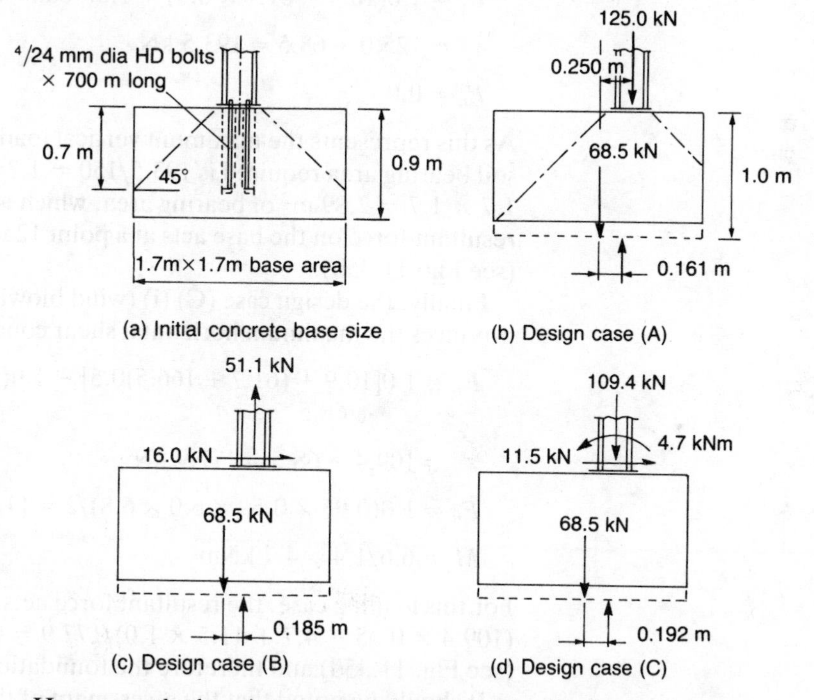

Fig. 11.35 Foundation block detail

However, the bearing pressure has to be checked for the complete design case (B), i.e. uplift, together with the moment generated by the horizontal shear (16 kN) at the bottom of the block. If the resultant soil bearing force lies within the middle third of the block, i.e. not more than $L/6(= 0.283 \text{ m})$ from the centreline of the base, then tension will not occur at the concrete/soil interface. In order to achieve this condition, the column member needs to be offset a horizontal distance y from the centreline of the base, i.e.

$$\frac{16.0 \times 0.9 - 51.1 \times y}{61.6 - 51.1} \leqslant 0.283$$

$$y \geqslant 0.224 \text{ m, say } 0.250 \text{ m}$$

This means that the 45° spread does not cut one of the vertical sides of block, i.e. $(0.85 + 0.25 - 0.14) = 0.96 \text{ m} > 0.9 \text{ m}$. Therefore, the depth of the block is increased to 1.0 m, which results in a block weight of 68.5 kN.

By taking moments about the centreline of the base, the point at which the resultant force acts at the concrete/soil interface can be determined, i.e:

$$(16.0 \times 1.0 - 51.1 \times 0.25)/(68.5 - 51.1) = 0.185 \text{ m}$$

from the base centreline (see Fig. 11.35c) and therefore the base is adequate for this loading case. Figures 11.35b and 11.35d show the loading for the other two cases (A) and (C).

Check that the proposed size is satisfactory with respect to these other design cases. First, examine the design case (A), the loads for which are:

$$F_v = 1.0(10.9 + 61.7 \times 0.5) + 1.0(166.5 \times 0.5) + 68.5$$

$$= 125.0 + 68.5 = 193.5 \text{ kN}$$

$$F_h = 0.0$$

As this represents the maximum vertical load condition, the minimum soil bearing area required is $193.5/150 = 1.24 \text{ m}^2$. The block provides $1.7 \times 1.7 = 2.89 \text{ m}^2$ of bearing area, which is more than adequate. The resultant force on the base acts at a point $125.0 \times 0.25/193.5 = 0.161$ m (see Fig. 11.35b).

Finally, the design case (C) (i) (wind blowing on the side walls), which produces the maximum horizontal shear condition, has to be checked:

$$F_v = 1.0[10.9 + (61.7 + 166.5)0.5] - 1.0(42.6 \times 0.25 + 6.6 \times 0.75) + 68.5$$

$$= 109.4 + 68.5 = 177.9 \text{ kN}$$

$$F_h = 1.0(0.95 \times 0.59 \times 6.0 \times 6.8)/2 = 11.5 \text{ kN per column}$$

$$M = 6.6/1.4 = 4.7 \text{ kNm}$$

For this loading case, the resultant force acts through a point $(109.4 \times 0.25 - 4.7 + 11.5 \times 1.0)/177.9 = 0.192$ m from the centreline (see Fig. 11.35d) and therefore the foundation block is satisfactory.

It should be noted that the assessment of the forces in the vertical bracing system (Section 11.8.3.3) indicates an additional factored uplift force in the penultimate columns (see Fig. 11.19a), i.e. −50.6 kN. The foundation blocks for these particular columns have to be made larger than that just calculated for the normal conditions, but again the uplift condition controls the design, i.e. $-(51.1 + 50.6) = -101.7$ kN. Use a 2.2 m × 2.0 m × 1.0 m base, which weighs 104.3 kN. However, in this case the column need only be offset 0.15 m from the base centreline (see Fig. 11.36). Also, this non-standard block has to be checked for the normal conditions (excluding bracing force) in anticipation of the building being extended. Further calculations would show that all the design conditions are satisfied.

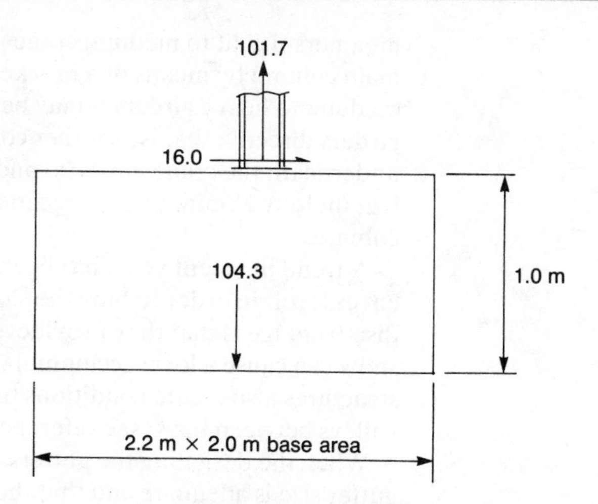

Fig. 11.36 Foundation block – penultimate frame

The foundation pads for the gable posts can be designed in a similar manner.

11.12 OTHER CONSIDERATIONS

Both examples of single-storey construction reviewed in detail deal essentially with the basic design of the structural members for standard arrangements, i.e. lattice girder and column construction in this chapter and portal frame construction in Chapter 12. In practice, it can be said that generally there are no 'standard' buildings, as each client has different requirements. Though they will not be discussed in detail, the reader should be aware of other considerations which might affect the basic design.

For example, many single-storey buildings serve to house goods/materials or enclose large stores requiring the movement of large volumes of goods, being both delivered and dispatched. The design examples do not specifically deal with framing of doorways, large access openings, effect of loading bays (with or without canopies). It is essential that the client defines his requirements as early as possible, preferably no later than the design stage, otherwise late alterations can radically affect the structural arrangements and therefore costs.

11.12.1 Loading

Common types of loads which can be imposed on to a single-storey building and which could significantly influence the design, include ventilation extractors, crane loading and drifted snow. Extractor units are usually located on the roof, near the ridge. Generally, these extractor units require a special framing (trimmer) detail in order to transfer their own local load to the purlins and hence to the main frame. The main frame itself is rarely affected.

Occasionally monorails are required to be hung from the roof members, thereby imposing local concentrated loads on particular members. Light to medium crane girders can be supported from the main column by means of a bracket (see Section 8.6). However, for medium to heavy girders it may be more economic to support the girders directly; that is, another column is positioned immediately underneath the crane girder(s) and battened back to the main column leg, thereby eliminating a large moment from acting on the main column.

A trend in recent years has been the use of a facade or parapet at eaves level, in order to hide the sloping roof and gutter detail. The disadvantage is that they provide an effective barrier whereby drifting snow can cause a local accummulation of snow in that region. Multi-bay structures also create conditions that lead to snow drifting into the valleys between bays; see reference (6).

When the designing the gutters, the engineer must ensure that the gutter size is adequate and that the outlets can cope with the expected volume of water. Badly designed outlets have been known to cause overflow problems; see references (10) and (11) for guidance.

11.12.2 Deflections

BS 5950 recommends limitations on deflections for most buildings, but specifically does not give guidance for portal frames. Though deflections for the normal pitched portal frame used in practice do not usually cause any problems, the designer should be wary of non-standard frames, especially if they are relatively tall. The reason for deflection limits is to prevent damage to the cladding or other secondary members and to avoid psychological unease of the client or employees. 'Excessive' deflection does not necessarily indicate that the building is unsafe in a structural sense, i.e. deflections not acceptable to the client might be acceptable to the structural engineer. The designer can always offset a predicted downward deflection due to the effect of dead plus imposed loads, by deliberately introducing an upward camber at the fabrication stage. The deflection itself is, of course, unchanged.

Another problem that possibly could arise is ponding. Ponding is associated with shallow flexible roofs, where even 'dead load' deflections are sufficient to create a hollow on the roof surface, increasing in size as rainwater is collected. This can lead to leakage of the water into the building, which might prove costly. Also, if a building is subjected to constant buffeting by wind forces, then the cladding might suffer from low cycle failure resulting from cracking of the cladding along its crests or troughs. Asbestos corrugated sheets in existing buildings are more prone to this form of damage, as the sheets become brittle with age.

11.12.3 Fire

Normally fire protection is not required by Building Regulations for single-storey buildings, unless there is a potential fire risk arising from a particular use of the building or the building is located within the fire boundary of another person's property. If it becomes necessary to control the spread of fire to structural members, then the SCI guidance regarding fire boundary and fire protection should be read[13], being particularly relevant to portal frames. The function of any fire protection is to allow the occupants to escape within a specified period of time, e.g. one hour. There are various methods of producing the required fire rating for a building, which is dependent on use of building. Most of these methods are passive, i.e. they delay the effect of fire on structural members. However, one method which might have long-term financial benefits is the water sprinkler system. As this method is an active system (attempts to control the fire), lower insurance premiums might be negotiated. The disadvantage is that the building has to be designed to accommodate the additional loading from the pipe network supplying the water to the individual sprinklers in the roof zone. Also, if the pipes have to be supported eccentric to the plane of the supporting members then the torsional moments acting on the supporting member must be taken into account.

11.12.4 Corrosion

Generally, for most single-storey structures, corrosion is not a problem as the steelwork is contained within the cladding envelope. The minimum statutory temperature requirements are such that the ambient conditions inside a modern building tend to be dry and warm, which are not conducive to the propagation of corrosion and therefore any internal steelwork only needs a nominal paint specification, unless the function of the building is such as to generate a corrosive atmosphere[14].

11.12.5 General

The designer should always bear in mind the erection process and eliminate at the design stage any difficulties which could arise[15]. Also, he should be aware of the limitations on length, width and height of structural steelwork which has to be transported by sea, rail or road[16].

Finally, the reader is advised to consult copies of references (7) and (17), in which numerous construction details for different forms of single-storey buildings are graphically illustrated.

STUDY REFERENCES

Topic	*Reference*
1. Comparative costs	**Horridge, J.F. & Morris, L.J.** (1986) Comparative costs of single-storey steel framed structures, *Structural Engineer,* vol. 64A (no. 7), pp. 177–81
2. Loading	BS 6399 *Loading for Buildings* Part 1: *Dead and Imposed Loads* (1984)

3. **Wind loading** — BS 6399 *Loading for buildings* Part 2: *Wind loads* (to be published; presently CP3, Ch. V Part 2)
4. **Wind loading** — **Newberry & Eaton, K.** (19??) *Wind Loading on Buildings.* Building Research Establishment
5. **Cold formed sections** — BS 5950 *Structural Use of Steelwork in Buildings* Part 5: *Design of Cold-formed Sections* (1987)
6. **Snow drifting** — Building Research Establishment (1988) *Loads on Roofs from Snow Drifting against Vertical Obstructions and in Valleys,* Digest 332
7. **Multibeam purlins** — Ward Building Components (1986) *Multibeam – Purlin and Cladding Rail System.* Sherburn, North Yorkshire
8. **Truss analysis** — **Coates, R.C., Coutie, M.G. & Kong, F.K.** (1988) Analysis of plane trusses, *Structural Analysis,* pp. 30–40. Van Nostrand Reinhold
9. **Section properties** — (1985) *Steelwork Design,* vol. 1, Section properties, member properties. Steel Construction Institute
10. **Roof drainage** — Building Research Establishment (1976) *Roof Drainage – Part 1,* Digest 186
11. **Roof drainage** — Building Research Establishment (1976) *Roof Drainage – Part 2,* Digest 187
12. **Holding down bolts** — (1980) *Holding Down Systems for Steel Stanchions.* Concrete Society/BCSA/Steel Construction Institute
13. **Fire boundary** — (1980) *The Behaviour of Steel Portal Frames in Boundary Conditions.* Steel Construction Institute
14. **Painting** — **Haigh, I.P.** (1982) *Painting Steelwork,* CIRIA Report no. 93
15. **Lack of fit** — **Mann, A.P. & Morris, L.J.** (1981) *Lack of Fit in Steelwork,* CIRIA Report no. 89
16. **Transportation** — (1980) *Construction Guide.* British Steel Corporation
17. **Design details** — (1984) *Design Manual – Single-Storey Steel Framed Buildings.* British Steel Corporation/Conder, Steel House, Redcar

12

DESIGN OF SINGLE-STOREY BUILDING – PORTAL FRAME CONSTRUCTION

12.1 INTRODUCTION

An alternative, economic solution to the design of a single-storey building is to use portal frame construction[1] (see Fig. 12.1). Therefore, the building designed in the previous chapter will be redesigned, replacing the lattice girder and column framework by a portal frame. However, the selection of cladding, purlins, sheeting rails, together with the design of bracing, will not be undertaken in detail as these member sizes will be similar to the corresponding members for the previous solution based on lattice girder construction. The only fundamental change to the structural arrangement is the direct substitution of the lattice girder and column by a portal frame. This chapter deals basically only with the design of the portal frame together with the gable framing, which is different from that used in Chapter 11. Therefore, detailed design of other structural members in the building can be obtained by referring to the appropriate section in the previous chapter.

Though the steel portal frame is one of the simplest structural arrangements for covering a given area, the designer probably has to satisfy at least as many different structural criteria as for more complex structures. In essence, the portal frame is a rigid plane frame with assumed full continuity at the intersections of the column and roof (rafter) members. It will be assumed that the columns are pinned at the bases. This is the normal practice as the cost of the concrete foundations for fixed bases (due to the effect of large fixing moments) more than offsets the savings in material costs that result from designing the frame with fixed feet. Indeed, the pinned-based portal frame represents the economic solution in virtually all practical design cases. However, fixed bases may become necessary with tall portal frames as a means of limiting deflections.

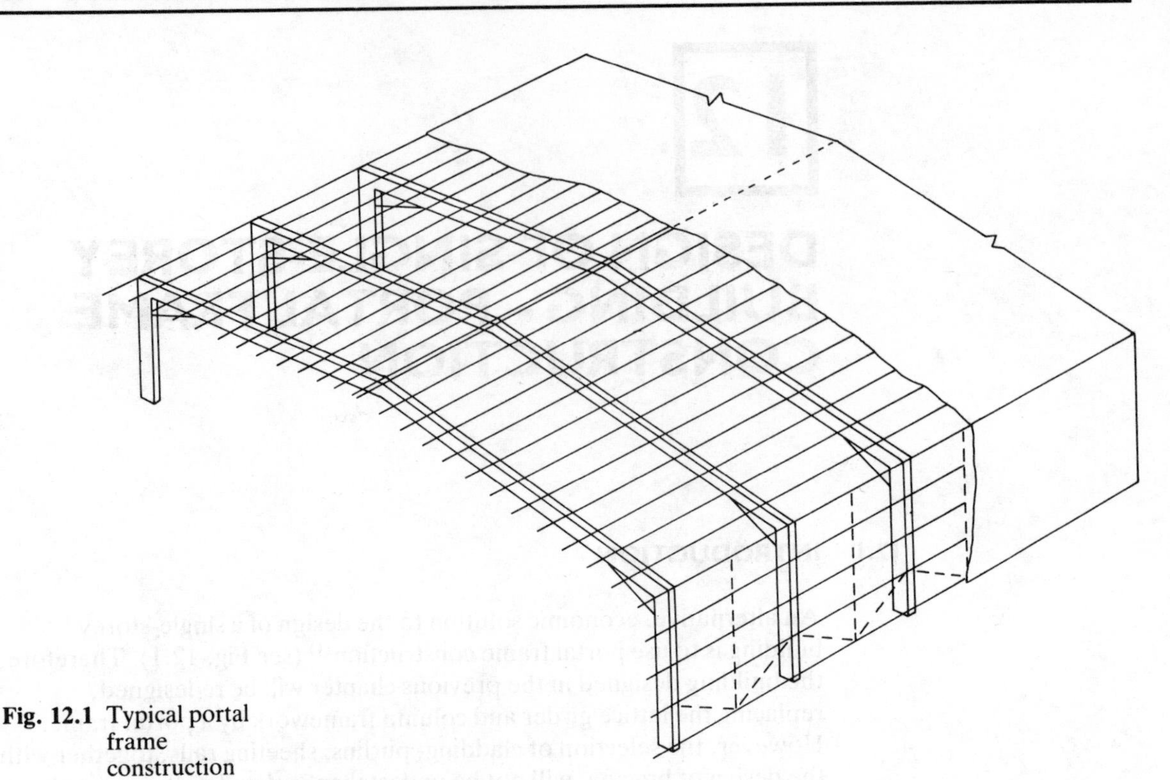

Fig. 12.1 Typical portal frame construction

12.2 DESIGN BRIEF

A client requires a similar single-storey building to that described in the previous chapter (Section 11.2), i.e. a clear floor area, 90 m × 36.4 m (see Fig. 11.2), with a clear height to underside of the roof steelwork of 4.8 m. However, in this case the client has decided that there will be no extensions of the building in the future. The building is also to be insulated and clad with BSC profiled metal sheeting, Long Rib 1000R, and the slope of the roof member is to be at least 6°. The site survey showed that the ground conditions can support a bearing pressure of 150 kN/m² at 0.8 m below existing ground level.

12.3 DESIGN INFORMATION

Though the design brief specified at least 6° it has been decided to make the slope at least 10° so that the sheeting can be laid without special strip mastic lap sealers, which are necessary for shallower slopes in order to prevent capillary action and rain leakages into the building. If it is assumed that the depth of the column is 0.6 m and the distance from the column/rafter intersection to the underside of haunch is 0.7 m, then with reference to Fig. 12.2:

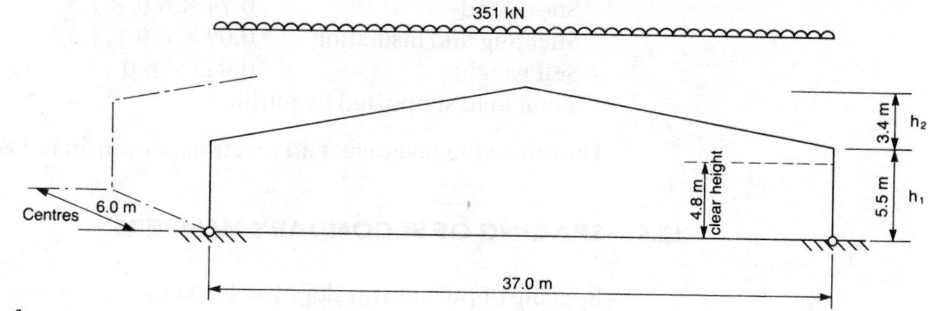

Fig. 12.2 Design information for proposed portal frame

Span of portal frame	$(L = 36.4 + 0.6)$	= 37.0 m
Centres of portal frames		= 6.0 m
Height to eaves intersection	$(h_1 = 4.8 + 0.7)$	= 5.5 m
Height of eaves to apex	$[h_2 = (37.0 \tan 10°)/2]$	= 3.4 m (say)
Actual slope of rafter	$[\theta = \tan^{-1}(3.4/18.5)]$	= 10.41°

The imposed load (snow) is 0.75 kN/m² on plan, which gives an equivalent load of $0.75 \times \cos\theta = 0.75 \times 0.984 = 0.74$ kN/m² on slope along the roof member. The use of an equivalent load makes due allowance for purlin spacing which is usually calculated as a slope distance. As this building is to be located on the same site as the building designed in Chapter 11, then the design wind pressure is identical, i.e. 0.59 kN/m². The self weight of the BSC profile Long Rib 1000R, plus insulation, is taken as 0.09 kN/m².

12.4 DESIGN OF PURLINS AND SHEETING RAILS

The design of the purlins and sheeting rails is virtually identical to the corresponding members in the lattice girder and column construction, apart from the equivalent snow load along the roof member (rafter) being 0.74 kN/mm², giving a total combined load (dead + snow) of 0.83 kN/mm². Therefore, the same cold formed section, Ward Multibeam A170/160 has been selected for the purlins and the B170/155 for the sheeting rails (Table 11.1), for the reasons noted in Section 11.5. However, a slightly reduced purlin spacing of 1.5 m from that used in the previous example, is assumed in anticipation of probable restraint required at haunch/rafter intersection (Fig. 12.12).

Again, the joints of the double spanning purlins/rails are assumed to be staggered across each frame, thereby ensuring that each intermediate portal frame receives approximately the same total loading via the purlins. The self weight of the selected purlin section is found to be 0.043 kN/m and hence the 'average' load (unfactored) being transferred by each purlin is:

Snow load	$0.74 \times 6.0 \times 1.5$	= 6.66 kN
Sheeting and insulation	$0.09 \times 6.0 \times 1.5$	= 0.81 kN
Self weight	0.043×6.0	= 0.26 kN
Total load supported by purlin		= 7.73 kN

Therefore the 'average' end reaction per purlin is 3.86 kN.

12.5 SPACING OF SECONDARY MEMBERS

Spacing of purlins (on slope) = 1.500 m
Spacing of purlins (on plan) = 1.475 m
Spacing of sheeting rails – see Section 12.8.1.1

12.6 DESIGN OF PORTAL FRAME

Since the mid 1950s portal frame construction in the United Kingdom has been widely based on the principles of plastic design[2,3] developed by Baker and his team at Cambridge[4]. By taking advantage of the ductility of steel, plastic design produces lighter and more slender structural proportions than similar rigid frames designed by elastic theory.

12.6.1 Gravity load condition

In applying plastic design to the modern haunched portal frame, the initial step is to estimate the plastic moment capacity of a uniform frame (no haunching) with pinned bases for the dead + snow loading condition. As most portal frame designs are governed by the gravity loading case, a general expression for M_p has been derived[3,5], i.e. with reference to Section 12.3 and Fig. 12.3:

$$M_p = \frac{W^* L}{8} \frac{1}{[(1 + k/2) + \sqrt{(1 + k)}]}$$

where $k = h_2/h_1 = 3.4/5.5 = 0.618$
w_d = self weight (i.e. rafter + cladding + purlins)
$= [0.90 \text{ (est)} + 0.09 \times 6.0 + 0.043 \times 6.0/1.5] = 1.60$ kN/m (on slo
w_i = imposed load (i.e. snow)
$= 0.74 \times 6.0 = 4.44$ kN/m (on slope)
$W^* = (1.4w_d + 1.6w_i)L/\cos 10.41°$ (on plan)
$= (1.4 \times 1.60 + 1.6 \times 4.44)37.0/0.984 = 351$ kN (on plan)

$$\text{Hence } M_p = \frac{351 \times 37.0}{8} \frac{1}{[(1 + 0.309) + \sqrt{(1 + 0.618)}]} = 629 \text{ kNm}$$

It is important to note that the value of M_p has been determined for a rigid-jointed plane frame without requiring any prior knowledge of section properties, unlike elastic design. The design of the structural steelwork is to be based on grade 43 steel being used throughout the project. (The use of higher grades of steel would produce more slender frames, which could lead to difficulties with respect to stability and

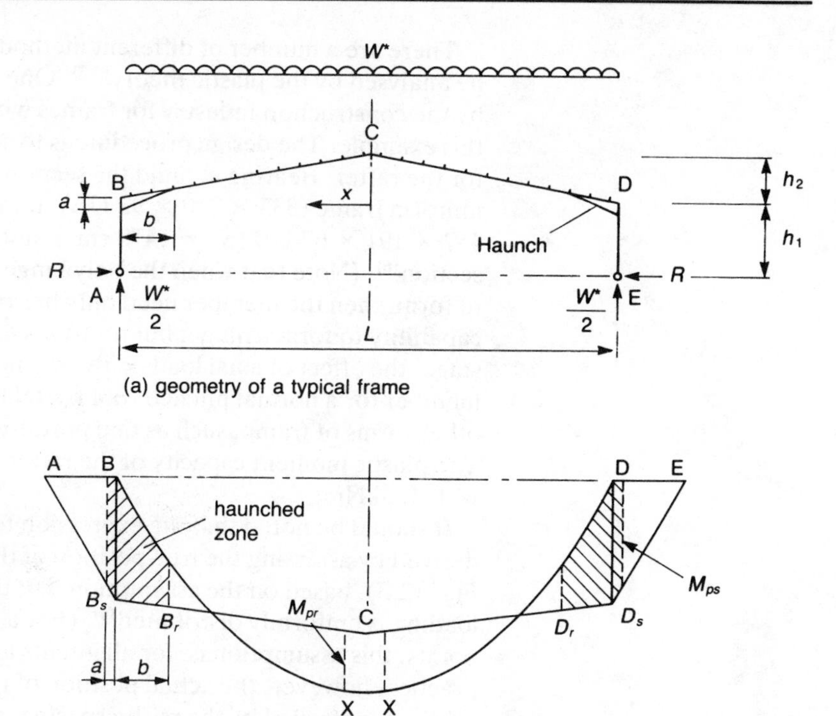

Fig. 12.3 Typical moment distribution for haunched portal frame

deflections.) Therefore, at this stage of calculation assume that the design strength (p_y) is 275 N/mm^2; then the plastic modulus required is:

$$S_x \text{ required} = M_p/p_y$$
$$= 629/0.275 = 2287 \text{ cm}^3$$

This would indicate that a 533 × 210 × 92 UB section (S_x = 2370 cm^3) could be chosen, assuming that the frame has a constant section throughout its length. However, the section must be checked to see whether or not it can sustain **plastic action**. Referring to the SCI guide, p.130[(6)], the section is designated as a 'plastic' section, i.e. the section is capable of adequate rotation at a plastic hinge position without local flange buckling developing.

However, the most cost-effective form for portal frames with spans greater than 15–20 m is to haunch the rafter member in the eaves region[(1)], thereby allowing a lighter (in weight) section to be used for the rafter than the column; that is, the rafter size, for the major part of the rafter, can be reduced below the section based on the frame having a constant cross-section, while the column section would probably need to be increased in size to compensate for the increased moment at the eaves. As the rafter member is generally much longer than the combined length of the column members, the resulting design becomes more efficient and therefore more economic.

There are a number of different methods by which portal frames can be analysed by the plastic theory[2,3]. One method commonly adopted by the construction industry for frames with pinned bases[3] is used for this example. The design procedure is to select a suitable member size for the rafter. Bearing in mind the section previously obtained for a uniform frame (533 × 210 × 92 UB), try the section 457 × 191 × 67 UB (S_x = 1470 cm³), noting that this is also a 'plastic' section[6]. (Note that when the only hinge in a member is the last hinge to form, then the member need only be compact, i.e. requires the capability to form a hinge, but rotation capacity not essential.) At this stage, the effect of axial load on the moment capacity of a rafter member for a normal pitched roof portal frame can be neglected. (For other forms of frame, such as tied portal[2], it could become significant.) The plastic moment capacity of the rafter member is 1470 × 0.275 = 404.3 kNm.

It should be noted that the expression for M_p, given previously, was derived by assessing the true position of the 'apex' hinges (see Fig. 12.3), based on the assumption that the total factored vertical loading is uniformly distributed[5]. (For a frame with at least 16 purlin points, this assumption is, for all intents and purposes, accurate.) In practice, however, the actual position of the 'apex' plastic hinges in the rafter is controlled by the purlin spacing, since hinges coincide with particular purlin supports, i.e. where the point loads are imposed on the rafter member. If the haunch is to remain elastic (by doing so, it helps with member stability in that region), assume that the hinge positions occur at the column/haunch intersection (in the column) and near the apex.

For gravity loading condition (dead + snow) experience has found that the 'apex' hinge usually occurs at the first or second purlin support down from the apex purlin (cf. wind condition, Section 12.6.2). Assuming that the hinge will form at the second purlin support from the apex (point X), then take moments about, and to the left of the left-hand X. As the plastic moment at X is known (404.3 kNm), then with reference to Fig. 12.3, the value of R (the horizontal thrust at the pinned base) can be calculated from the resulting equilibrium equation:

$$-M_{pr} = -\frac{w^*}{2}[(L/2)^2 - x^2] + [h_1 + (1 - 2x/L)h_2]R$$

$$L = 37.0 \text{ m}$$
$$h_1 = 5.5 \text{ m}$$
$$h_2 = 3.4 \text{ m};\ x = 2.95 \text{ m}$$
$$w^* = 351/37.0 = 9.50 \text{ kN/m}$$
$$-404.3 = -\frac{9.50}{2}[18.5^2 - 2.95^2] + [5.5 + (1 - 5.9/37.0)3.4]R$$
$$= -1584.4 + 8.358R$$
$$R = (1584.4 - 404.3)/8.358 = 141.2 \text{ kN}$$

Based on the assumption that a plastic hinge will occur at position B_s (which in fact is the clear height to the underside of the haunch; see

Fig. 12.3), then the required plastic modulus for the column member can be readily determined by taking moments about and below B_s, i.e:

$$M_{ps} = (h_1 - a)R$$

where a = distance from eaves intersection, B, to B_s

$$M_{ps} = (5.5 - 0.7)141.2 = 667.8 \text{ kNm}$$

$$S_{xs} = 677.8/0.275 = 2465 \text{ cm}^3$$

Though there seems to be a choice between a 533 × 210 × 101 UB or a 610 × 229 × 101 UB (both are 'plastic' sections[6]), the flange thickness of the former section (17.4 mm) is such that the design strength would have to be reduced to 265 N/mm² (see BS table 6). In any case, the second choice gives a better plastic modulus and increased stiffness for an identical weight, which could prove useful in controlling eaves deflections. Now check the adequacy of the frame by using a **457 × 191 × 67 UB** for the rafter and a **610 × 229 × 101 UB** for the column member.

First the full plastic modulus of the column section needs to be reduced to take account of the axial load in the column member, i.e:

$$S_{xs} = 2880 - 3950n^2 \text{ (reference 6)}$$

$$n = \alpha f_c/p_y$$

α = adequacy ratio (see below) – estimate 1.1

As the weight of side cladding and rails generally affects the lower portions of the column, it is therefore not included in F.

$$f_c = F/A$$

$$= 351 \times 10/(2 \times 129) = 13.6 \text{ N/mm}^2$$

$$n = 1.1 \times 13.6/275 = 0.05$$

$$S_{xs} = 2880 - 3950(0.05)^2 = 2870 \text{ cm}^3$$

$$M_{ps} = 2870 \times 0.275 = 789.3 \text{ kNm}$$

The adequacy ratio (α) indicates the reserve of strength of the frame, over and above that necessary to produce a safe design, and hence must not be less than unity. Therefore it is essential to check the value of the adequacy ratio. This is done by setting up the equilibrium equations by the method outlined in references (2) and (3), i.e. the frame is cut at the apex to give three internal releases M, R and S (see Fig. 12.4). Each half frame acts as a cantilever with the free end at the apex. It can be shown that due to the symmetry of both the loading on, and the geometry of, the frame, the internal vertical force S is zero[2].

This leaves three unknowns, M, R and α to be determined; therefore three independent equilibrium equations are necessary. As the net moments at the positions A, B_s and X are known, equate them to the combined effect of the internal and external moments acting at those points, remembering that X is the second purlin from apex purlin.

$$\text{A:} \quad 0.0 = \frac{351 \times 37.0}{8}\alpha - M - 8.90R$$

$$B_s\text{:} \quad 789.3 = \frac{351 \times 37.0}{8}\alpha - M - 4.10R$$

$$\text{X:} \quad -404.3 = \frac{351 \times 2.95^2}{37.0 \times 2}\alpha - M - 0.54R$$

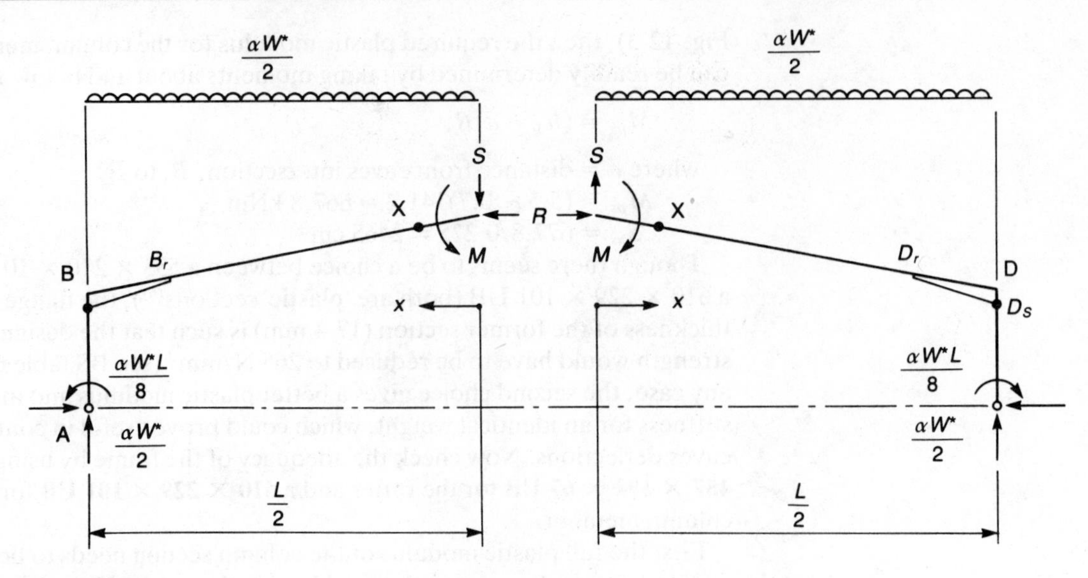

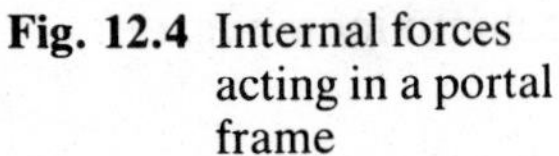
Fig. 12.4 Internal forces acting in a portal frame

Solving these equations gives:

$$\alpha = 1.124$$
$$M = 361.9 \text{ kNm}$$
$$R = 164.44 \text{ kN}$$

Note that the values of M and R include the adequacy factor.

This particular design produces a 12.4% overstrength in the frame; this overstrength arises due to the range of discrete sizes of sections available to the designer. This enhanced capacity of the frame will be used throughout the remaining calculations, i.e. the design will include the factor 1.124 so that the frame will effectively be capable of carrying $1.124 \times 351 = 395$ kN. This allows the client to take advantage of this overstrength in the future without any additional cost.

The general expression for calculating the net moment at any position along the rafter for the half frame, shown in Fig. 12.4, is:

$$M_x = \frac{1.124 \times 351}{37.0 \times 2} x^2 - 361.9 - \frac{3.4 \times 164.44}{18.5} x$$
$$= 5.331x^2 - 30.221x - 361.9$$

Now the assumed position of the 'apex' hinge has to be checked by evaluating the net moments at the purlin supports immediately adjacent to, and on either side of, the second purlin (hinge position), using the expression for M_x, i.e.:

$$M_2 = 5.331 \times 1.475^2 - 30.221 \times 1.475 - 361.9 = -394.9 \text{ kNm}$$
$$M_4 = 5.331 \times 4.425^2 - 30.221 \times 4.425 - 361.9 = -391.2 \text{ kNm}$$

As the moments at the purlin supports on either side of the second

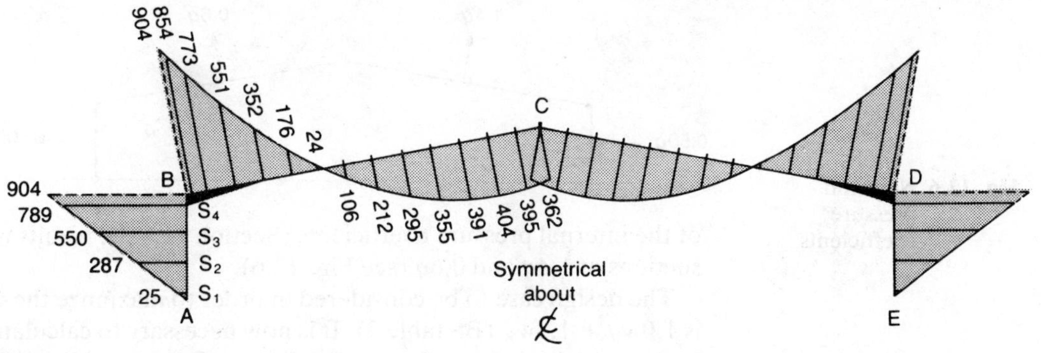

Fig. 12.5 Moment distribution for dead plus imposed load condition

purlin are less than the moment (404.3 kNm) at the assumed plastic hinge position near the apex, then the assumption is correct.

Next, the length of the haunch needs to be determined so that the haunch/rafter intersection remains just elastic, i.e. moment at intersection to be less than or equal to the **yield moment** M_y $(= p_y Z)$, which gives:

$$0.275 \times 1300.0 = 5.331x^2 - 30.221x - 361.9$$

Hence $x = 14.79$ m

Therefore the minimum length of haunch from the column centre-line is $(18.50 - 14.79) = 3.71$ m. By increasing the required haunch length by 0.04 m, it allows the haunch/rafter intersection to coincide with a purlin support position (some 14.75 m on plan from apex) which might prove useful when checking the adequacy of the haunched rafter against member instability. This means the actual haunch length, as measured from the column flange/end-plate interface is $3.75 - 0.30 = 3.45$ m.

Now draw the bending moment diagram to check that nowhere does any moment exceed the moment capacity of the frame. Also, the resulting information will be required to enable checks on member stability to be undertaken. The resulting net bending moment diagram for the gravity loading condition (dead + snow) is plotted on the tension side of the frame in Fig. 12.5.

12.6.2 Wind load condition

In elastic analysis, the analyses from different loading cases can be added together. However, because plastic analysis deals with the final collapse state of a structure, then each loading pattern generates its own unique failure mode. Therefore moments and forces from individual analyses cannot be added together. As a result, the dead + wind loading condition has to be analysed separately.

The basic wind pressure (q) of 0.59 kN/m^2 has already been established (Section 11.4.3). Knowing the slope of the rafter is 10.41° and that $h/w = 8.9/37.0 = 0.24$ and $l/w = 90.0/37.0 = 0.43$, then the external coefficients for the roof are -1.2 (windward) and -0.4 (leeward)[7]. These coefficients, when combined with the more onerous

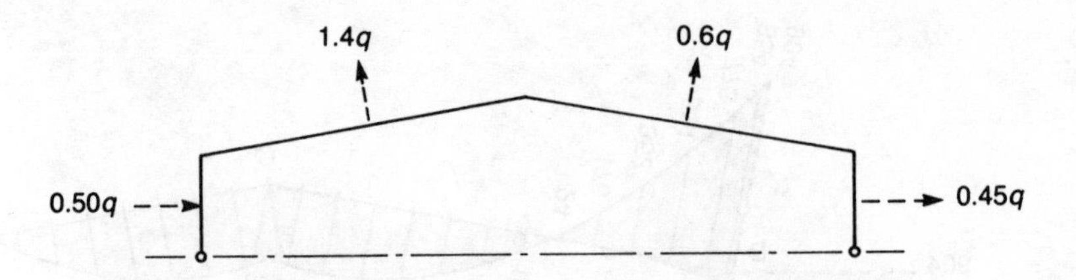

Fig. 12.6 Net wind pressure coefficients

of the internal pressure coefficients (Section 11.4.3), results in roof wind suctions of $1.4q$ and $0.6q$ (see Fig. 12.6).

The design case to be considered in order to maximize the wind uplift is $1.0w_d + 1.4w_w$ (BS table 2). It is now necessary to calculate the various purlin and sheeting rail loads, in order to establish the total loading pattern acting on the portal frame. For example, the vertical and horizontal components of load for typical purlins on the roof are:

Dead load vert. load $= 1.0 \times 1.60 \times 1.475/\cos 10.41° = 2.40$ kN

Wind load (windward):

vert. component $= 1.4 \times 1.4 \times 0.59 \times 1.475 \times 6.0 = 10.23$ kN

horiz. component $= 10.23 \times 3.4/18.5 = 1.88$ kN

Wind load (leeward):

vert. component $= 1.4 \times 0.6 \times 0.59 \times 1.475 \times 6.0 = 4.39$ kN

horiz. component $= 4.39 \times 3.4/18.5 = 0.81$ kN

The wind loads for other purlins and sheeting rails can be determined in a similar manner, using the net pressure coefficients given in Fig. 12.6. The factored loads (dead + wind) for the complete frame are shown in Fig. 12.7.

Assume the collapse mechanism shown in Fig. 12.8; that is, hinges are deemed to occur at the fourth purlin (X) down from the apex purlin on the left-hand side, and at the rafter/haunch intersection (D_r) on the right-hand side of the frame.

Unlike the previous analysis, four independent equilibrium equations are required, as the loading is non-symmetrical and therefore S will not be zero:

$$\begin{aligned}
A:&\quad 0.0 = -1039.1\alpha + M + 8.900R - 18.50S \\
X:&\quad -404.3 = -96.5\alpha + M + 1.084R - 5.90S \\
D_r:&\quad 404.3 = -157.8\alpha + M + 2.710R + 14.75S \\
E:&\quad 0.0 = -339.8\alpha + M + 8.900R + 18.50S
\end{aligned}$$

Solving these equations gives:

$$\begin{aligned}
\alpha &= 2.375 \\
M &= -192.68 \text{ kNm} \\
R &= -162.31 \text{ kN} \\
S &= 44.89 \text{ kN}
\end{aligned}$$

Clearly the adequacy ratio or this load condition is higher than that of the dead + snow load condition. This means that, in order to be consistent with the load level achieved with dead + snow loading, checking the portal frame design for the wind case need only be

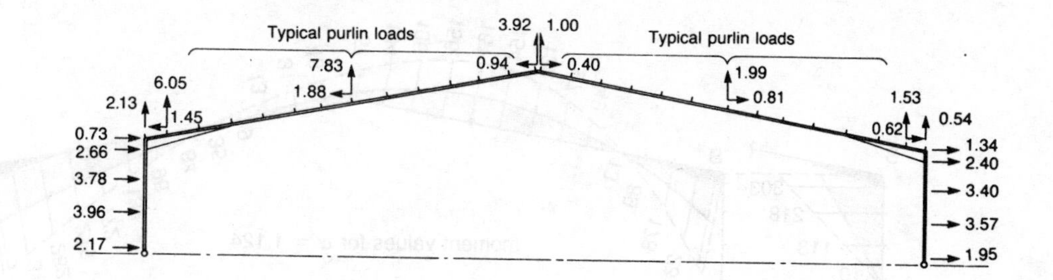

Fig. 12.7 Loading on frame for dead plus wind load condition

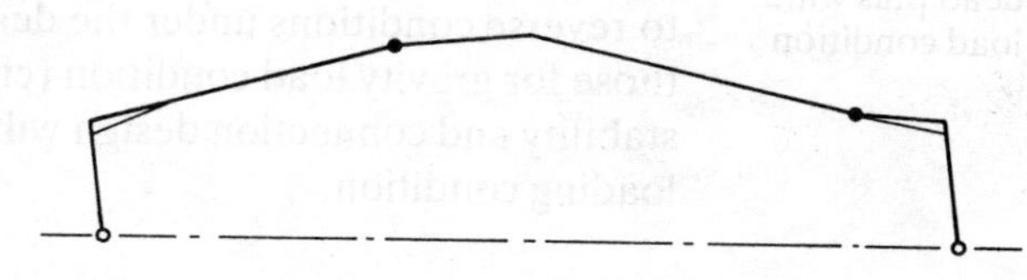

Fig. 12.8 Collapse mechanism for dead plus wind load condition

executed at a load level of 1.124. Figure 12.9 indicates the load levels at which hinges develop for the mechanisms associated with the two loading cases being discussed. It is clear from the diagram that at a load level of 1.124, no hinges will have formed in the frame for the dead + wind case, i.e. frame remains elastic at that load level.

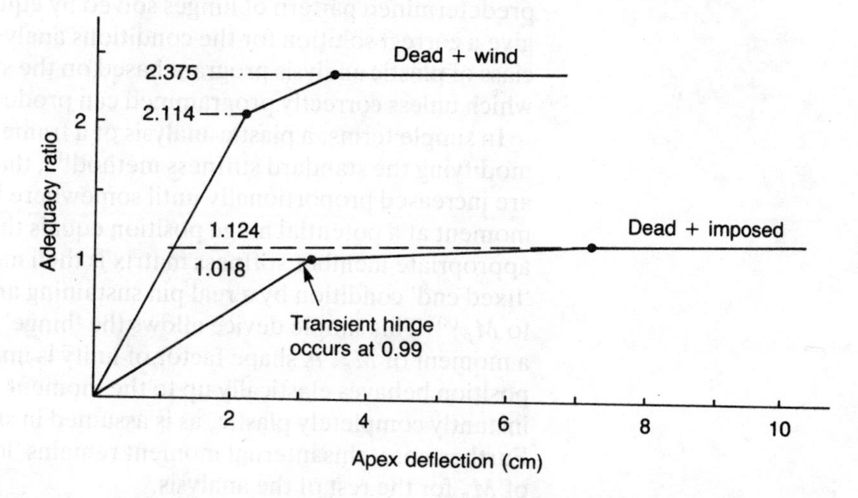

Fig. 12.9 Load-apex deflection for both load cases

Hence the moments and forces which are required for the purpose of checking member stability and connection design, can be obtained from an elastic analysis, using a design loading equal to 1.124 × factored loads. The bending moment diagram from such an analysis is represented by the full line in Fig. 12.10, as well as the bending moment diagram at collapse ($\alpha = 2.375$) – see dashed line.

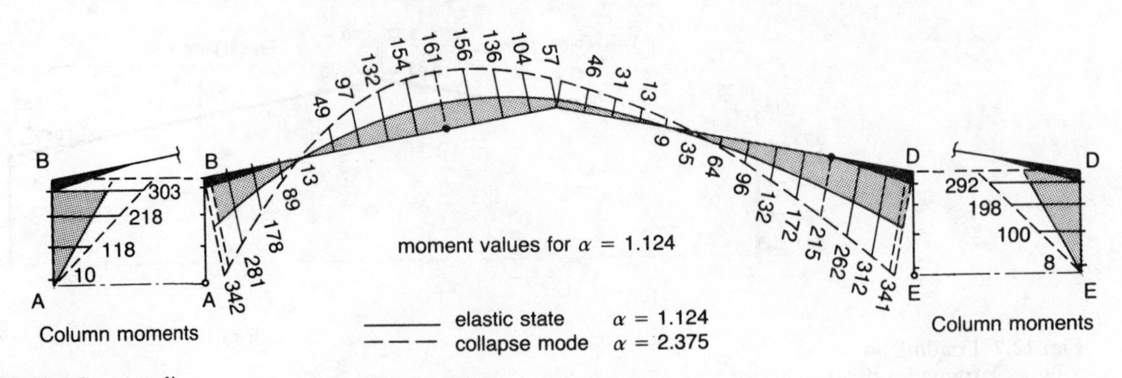

Fig. 12.10 Moment distribution for dead plus wind load condition

The bending moment diagram illustrates that the frame is subjected to reverse conditions under the dead + wind load case, compared with those for gravity load condition (cf. Fig. 12.5). Therefore, member stability and connection design will also need to be checked for the wind loading condition.

12.6.3 Computer analysis

The plastic analysis of a portal frame can easily be executed by a suitable computer program, of which there are several commercial systems on the market, as well as numerous versions developed by individuals for their own use. Those programs that analyse the final collapse state directly, i.e. linear programming (optimization) methods or predetermined pattern of hinges solved by equilibrium equations, will give a correct solution for the conditions analysed. However, there is a class of plastic analysis program based on the stiffness matrix method, which unless correctly programmed can produce false solutions.

In simple terms, a plastic analysis of a frame can be obtained by modifying the standard stiffness method[8]; that is, the factored loads are increased proportionally until somewhere in the frame the elastic moment at a potential hinge position equals the local M_p value. The appropriate member stiffness matrix is then modified by replacing the 'fixed end' condition by a real pin sustaining an internal moment equal to M_p[8]. This simple device allows the 'hinge' to rotate while sustaining a moment of M_p. A shape factor of unity is implied, i.e. the hinge position behaves elastically up to the moment M_p, when it becomes instantly completely plastic, as is assumed in simple plastic theory[2,3]. Furthermore, this internal moment remains 'locked' at this local value of M_p for the rest of the analysis.

The analysis continues with increasing load, modifying the stiffness of appropriate members every time a new hinge forms, until such time as there are sufficient hinges to produce a mechanism. The advantage of this method is that the analysis provides the sequence of hinge formation, while indicating the load level at which the individual hinges occur. The load level at which a mechanism is produced is in fact equal to the adequacy ratio, α.

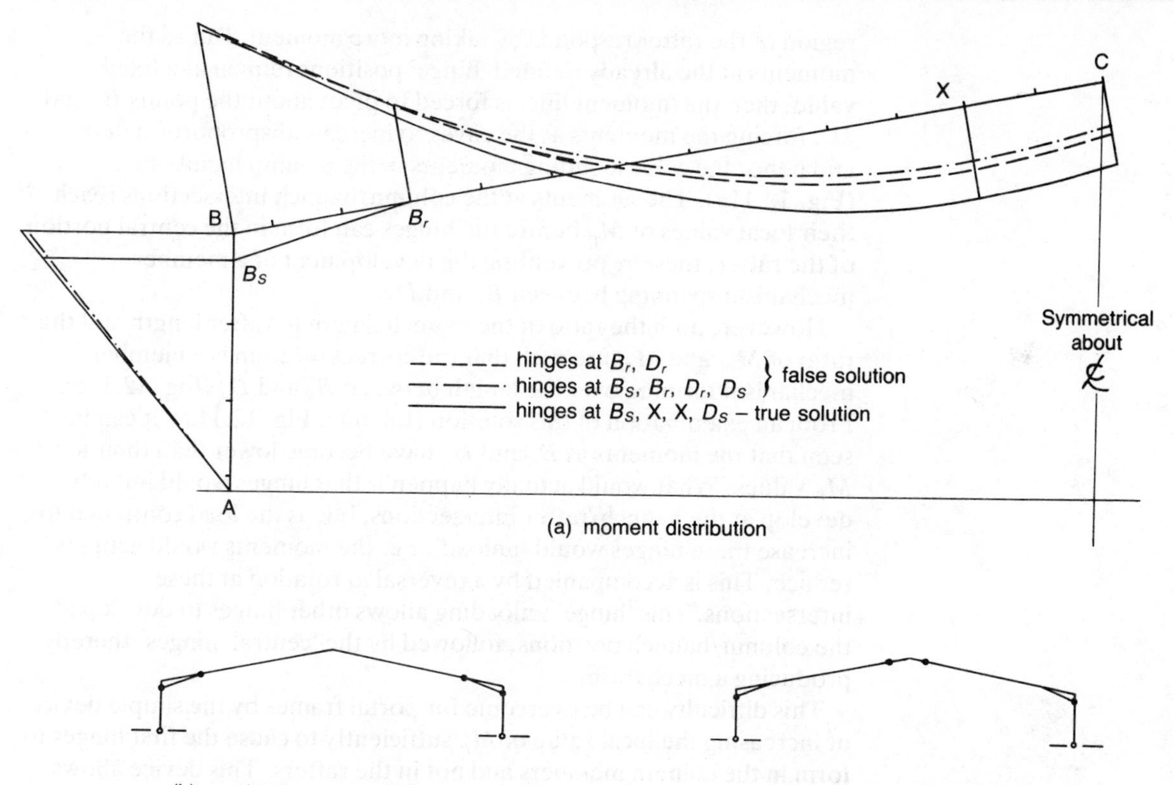

Fig. 12.11 Transient plastic hinge condition

However, unless there is a facility in the program which allows the moment at any defined hinge to unload, i.e. as the loading is increased, the moment at the 'hinge' position might reduce below the 'locked-in' value of M_p, then false solutions can be obtained, which might give rise to a potential problem.

The dead plus imposed loading condition for this design example affords a good illustration of the problem. The proportions of the frame, particularly the haunch length, are such that the first hinges to 'form' during an analysis, using the modified stiffness method, occur at the haunch/rafter intersections. The two hinges at B_r and D_r form simultaneously due to the symmetry of the frame and loading (Fig. 12.11b). Loading continues to increase until further hinges develop in the column members at the column/haunch intersections and the analysis stops, indicating a mechanism. However, inspection of the 'signs' of the moments at the defined hinge positions show that all hinges are rotating in the same sense (Fig. 12.11a). This is a physical impossibility as a mechanism cannot form under these conditions, i.e. a false solution.

This situation arises because the moments at the first hinge positions (in the rafter) are 'locked' and remain constant in spite of the effect of the additional load. As the loading increases the frame tries to fail as a member mechanism between B_r and D_r and the moment in the central region of the rafter responds by taking more moment. But as the moments at the already defined 'hinge' positions remain at a fixed value, then the 'moment line' is forced to pivot about the points B_r and D_r, forcing the moments at the eaves to increase disproportionately, and hence the corresponding moments in the column members (Fig. 12.11a). The moments at the column/haunch intersections reach their local values of M_p before the hinges can form in the central portion of the rafter, thereby preventing the development of a member mechanism spanning between B_r and D_r.

However, both the ratio of the haunch length to rafter length and the ratio of M_{pr} and M_{ps} are such that the correct solution is a member mechanism forming over the length between B_s and D_s (Fig. 12.11c). From an examination of this solution (full line, Fig. 12.11a), it can be seen that the moments at B_r and D_r have become lower than their local M_p values. What would actually happen is that hinges would initially develop at the haunch/rafter intersections, but as the load continued to increase these hinges would 'unload', i.e. the moments would actually reduce. This is accompanied by a reversal in rotation at these intersections. This 'hinge' unloading allows other hinges to develop at the column/haunch positions, followed by the 'central' hinges, thereby producing a mechanism.

This difficulty can be overcome for portal frames by the simple device of increasing the local value of M_p sufficiently to cause the first hinges to form in the column members and not in the rafters. This device allows the correct mechanism to develop, but the solution is only acceptable if the final moments at the haunch/rafter intersections are below the correct M_p values for the rafter.

Nevertheless, this example does highlight the fact that transient hinges can develop at an intermediate load level and subsequently disappear from the final mechanism. It might be advisable under these circumstances to restrain the transient hinge positions. Also, analysis by direct methods, though giving the correct solution, would not indicate transient hinges and alert the designer to the potential problem of member instability.

Clearly, this example, based on the dead plus imposed loading condition, could be used as a test case for checking plastic analysis programs.

Note that, as with the lattice girder analysis executed by computer program, the axial load in the rafter changes at the purlin positions as each component of purlin loading parallel to the rafter member takes effect. In this case, the variation of axial load along the rafter in the vicinity of the hinge position is of the order 1–2% which would not affect significantly the local rafter capacity.

12.7 FRAME STABILITY

Before checking the frame for member stability and designing the connections, it is prudent to check the selected member sizes against the possibility of frame instability (clause 5.5.3), as new sections would need to be chosen if these criteria are not satisfied.

12.7.1 Sway stability of frame

First, the member sizes have to be checked against frame sway instability (clause 5.5.3.2). This condition, modified to take effect of the haunch into account, has to be satisfied in the absence of a rigorous analysis of frame stability[3], i.e:

$$\frac{L-b}{D} \leqslant \frac{44}{\Omega}\frac{L}{h}\frac{\rho}{(4+\rho L_r/L)}\frac{275}{p_{yr}}$$

$$\rho = \frac{2I_c}{I_r}\frac{L}{h} \text{ for single bay frame}$$

$\Omega = \alpha W^* L/(16 M_{pr})$

$L = 37.0$ m

$L_r = 37.0/\cos 10.41° = 37.62$ m

$D = 0.454$ m

$h = h_1 = 5.50$ m

$b =$ average haunch length $= 3.75$ m

$I_c = 75\,700$ cm^4

$I_r = 29\,400$ cm^4

$\rho = 2 \times 75\,700 \times 37.0/(29\,400 \times 5.5) = 34.6$

$\Omega = 1.124 \times 351 \times 37.0/(16 \times 404.3) = 2.26$

$$\frac{37.0-3.75}{0.454} \leqslant \frac{44}{2.26}\frac{37.0}{5.5}\frac{34.6}{(4+34.6\times 37.62/37.0)}\frac{275}{275}$$

$73 < 116$

Frame stable

12.7.2 Snap-through buckling of rafter

The condition of snap-through buckling of the rafters (clause 5.5.3.3) is unlikely to apply to single-bay frames unless roof slope is shallow. Such a possibility increases with multi-bay frames due to the accumulative spread at the eaves which may be sufficient to allow the rafters of the internal bays to snap-through.

12.8 MEMBER STABILITY – LATERAL-TORSIONAL BUCKLING

BS 5950 caters specifically for different moment gradients and geometry of member for portal frames and the appropriate methods will be adequately illustrated when checking the lateral-torsional resistances of both column and haunched rafter members. Guidance is given in those situations which are not specifically covered by the code or where it is felt the code recommendations are not appropriate. First, the various parts of the frame are checked for stability under gravity loading condition and then rechecked for the dead plus wind loading condition.

12.8.1 Stability checks for dead plus snow loading condition

As noted in Section 12.6.2, member stability is to be checked at a load level 1.124 × factored loads; see Fig. 12.5 for the relevant bending moments.

12.8.1.1 CHECK COLUMN MEMBER BUCKLING

The design of the frame assumes a plastic hinge forms at the top of the column member (610 × 229 × 101 UB), immediately below the haunch level. This plastic hinge position must be torsionally restrained in position by diagonal stays; see later comment regarding size of stays (Section 12.8.3). With the hinge position restrained, check the plastic stability of the length of column from the hinge position, between the two sheeting rails, S_3 and S_4 (see Fig. 12.12). This is done by using the expression for uniform members, with no tension flange restraint between effective end torsional restraints, given in clause 5.3.5(a), which is based on the stability curves given in reference[9].

$$L_m = \frac{38r_y}{\sqrt{\left[\frac{f_c}{130} + \left(\frac{p_y}{275}\right)^2\left(\frac{x}{36}\right)^2\right]}}$$

$$f_c = 1.124 \times 351 \times 10/(2 \times 129) = 15 \text{ N/mm}^2$$
$$x = 43.0 \text{ (reference 6)}$$

Note that the value 1.124 is the adequacy ratio.

$$L_m = \frac{38 \times 47.5 \times 10^{-3}}{\sqrt{\left[\frac{15}{130} + \left(\frac{275}{275}\right)^2\left(\frac{43.0}{36}\right)^2\right]}} = 1.454 \text{ m}$$

This indicates that the sheeting rail S_3, immediately below the rail at the hinge position S_4, is required to restrain the column member on both flanges, i.e. by means of diagonal stays which must be placed not more than 1.454 m from the stay at the hinge position. As a consequence, the side rails are positioned from ground level at 0.15 m (S_1), 1.75 m (S_2), 3.35 m (S_3) and 4.80 m (S_4); see Fig. 12.12.

Now check the elastic stability of the column length between sheeting rail S_3 and the pinned base A. Under the action of gravity loading (dead + snow) the inner flanges of the column members are in compression. Now:

$$\frac{F}{P_c} + \frac{\bar{M}}{M_b} \leqslant 1$$

$$F = 1.124 \times 351/2 = 197.3 \text{ kN}$$
$$M_A = 550.3 \text{ kNm (max. moment occurs at } S_3)$$

As the moment at base A is zero then $\beta = 0$, hence $m = 0.57$ (BS table 18) and:

$$\bar{M} = mM_A = 0.57 \times 550.3 = 313.7 \text{ kNm}$$

It is assumed that the effective length for the column member buckling about the x–x axis is the distance A–S_4, i.e. 4.80 m, hence:

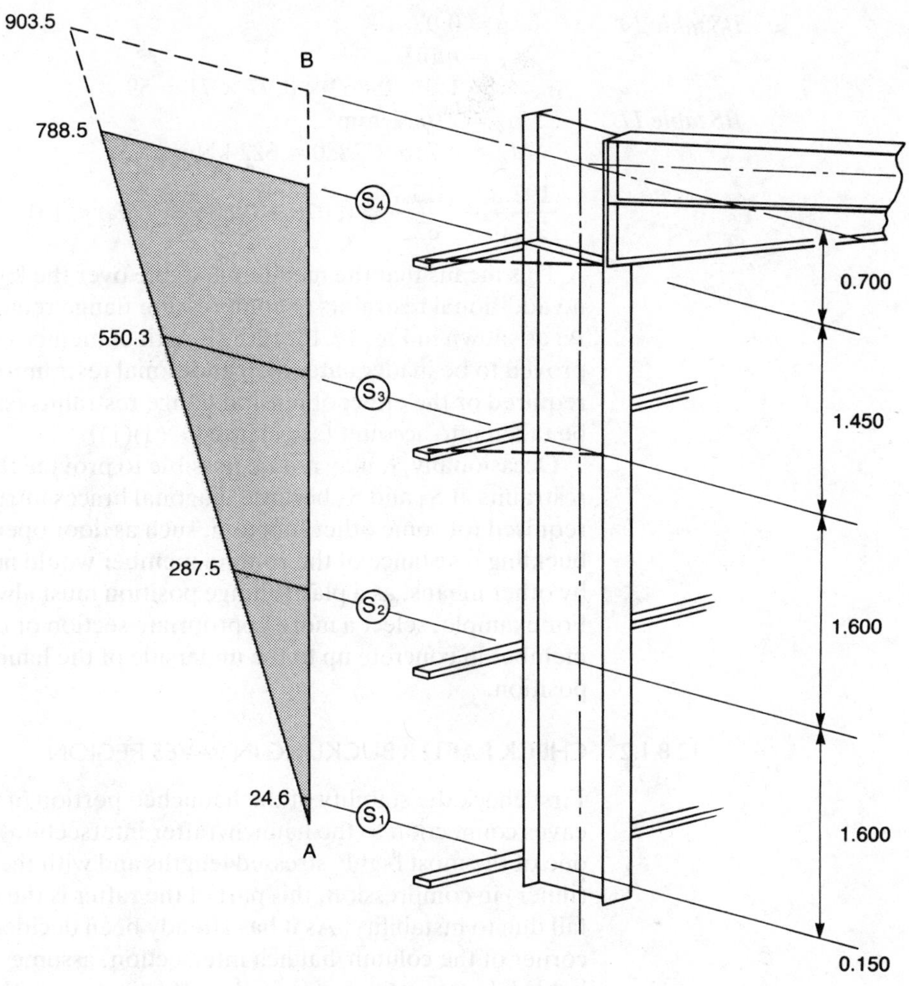

Fig. 12.12 Member stability – column member

BS table 27(a) $L/r_x = 4800/242 = 20$

Hence $p_c = 273\ \text{N/mm}^2$

At this stage, it is assumed that there are no further restraints to the inner (compression) flange below S_3 and therefore the effective length of the compression flange about the y–y axis is A–S_3, i.e. 3.35 m.

BS table 27(b) $L/r_y = 3350/47.5 = 71$

$p_c = 200\ \text{N/mm}^2$

$P_c = 200 \times 129/10 = 2580\ \text{kN}$

BS table 13 $n = 1.0$

$u = 0.863$ (reference 6)

$x = 43.0$ (reference 6)

$\lambda = 71$

$\lambda/x = 71/43.0 = 1.65$

As the universal section has equal flanges then $N = 0.5$ and with $\lambda/x = 1.65$ hence:

BS table 14 $v = 0.97$

$$\lambda_{LT} = nuv\lambda$$
$$= 1.0 \times 0.863 \times 0.97 \times 71 = 59$$

BS table 11 $p_b = 216 \text{ N/mm}^2$

$$M_b = 0.216 \times 2880 = 622 \text{ kNm}$$

$$\frac{197.3}{2580} + \frac{313.7}{622} = 0.076 + 0.504 = 0.580 < 1.0$$

This means that the member is stable over the length considered, i.e. no additional restraints to compression flange required between S_3 and A, as shown in Fig. 12.12. If the length of member being considered had proved to be inadequate, then additional restraint(s) would have been required or the effect of tension flange restraints (sheeting rails) could be taken into account (see clause G2(a)(1)).

Occasionally, it may not be possible to provide the necessary torsional restraints at S_4 and S_3 because diagonal braces intrude into space required for some other function, such as door openings. Then the buckling resistance of the column member would need to be increased by other means, as a plastic hinge position must always be restrained. For example, select a more appropriate section or encase the column member in concrete up to the underside of the haunch, i.e. hinge position.

12.8.1.2 CHECK RAFTER BUCKLING IN EAVES REGION

First check the stability of the haunched portion of the rafter (from the eaves connection to the haunch/rafter intersection) as this represents one of the most highly stressed lengths and with the outstand flange (inner) in compression, this part of the rafter is the region most likely to fail due to instability. As it has already been decided to stay the inside corner of the column/haunch intersection, assume that the haunch/rafter intersection is also effectively stayed by diagonal braces (see Fig. 12.13), giving a length of 3.450 m between restraints.

The depth of a haunch is usually made approximately twice the depth of the basic rafter section, as it is the normal practice to use a UB cutting of the same serial size as that of the rafter section for the haunch, which is welded to the underside of the basic rafter (457 × 191 × 67 UB). Therefore, assume haunch has the same section as the rafter member and therefore the overall depth at the connection is 0.90 m. It would appear that clause G2(b)(2) is the most appropriate criterion to check stability of the haunched portion, as there are tension flange restraints between the effective end restraints, i.e:

$$L_t \leqslant L^* = \frac{L_k}{cn_t}$$

However, the formula for L_k is only appropriate if there is a plastic hinge associated with the length being considered, i.e. a severe condition. In this particular design exercise, the section sizes were chosen so that the stresses in the haunched portion and the majority of

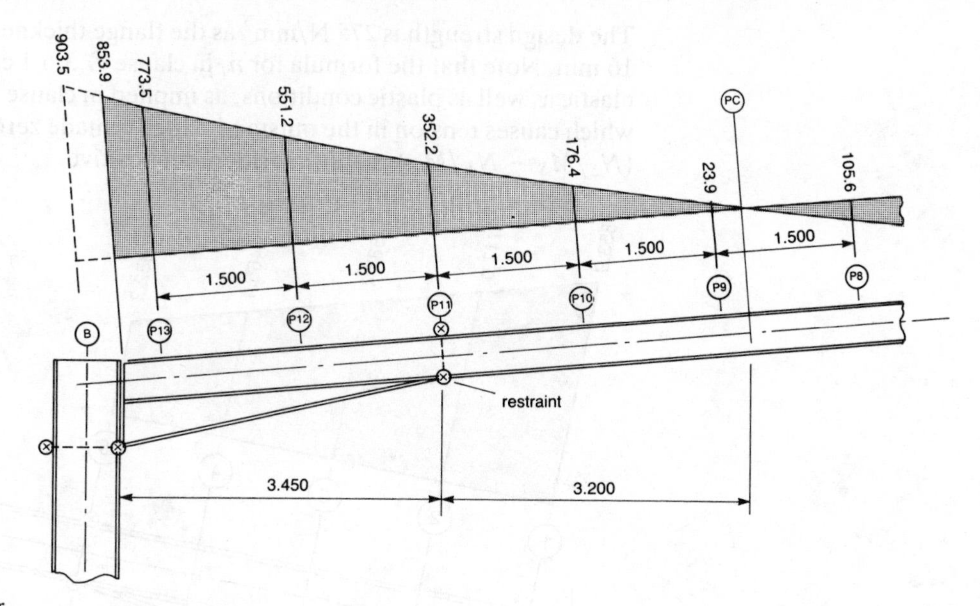

Fig. 12.13 Member stability – haunched rafter region

the basic rafter member remain elastic. A more suitable formula for L_k (limiting length for uniform member) under elastic conditions is[10]:

$$L_k = \frac{[8.0 + 150(p_y/E)]xr_y}{\sqrt{[4.4(p_y/E)x^2 - 1]}}$$

$$= \frac{[8.0 + 150 \times 275/205\,000]41.2 \times 37.9 \times 10^{-3}}{\sqrt{[4.4(275/205\,000)37.9^2 - 1]}} = 4.683 \text{ m}$$

This value of L_k has to be modified by the shape factor (c) which accounts for the haunch and n_t which allows for the moment gradient.

$$c = 1 + \frac{3}{(x-9)}(R-1)^{2/3}\,q^{1/2}$$

R = greater depth/lesser depth

q = haunched length/total length

Over the length being considered (3.450 m), then:

$$R = 900/454 = 1.98$$

$$q = 3.450/3.450 = 1.00$$

$$c = 1 + 3(1.98 - 1.0)^{2/3}\,1.0^{1/2}/(37.9 - 9) = 1.102$$

$$n_t = \sqrt{\left[\frac{1}{12}\left\{\frac{N_1}{M_1} + \frac{3N_2}{M_2} + \frac{4N_3}{M_3} + \frac{3N_4}{M_4} + \frac{N_5}{M_5} + 2\left(\frac{N_S}{M_S} - \frac{N_E}{M_E}\right)\right\}\right]}$$

N_i = applied factored moments at the ends, quarter points and mid-length of the length considered

M_i = moment capacity at section $i = p_yS_i$ or p_yZ_i

N_S/M_S = greatest of N_2/M_2, N_3/M_3, N_4/M_4

N_E/M_E = greater of N_1/M_1, N_5/M_5

The design strength is 275 N/mm^2 as the flange thickness is less than 16 mm. Note that the formula for n_t in clause G.3.6.1 can be used for elastic as well as plastic conditions, as implied in clause G.3.6.3. Any N_i which causes tension in the outstand flange is made zero. Also, the term $(N_S/M_S - N_E/M_E)$ is only considered if positive.

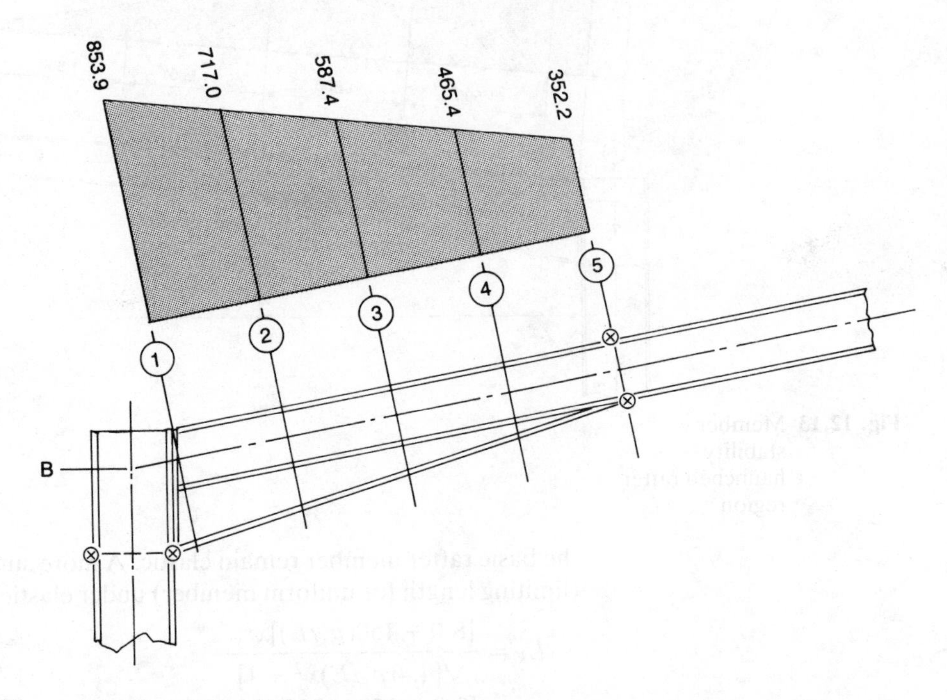

Fig. 12.14 Member stability – haunch region

The elastic moduli are determined for the five cross-sections, indicated on Fig. 12.14; the actual cross-sections considered are taken as being normal to the axis of the basic rafter (unhaunched) member. The worse stress condition at the haunch/rafter intersection (location 5) is taken, i.e. that in the basic rafter immediately adjacent to the intersection.

		1	2	3	4	5
Distance from apex	(m)	18.20	17.34	16.48	15.61	14.75
Factored moment	(kNm)	853.9	717.0	587.4	465.4	352.2
Elastic modulus	(cm^3)	3100	2692	2359	2134	1300
Moment capacity	(kNm)	852.5	740.3	648.7	586.9	357.5
Ratio	N/M	1.000	0.969	0.905	0.793	0.985

$$n_t = \sqrt{[1.000 + 3 \times 0.969 + 4 \times 0.905 + 3 \times 0.793 + 0.985 + 2(0.969 - 1.000)/12]}$$
$$= \sqrt{[0.908]} = 0.953$$

$$L^* = \frac{L_k}{cn_t} = 4.683/(1.102 \times 0.953) = 4.460 \text{ m} > 3.450 \text{ m}$$

The haunched portion of the rafter is therefore stable over the assumed effective length of 3.450 m, with the proviso that both flanges are torsionally restrained at the column/haunch intersection and haunch/rafter intersection, as indicated in Fig. 12.14. See also p. 238.

An alternative method of assessing n_t

In Appendix B, a rapid method for giving an approximate assessment of n_t is outlined. The method enables the designer to establish quickly whether or not the more exact method just used in this example needs to be undertaken. If the approximate assessment indicates that the resulting permissible length is at least **5% longer** than the length being checked, then there is no need for an exact check. Taking the design example, calculate the approximate value of n_t (with reference to Appendix B):

$$n_t = \sqrt{\left[\frac{1}{12M_y}\left\{\frac{N_1}{R_1} + \frac{3N_2}{R_2} + \frac{4N_3}{R_3} + \frac{3N_4}{R_4} + \frac{N_5}{R_5} + 2\left(\frac{N_S}{R_S} - \frac{N_E}{R_E}\right)\right\}\right]}$$

$M_y = 0.275 \times 1300 = 357.5$ kNm

From Appendix B:

$R_1 = 2.45$; $R_2 = 2.14$; $R_3 = 1.83$; $R_4 = 1.63$; $R_5 = 1.00$

$$n_t = \sqrt{\left[\frac{1}{12 \times 357.5}\left\{\frac{853.9}{2.45} + \frac{3 \times 717.0}{2.14} + \frac{4 \times 587.4}{1.83} + \frac{3 \times 465.4}{1.63} + \frac{352.2}{1.00} + 2\left(\frac{717}{2.14} - \frac{853.9}{2.45}\right)\right\}\right]}$$

$= 0.947$

$L^* = 4.683/(1.102 \times 0.947) = 4.488$ m (cf. 4.460 m by exact method)
and 95% of $4.488 = 4.264 > 3.450$ m
No need for exact method

Note that the rapid method is only applicable for the typical British haunch (with middle flange) where the haunch depth is approximately twice that of the basic rafter section and the haunch cutting is from the same section size as the rafter member.

The uniform rafter from the haunch/rafter intersection to the point of contraflexure (where outstand flange is still in compression) now needs to be checked for elastic stability by the method given in section 4, BS 5950. It is accepted generally that the length of rafter beyond the point of contraflexure acts as an effective end restraint, as its outstand flange is in tension. The point of contraflexure, being the position of zero moment in the rafter, can rapidly be determined from:

$$0 = 5.331x^2 - 30.221x - 361.9$$

Hence $x = 11.550$ m

Therefore, the effective length of this part of the rafter is 14.750 − 11.550 = 3.200 m (see Fig. 12.13). Note that this portion of the rafter has two tension flange restraints (purlins) between the effective end restraints and therefore clause G2(a)(1) can be used. However, in this case, the length between effective full restraint can be justified for checking the elastic stability of the uniform rafter, i.e. using the simpler procedure of clause 4.8.3.3.1 without the assistance of the intermediate partial restraints:

$$\frac{F}{P_c} + \frac{\overline{M}}{M_b} \leqslant 1$$

$F = 164.4$ kN (thrust in rafter − R)

$M_A = 352.2$ kNm (max. moment occurs at intersection)

As the point of contraflexure represents the position of zero moment then $\beta = 0$, hence:

BS table 18 $m = 0.57$, and

$\overline{M} = mM_A = 0.57 \times 352.2 = 200.8$ kNm

$L/r_y = 3200/41.2 = 78$

BS table 27(b) $p_c = 186$ N/mm^2

$P_c = 186 \times 85.4/10 = 1588$ kN

BS table 13 $n = 1.0$

$u = 0.873$ (reference 6)

$x = 37.9$ (reference 6)

$\lambda = 78$

$\lambda/x = 78/37.9 = 2.06$

As the universal section has equal flanges then $N = 0.5$ and with $\lambda/x = 2.06$, hence:

BS table 14 $v = 0.96$

$\lambda_{LT} = 1.0 \times 0.873 \times 0.96 \times 78 = 65$

BS table 11 $p_b = 201$ N/mm^2

$M_b = 0.201 \times 1470 = 295$ kNm

$$\frac{164.4}{1588} + \frac{200.8}{295} = 0.104 + 0.681 = 0.785 < 1.0$$

The member is stable over the length P_{11}–PC and no further restraints are necessary within this length of member.

12.8.1.3 CHECK RAFTER BUCKLING IN APEX REGION

Another highly stressed region is the length of rafter in which the 'apex' hinge occurs (see Fig. 12.15). Under dead + snow loading, the outstand flange is in tension, while the compression flange is restrained by the purlin/rafter connections.

Therefore, the buckling resistance of the rafter member between purlins in the apex region needs to be checked. Provided that the 'apex' hinge is the **last** hinge to form in order to produce a mechanism (which is true for low pitched portal frames under dead + snow loading), then adequate rotation capacity is not a design requirement, i.e. the hinge is only required to develop M_p, not rotate – purely a strength condition.

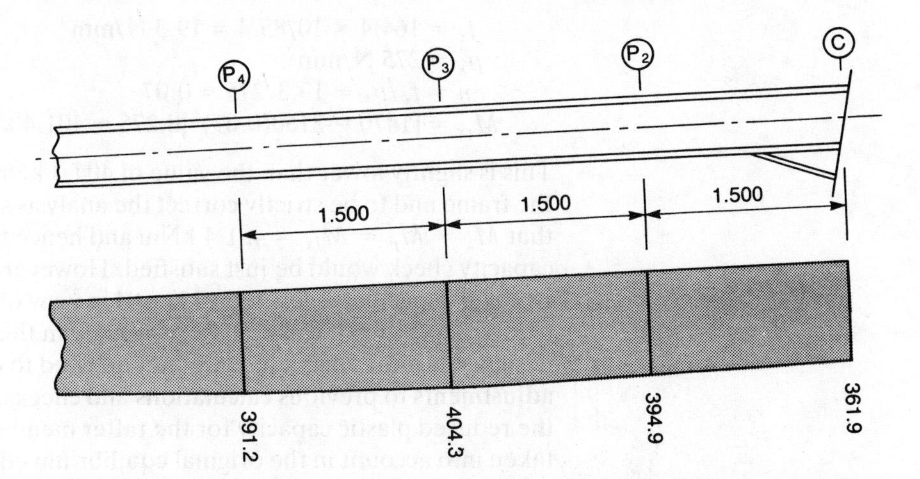

Fig. 12.15 Member stability – apex region

This situation is not specifically covered by BS 5950 (1985), apart from a sentence in clause 5.5.3.1 which states that clause 5.3.5 (dealing with the spacing of restraints) need not be met. However, it is suggested, based on research evidence, that if the value of L_m (as defined by clause 5.3.5) is factored by 1.5 then this would represent a safe criterion for restraint spacing in the region of the 'last hinge' position. That is, assuming that the purlins act as restraints because of their direct attachment to the compression flanges in the apex hinge region, then the purlin spacing should not exceed:

$$L_m = \frac{57r_y}{\sqrt{\left[\frac{f_c}{130} + \left(\frac{p_y}{275}\right)^2\left(\frac{x}{36}\right)^2\right]}}$$

$$f_c = 164.4 \times 10/85.4 = 19.3\ \text{N/mm}^2$$

$$x = 37.9\ \text{(reference 6)}$$

$$L_m = \frac{57 \times 41.2 \times 10^{-3}}{\sqrt{\left[\frac{19.3}{130} + \left(\frac{275}{275}\right)^2\left(\frac{37.9}{36}\right)^2\right]}} = 2.095\ \text{m}$$

As the purlin spacing is 1.500 m (on slope), then no additional restraints are required. (Note that if the 'apex' hinge is not the last hinge to form, then L_m reverts to $2.095/1.5 = 1.307\ \text{m} < 1.500\ \text{m}$. This means that the purlin spacing would have to be reduced to satisfy this limitation.)

Now check the strength capacity of rafter in the 'apex' region, using clause 4.8.3.2(b), i.e:

$$\left(\frac{M_x}{m_{rx}}\right)^2 + \frac{M_y}{M_{ry}} \leqslant 1$$

There is no minor axis moment, M_y, and M_{rx} is the reduced capacity of the rafter section about the major axis in the presence of axial load. Remembering that the axial thrust of 164.4 kN includes the adequacy ratio, then:

$$M_{rx} = (1470 - 2150n^2)p_y\ \text{(reference 6)}$$

$$f_c = 164.4 \times 10/85.4 = 19.3\ \text{N/mm}^2$$
$$p_y = 275\ \text{N/mm}^2$$
$$n = f_c/p_y = 19.3/275 = 0.07$$
$$M_{rx} = [1470 - 2150(0.07)^2]0.275 = 401.4\ \text{kNm}$$

This is slightly lower than the value of 404.3 kNm used in the analysis of the frame and to be strictly correct the analysis should be modified, such that $M_x = M_p = M_{rx} = 401.4$ kNm and hence the above strength capacity check would be just satisfied. However, as the difference in strength capacities is less than 1% and in view of the adequacy ratio being 12.4% then it can be safely assumed in this case that the frame as designed is more than adequate, i.e. no need to carry out the minor adjustments to previous calculations and checks. It can be argued that the reduced plastic capacity for the rafter member should have been taken into account in the original equilibrium equations. The actual reduction in the adequacy ratio is dependent on the geometry and the relative strengths of the column and rafter sections, i.e. in this example the ratio is 0.2%.

This completes the member stability checks of the design frame for the dead plus snow load condition, but this set of checks must be repeated for the dead plus wind load case.

12.8.2 Stability checks for dead plus wind loading condition

Figure 12.10 shows clearly that the dead plus wind loading condition causes a reversal of moments in the frame compared with those obtained for the dead plus imposed condition (Fig. 12.5) and therefore the series of checks undertaken in Sections 12.8.1.1 to 12.8.1.3 need to be repeated. Note that only one dead plus wind condition is being considered in this analysis; nevertheless, when the wind direction is normal to the gables, a worse reversed moment condition might arise for the apex region.

12.8.2.1 CHECK COLUMN MEMBER BUCKLING

Under the dead plus wind loading (Fig. 12.10), the elastic bending moments in the right-hand column vary from 292 kN, at S_4 to 0 at the base, causing the outer flange to go into compression. This moment of 292 kNm compared with the corresponding value of 788.5 kNm for the dead plus imposed loading (Fig. 12.5), coupled with the fact that the outer flanges of the columns are restrained by the sheeting rails (see Fig. 12.12), indicates that this loading condition is not as severe as that investigated in Section 12.8.1.1. A check using section 4, BS 5950 (elastic condition) indicates the column member requires no further restraints.

12.8.2.2 CHECK RAFTER BUCKLING IN EAVES REGION

The wind loading condition causes the upper flanges in the eaves regions to sustain compression. These flanges are restrained by purlin cleats at

1.5 m intervals (see Fig. 12.13). As the magnitude of the moments are significantly less than those for gravity loading, the application of section 4, BS 5950 would prove that the haunched rafters in these regions are more than adequate. No futher restraints are necessary.

12.8.2.3 CHECK RAFTER BUCKLING IN APEX REGION

In the apex region, due to the stress reversal conditions arising from the dead plus wind case, the outstand (lower) flange is in compression and at present there are no restraints to that flange, though the 'tension' flange is restrained by the purlins. Therefore, this part of the rafter needs to checked in detail as member buckling may prove to be more severe than that checked in Section 12.8.1.3, despite the moments being appreciably smaller. The unrestrained length between the two points of contraflexure is about 12.4 × 1.5 = 18.6 m (see Fig. 12.10), hence:

$$\lambda = \frac{18600}{41.2} = 451$$

Clearly this unrestrained length of 18.6 m is too slender, exceeding the limitation of 350 for slenderness for wind reversal (clause 4.7.3.2). Therefore, restraints are required in the apex region. In considering the best location for the restraints the moment distribution for the dead plus imposed load condition has a bearing on the decision. Also, it has to be remembered that the wind can blow in the opposite direction, i.e. right to left, therefore the restraints should be arranged symmetrically about the apex of the frame.

As the moment distribution for dead plus imposed case in the apex region is fairly constant, there is a choice for the location of the restraints, i.e. at either first, second or third purlin position down for apex purlin. If restraints are placed at the third purlin down for the apex purlin on either side of the apex, then the unrestrained lengths become 8.7 m, 9.0 m and 0.9 m from left to right in Fig. 12.16.

Consider the left-hand portion, 8.7 m long. Clause G2.a(1), BS 5950, states that for checking the elastic stability of a uniform member which is restrained by intermediate restraints on the tension flange between effective torsional restraints; then:

$$\frac{F}{p_c} + \frac{\bar{M}}{M_b} \leqslant 1$$

However, the axial load (F) for the dead plus wind condition is in tension and although axial tension should improve a member's buckling resistance, the code does not allow any benefit for this condition. Conversely, clause 4.8.2 gives an erroneous impression that any axial tension would obviate a member buckling check, whereas this would depend on the relative magnitude of the bending moment and axial tension. In the absence of clear guidance, a member buckling resistance check should be based on:

$$\frac{\bar{M}}{M_b} \leqslant 1$$

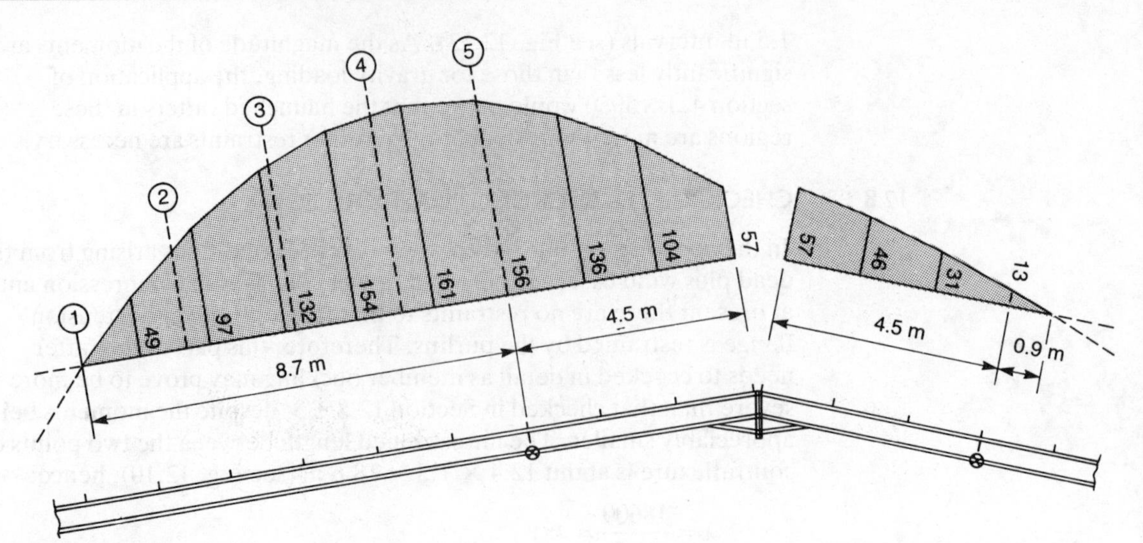

Fig. 12.16 Member stability – apex region – reversed loading

$$M_A = 161 \text{ kNm}$$

$$m_t = 1.0$$

clause G3.4

$$\bar{M} = m_t M_A = 161 \text{ kNm}$$

The minor axis slenderness ratio for this particular stability check is defined by clause G3.3, i.e:

$$\lambda_{TB} = n_t u v_t c \lambda$$

$$\lambda = \frac{8700}{41.2} = 209$$

The local moment capacity of a uniform member under elastic condition (load level = 1.124) is the yield moment ($p_y Z_x$), hence the expression for n_t (clause G3.6) becomes:

$$n_t = \sqrt{\left[\frac{N_1 + 3N_2 + 4N_3 + 3N_4 + N_5 + 2(N_S - N_E)}{12\text{M}_\text{y}}\right]}$$

The values of N_i can be evaluated from Fig. 12.16 and hence:

$$n_t = \sqrt{\left[\frac{0 + 3 \times 80 + 4 \times 134 + 3 \times 158 + 156 + 2(158 - 156)}{12 \times 0.275 \times 1300}\right]}$$

$$= 0.573$$

$$u = 0.873 \text{ (reference 6)}$$

$$x = 37.9 \text{ (reference 6)}$$

$$\lambda/x = 209/37.9 = 5.51$$

clause G3.3

$$v_t = \sqrt{\left[\frac{(4a/h_s)}{1 + (2a/h_s)^2 + (1/20)(\lambda/x)^2}\right]}$$

a = half depth of purlin + half depth of member = 85 + 227 = 312 mm

h_s = distance between shear centres of flanges = 453.6 − 12.7 = 441 mm

$$\frac{a}{h_s} = \frac{312}{441} = 0.707$$

$$v_t = \sqrt{\left[\frac{4 \times 0.707}{1 + (2 \times 0.707)^2 + \frac{1}{20}(5.51)^2}\right]} = 0.791$$

$$c = 1.0 \quad \text{uniform member}$$
$$\lambda_{TB} = 0.573 \times 0.873 \times 0.791 \times 1.0 \times 209 = 82$$
$$p_b = 161 \text{ N/mm}^2$$
$$M_b = 0.161 \times 1470 = 237 \text{ kNm} < p_y Z_x$$
$$\frac{\bar{M}}{M_b} = \frac{156}{237} = 0.658 < 1.0$$

There still remains the checking of the rafter between the new restraints 5 to 13 (see Fig. 12.16). A check would indicate that this part-member is stable for the dead plus wind case.

12.8.3 Design of lateral restraints

During the various checks on member buckling undertaken in the previous sections, several positions along the frame have been assumed to be effectively restrained against both lateral and torsional displacements. Such restraints must be capable of carrying the lateral forces while being sufficiently stiff so that the member being braced is induced to buckle between braces.

Research evidence[(11)] has indicated that the magnitude of the restraining force in any **one** restraining element/brace before instability occurs is of the order of 2% of the squash load of the compression flange, i.e. $0.02BTp_y$. Though the restraining force is relatively small, it is **essential** that such a force (in the form of a brace) be supplied. Adequate stiffness is at least as important as the strength criterion of 2% squash load. In view of lack of sufficient experimental evidence, a limiting slenderness ratio of 100 is recommended for diagonal braces, illustrated in Fig. 12.17.

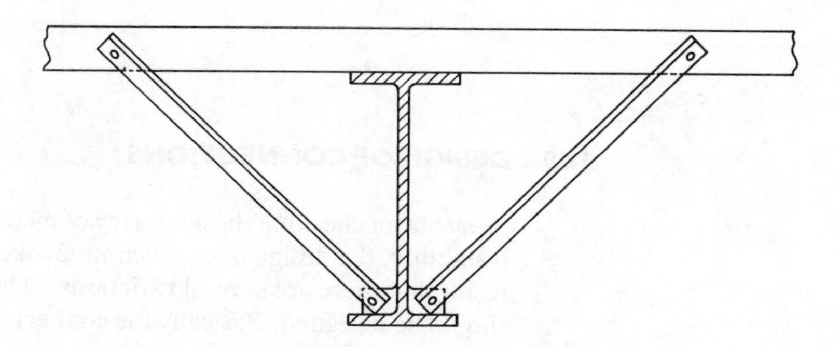

Fig. 12.17 Effective torsional restraints

Though individual design of the various column and rafter restraints would probably result in different sizes, it is more economic to design for the worst case and standardize on one size for all restraints. As the area of column flange is larger than that of the rafter, the design of the braces will be based on the conditions appropriate to the restraint at the column/haunch intersection. Therefore, assuming that the diagonal stays are approximately at 45° to the braced member, then:

length of brace = $\sqrt{2}$ × (depth of column section)

$$L = 1.41 \times 602 = 850 \text{ mm}$$

As an angle section is more effective as a strut (on a weight to weight basis) than a flat (thin rectangular section), an angle with a r_{vv} of at least 850/100 = 8.5 mm will be selected. From the SCI guide[6] an appropriate angle, say a **45 × 45 × 4 Angle**, is chosen and is then checked against both strength and stiffness requirements.

The slenderness ratio of a discontinuous single angle strut with a single bolt at each end (clause 4.7.10.2(b)) is either

$\lambda = 1.0L/r_{vv}$ or $0.7L/r_{xx} + 30$

$= 850/8.76$ or $0.7 \times 850/13.6 + 30$

$= 97$ or $74 < 100$ Stiffness satisfied

Lateral force $= 0.02BTp_y$

$F = 0.02 \times 227.6 \times 14.8 \times 0.275 = 18.5$ kN

BS table 27(c) $p_c = 130 \text{ N/mm}^2$

Clause 4.7.10.2(b) further states that for a single angle with a single fastener at each end, then the compression resistance must not be greater than 80% of the compression resistance of the angle when treated as an axially loaded strut; hence for the lateral stays:

$P_c = 0.80(0.130 \times 349) = 36.3 \text{ kN} > 18.5 \text{ kN}$ Strength satisfied

Use 45 × 45 × 4 Angle.

There are alternative forms of restraint and the reader is directed to reference (12). Unlike this design example, difficulty may be experienced in giving lateral support exactly at a plastic hinge position. Should this situation arise then the hinge position may be regarded as being laterally restrained, provided the point of attachment of the brace to the **compression flange** is not more than $D/2$ from the assumed hinge position.

12.9 DESIGN OF CONNECTIONS

Apart from checking the adequacy of members against lateral-torsional instability, the design of connections evokes much discussion with the result that there are several variations on how portal frame connections should be designed. Basically the connections have to perform as an elastic unit joining two main structural members together without loss of strength and undue distress such as gross deformations or plasticity. The method adopted for this design exercise[3] has been developed from theoretical considerations and experimental evidence. Connections based on this approach have been shown to perform satisfactorily; more details of the method can be found in reference (3).

In proportioning both the eaves and apex connections for the dead plus imposed case it should be remembered that the design is being undertaken at ultimate load level. Basically the eaves and apex joints are flush end-plate connections, in which the bolt lever arms are increased by means of a haunch (see Figs 12.18 and 12.19), thereby enhancing the moment capacity of the bolted connections. There are a number of different design criteria which need to be satisfied[3]. These will be explained during the process of designing the connections for the portal frame in this section.

As mentioned, the design of portal frame connections is generally governed by the moments and forces resulting from the dead plus imposed loading condition. However, if there is moment reversal due to another loading case (as in this example), then the connections have to be rechecked.

12.9.1 Design of eaves connection

From the portal frame analysis for the dead plus imposed loading condition (Section 12.6), it can be seen that the factored moment and vertical shear, acting on the eaves connection, are 853.9 kNm (Fig. 12.14) and 197.3 kN (i.e. 1.124 × 351/2) respectively. In addition, there is a reverse moment condition to be checked, see Section 12.9.1.8. The initial design decisions to be made are the geometry of the end-plate and the size of bolts to be used. From practical considerations (such as width of column and rafter flanges, discrete sizes of rolled plate sections), the end-plate is made 220 mm. As the depth of the haunch is approximately twice that of the basic rafter member, the end-plate is made 940 mm long (allowing for the top and bottom welds). The end-plate thickness is determined from consideration of the flexural action imposed on the plate by the forces in the bolts. Therefore, the first step is to evaluate the maximum force that can occur in any one bolt.

12.9.1.1 SIZE OF BOLTS

Assume that the vertical pitch of the bolt rows is 90 mm, with the top row being positioned at 50 mm from top surface at the tension flange of the rafter (see Fig. 12.18). By minimizing the clearance between the top row of bolts next to the flange, the bending of the end-plate is reduced. The appropriate load distribution in the bolt group to be used[3] is dependent on the ratio of the distance from the compression flange of the rafter to the penultimate row of the 'tension' bolts to that from the compression flange to the top row of bolts, i.e. 0.760/0.850 = 0.89.

As the ratio is about 0.9 then the two top rows of tension bolts can be assumed to carry equal load[3], with the remainder of the bolt loads varying linearly with their bolt distances (y_i) from the compression flange. With this distribution, the maximum bolt load (F_b) occurs in the four bolts furthest away from the compression flange, about which it is

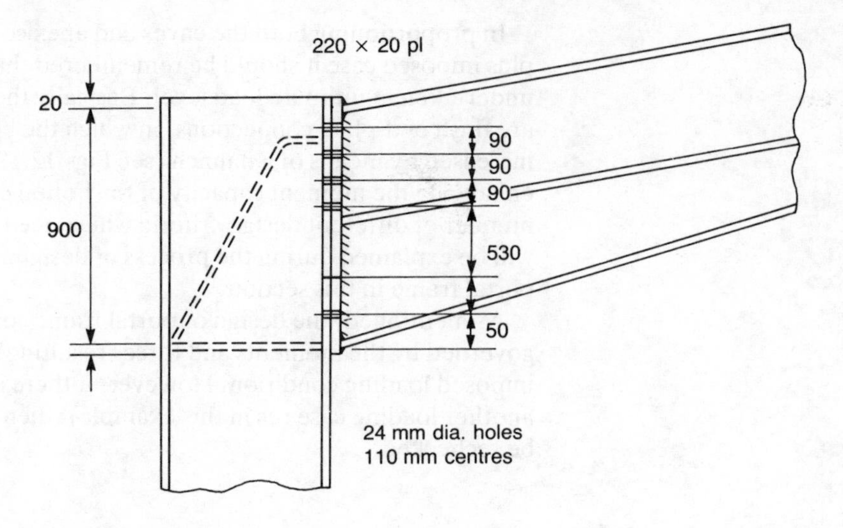

Fig. 12.18 Details of eaves connection

assumed that the connection rotates; hence:

$$F_b = M/[4d_e + 2(\Sigma y_i^2/d_e)]$$

The dimension d_e is the distance from the compression flange to a point midway between the two top rows of bolts, i.e:

$$d_e = (0.850 + 0.760)/2 = 0.805$$

Hence $F_b = 853.9/[4 \times 0.805 + 2(0.670^2 + 0.580^2)/0.805] = 165.1$ kN

Nowadays, high tensile bolts (grade 8.8) are used in moment connections. However, if conditions are such that wind or crane vibrations might be a problem, resulting in bolt loosening or fatigue, then it would be advisable to use preloaded HSFG bolts, though other devices are available which prevent nut loosening. Assuming that no such conditions exist, then the ultimate load capacity of a grade 8.8, 22 mm diameter bolt (based on the proof load of the bolt, assuming that the 'proof' load $\leqslant 0.7P_{ult}$) is:

$$P_L = 0.7 \times 785 \times 303/10^3 = 166.5 \text{ kN} > 165.1 \text{ kN}$$

The use of the ultimate capacity of these bolts is conditional on grade 10 nuts being used with the bolts[(13)] to prevent premature failure by thread stripping. Alternatively, clause 3.2.2 allows the use of non-preloaded HSFG bolts, which have a larger capacity than a grade 8.8 bolt of the same diameter. Otherwise, use the lower allowable stresses stipulated in BS 5950.

Use eight 22 mm diameter bolts (grade 8.8) (tension region).

A commonly accepted assumption is that the vertical shear is taken by the bolts in the compression zone of a connection. The minimum number of bolts required to carry the vertical shear of 197.3 kN is:

$$197.3/(0.6 \times 166.5) = 1.97 \text{ (say, 2 bolts)}$$

Use at least two 22 mm diameter bolts (grade 8.8) (compression region).

12.9.1.2 DETERMINATION OF END-PLATE THICKNESS

The centres of the holes (A) can be fabricator dependent, but a good guide is to make the dimension equal to approximately fives times the bolt diameter, i.e. 110 mm. The end-plate thickness (t_p) is calculated by assuming double curvature bending of the plate, from which the following expression has been derived:

$$t_p = \sqrt{\left(\frac{4F_b m}{p_{yp} L_e}\right)}$$

m = effective span from centre of bolt to edge of weld
$= (A - t_b - 2 \times \text{weld size})/2$
$= [110 - 8.5 - 2 \times 6(\text{est})]/2 = 44.8$ mm
L_e = effective length, based on 30° dispersal
= lesser of $(C + 3.5m)$ or $7.0m$
$= (90 + 3.5 \times 44.8) = 246.8$ mm or $7 \times 44.8 = 313.6$ mm
$t_p = \sqrt{[(4 \times 165.1 \times 44.8)/(0.265 \times 246.8)]} = 21.3$ mm

Note that the design strength of plate sections is 265 N/mm^2. A simple rule is to make the end-plate thickness equal to the bolt diameter, i.e. 22 mm, which is supported in this instance by the preceding calculation. However, due to the discrete sizes of plate rolled by the steel producers, the designer has to choose between 20 or 25 mm plate. In this case, due to the value of 21.3 mm being less than 22 mm, use a 20 mm thick end-plate.

Use 220 mm × 20 mm plate × 940 mm long.

12.9.1.3 WELD SIZES FOR END-PLATE

The end-plate is usually connected to the rafter section by means of fillet welds. A simple rule for proportioning the weld sizes at ultimate load condition is to make the combined throat thicknesses of the welds equal to at least the thickness of the plate element being welded[3]. Therefore:

Flange weld $= T_b/\sqrt{2}$
$= 12.7/1.41 = 8.98$ **Use 10 mm FW.**

Web weld $= t_b/\sqrt{2}$
$= 8.5/1.41 = 6.03$ **Use 6 mm FW.**

The heavier flange weld must be continued down the web on the tension side of the connection for a minimum distance of 50 mm in order to avoid premature weld cracking in the vicinity of the root fillet of the rafter flange, due to potential stress concentrations.

12.9.1.4 CHECK THE TENSION REGION OF THE COLUMN

The column flanges needs to be checked for the effects of cross-bending due to the action of the bolts[3]. The design formulae given in reference (3) have been modified; that is, it is assumed that the effect of a hole has been compensated by the flexural action of the bolt which effectively replaces the missing plate material. These modified formulae are used in the following calculations. Stiffening is required if:

$$4F_b \geqslant T_c^2\left[\frac{C+w+w^*}{m}+\left(\frac{1}{w}+\frac{1}{w^*}\right)(m+n)\right]p_{yc}$$

$$w = \sqrt{[m(m+n)]}$$
$$n = (B-A)/2 = (220-110)/2 = 55.0 \text{ mm}$$
$$w = \sqrt{[44.8(44.8+55.0)]} = 66.9 \text{ mm}$$
$$w^* = 70 < 2w$$

$$4 \times 165.1 > 14.8^2\left[\frac{90+66.9+70}{44.8}+\left(\frac{1}{66.9}+\frac{1}{70}\right)(44.8+55.0)\right]0.275$$

$660.4 > 481$ Inadequate.

Therefore stiffening of the flange is required in the tension region. Check that the stiffened flange is adequate[3], i.e:

$$4F_b \leqslant T_c^2\left[\frac{2v+w+w^*}{m}+\left(\frac{2}{v}+\frac{1}{w}+\frac{1}{2w^*}\right)(m+n)\right]p_{yc}$$

$$660.4 < 14.8^2\left[\frac{68+66.9+70}{44.8}+\left(\frac{2}{34}+\frac{1}{66.9}+\frac{1}{140}\right)(44.8+5.0)\right]0.2$$

$660.4 < 762$ kN Adequate.

12.9.1.5 CHECK SHEAR IN THE COLUMN WEB PANEL

The factored moment acting on the connection produces a shearing action in the column web adjacent to the connection. Therefore, the shear capacity of the web (P_v) needs to be checked against the induced shear force (F_v) of:

$$F_v = M/d_e$$
$$= 853.9/0.805 = 1060.7 \text{ kN}$$

The following design rule, which is slightly more correct than the guidance given in BS 5950, is based on research evidence[14]:

$$P_v = 0.6t_c(D-2T_c)p_{yc}$$
$$= 0.6 \times 10.6(602.2 - 2 \times 14.8)0.275 = 1001.4 \text{ kN}$$

$$F_v > P_v$$

The column web has to be stiffened. As the inner column flange in the tension zone has also to be stiffened (Section 12.9.1.4), use the Morris stiffener, which combines both functions into one stiffening arrangement (see Fig. 12.18). This form of shear stiffening has been shown to be both economic and structurally efficient[14]. Make the horizontal portion equal to 100 mm, thereby allowing easy bolt access for the erectors. Design the Morris stiffeners like diagonal stiffeners – they have to carry the excess shear force not taken by the column web, i.e. 1060.7 − 1001.4 = 59.3 kN. The required area of stiffeners is obtained from:

$$A_s \geqslant (F_v - P_v)/(p_{ys} \cos\theta)$$

$$\tan\theta = d_e/(D_c - 2T_c - 100)$$
$$= 805/(602.2 - 2 \times 14.8 - 100) = 1.703$$
$$\cos\theta = 0.506$$
$$A_s \geqslant (1060.7 - 1001.4/(0.265 \times 0.506) = 442 \text{ mm}^2$$

Use nominal sized stiffeners, say two 90 mm × 10 mm flats, which provide some 1800 mm^2 of area.

Use two 90 mm × 10 mm flats, 6 mm FW.

12.9.1.6 CHECK COMPRESSION ZONE

Web buckling A simple rule based on experimental evidence indicates that stiffening to the web is required if:

$$d/t \geqslant 52\varepsilon \quad \text{(cf. BS 5950)}$$

i.e. $d/t = 602.2/10.6 = 56.8 > 52$

Therefore the web needs stiffening to prevent plate buckling. Generally, full depth web stiffeners are required in this position.

Web crushing The force being transmitted from the compression flange of the haunched rafter into the column web is:

$$F_c = M/d_e + R$$
$$= 853.9/0.805 + 164.4 = 1225.1 \text{ kN}$$

Stiffeners required if:

$$F_c \geqslant P_c = [T_b + 5(T_c + \text{root fillet}) + 2t_p]t_c p_{yc}$$
$$= [12.7 + 5(14.8 + 12.7) + 2 \times 20]10.6 \times 0.275 = 554.4 \text{ kN}$$

1225.1 > 554.4 Stiffener required

Web stiffeners (placed on either side of the column web, opposite the rafter compression flange) are required to prevent both web buckling and crushing. The capacity of the stiffened column web in the compression zone is given by:

$$P_{vs} = A_s p_{ys} + 1.63 T_c (B_c t_c)^{1/2} p_{yc}$$

Hence $A_s \geqslant [F_v - 1.63 T_c (B_c t_c)^{1/2} p_{yc}]/p_{ys}$

$$= [1225.1 - 1.63 \times 14.8(227.6 \times 10.6)^{1/2} 0.275]/0.265$$
$$= 3393 \text{ mm}^2$$

Use two 100 mm × 20 mm flats, 6 mm FW.

Check the outstand edge of these compression stiffeners for buckling; stiffeners are adequate if:

$$b/T \leqslant 7.5\varepsilon$$

$b/T = 100/20 = 5.0 < 7.5$ Stiffeners adequate

12.9.1.7 LOCAL PLASTICITY ADJACENT TO END-PLATE

In advocating the design method outlined (Sections 12.9.1.1 to 12.9.1.6) it is accepted that as the connection approaches its ultimate capacity, local areas of plasticity would have developed adjacent to the end-plate, both on the tension and compression sides. On the tension side, there is a diffusion of load from the tension flange into the web and across to the end-plate. This diffusion, as well as residual stresses due to welding, are two of a number of factors which interact to produce large plastic strains in this region. It is certainly true that the load distribution in the bolts results from load emanating from the haunch web as well as from the tension flange via the end-plate. At present, there is no satisfactory criterion to check this zone, apart from imposing a more severe design restriction than is necessary. Similarly, the haunch flange and web in the compression zone would also exhibit plasticity. It is suggested, based on available research evidence, that these relatively small areas of plasticity are acceptable, when compared with the larger yielded zones associated with the formation of plastic hinges, in the later stages of loading.

12.9.1.8 CHECK FOR REVERSED MOMENT CONDITION

From Fig. 12.10 it can be seen that the eaves connection is also subjected to a reversed moment of 330 kNm and therefore the proposed connection details must be checked as to their suitability to sustain this moment. The following checks are based on the assumption that the size of end-plate and bolt diameter are not changed.

Moment capacity Assuming that the connection would rotate about the top flange, then the moment capacity of the connection as designed is:

$$M = 2 \times 166.5 \times 0.850 = 283 < 330 \text{ kNm}$$

i.e. moment capacity has to be increased. This is achieved by inserting two additional bolts in the bottom zone of the connection (see Fig. 12.18). Hence:

$$M = 2 \times 166.5(0.850 + 0.760) = 536 > 330 \text{ kNm}$$

Vertical shear The six bolts in the upper part of the connection are more than sufficient to cope with the vertical shear from the dead plus wind case.

'Tension' region of column This particular region of the column is already reinforced by full depth web stiffeners for a much larger force; it would be safe to assume that this region is adequate without a detailed check.

Shear in the column web panel As the reversed moment of 330 kNm is about 60% of the moment on which the original design was made, then the provisions for shear stiffening should prove adequate (see Fig. 12.18).

'Compression' zone of column The load to be transferred into the column web in the 'compression' zone at the top of the column member is:

$$F_c = M/d_e + R$$
$$= 330/[(0.850 + 0.760)/2] - 1.124 \times 62.1 = 340 \text{ kN}$$

Note that the axial load in the rafter for the dead plus wind case is tension.

It is known that the effectiveness of web stiffeners, when not placed in line with the application of the compression load, decreases rapidly as they are positioned further away from the load. In this case, it is decided to ignore any contribution that the horizontal part of the Morris stiffener may have on the bearing strength of the column web in the vicinity of the 'compression' zone.

With reference to Fig. 12.18, it can be seen that the compression force from the haunched rafter flange is only resisted by about half the web length compared with that used in Section 12.9.1.6, i.e:

$$P_c = [T_b + 2.5(T_c + \text{root fillet}) + 2t_p]t_c p_{yc}$$
$$= [12.7 + 2.5(14.8 + 12.7) + 2 \times 20]10.6 \times 0.275 = 354 \text{ kN}$$

Nevertheless, the bearing capacity of the column web is adequate.

This completes the design of the eaves connection.

12.9.2 DESIGN OF APEX CONNECTION

The design of the apex connection is simpler than that of the eaves connection in so far as only the depth of haunch, geometry of end-plate and bolt diameter need to be determined, i.e. no column involved (see Fig. 12.19).

The apex connection has to be designed for a moment of 394.9 kNm. (Note that if the apex moment has not been evaluated then a good estimate is to make it equal to the full plastic moment of the rafter

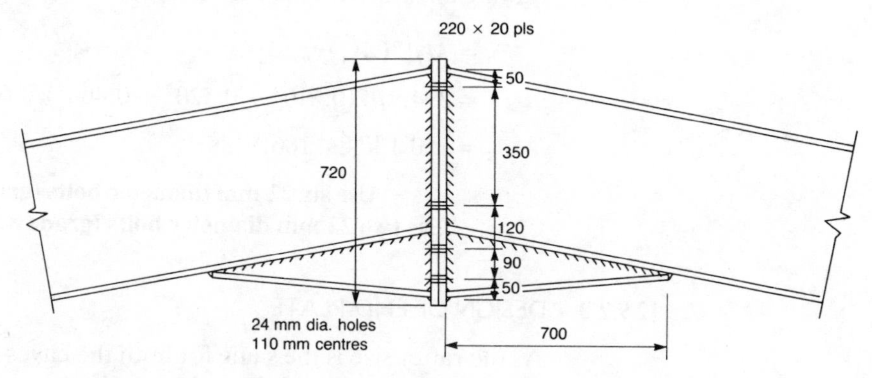

Fig. 12.19 Details of apex connection

section as the bending moment in this zone is virtually constant, i.e. for this exercise it would only be 2% in error and would not affect the outcome.)

The vertical shear is theoretically zero (due to symmetrical loading and frame). Nevertheless, it is advisable to have a minimum of two bolts in the compression region. The connection details need to be checked if there is moment reversal due to other loading conditions, which in this example is 57 kNm.

12.9.2.1 PROPORTIONS OF APEX HAUNCH

The apex connection has to be haunched in order to increase the tension bolt group lever arm so that the connection has sufficient moment capacity. The actual depth of haunch is chosen to accommodate sufficient bolts within its depth to sustain the moment. This is a trial and error process. In this design case, it has been decided to make the overall depth of the connection 680 mm deep (see Fig. 12.19). To allow for the dispersion of the tension flange load to the bolts the length of haunch should be at least 1.5 times depth of basic rafter section, i.e. $1.5 \times 454 = 681$ mm, or twice the depth of haunch cutting, i.e. $2(650 - 460) = 380$ mm. Therefore, make haunch length equal to 700 mm. Use minimum weld size (6 mm FW) to connect haunch cutting to underside of basic rafter member.

12.9.2.2 SIZING OF BOLTS

Having designed the eaves connection and established the bolt diameter to be 22 mm, it is economic to standardize on the size of bolts throughout the frame, i.e. check that 22 mm diameter bolts (grade 8.8) are suitable for the apex connection. Again, the ratio of the two largest lever arms of the tension bolts about the compression flange (see Fig. 12.19) is evaluated:

$$0.520/0.610 = 0.85$$

Therefore, the maximum bolt load (F_b) occurs in the extreme row of bolts, i.e. nearest tension flange, with the remainder of the bolt loads varying linearly with their distances from the compression flange, i.e:

$$F_b = M/[2(\Sigma y_i^2/y_{max})]$$

$$= 394.9/[2(0.610^2 + 0.520^2 + 0.400^2)/0.610]$$

$$= 150.1 \text{ kN} < 166.5 \text{ kN}$$

Use six 22 mm diameter bolts (grade 8.8) (tension region).
Use two 22 mm diameter bolts (grade 8.8) (compression region).

12.9.2.3 DESIGN OF END-PLATE

As the rafter size is the same for both the eaves and apex connections, then make the width of the end-plate the same as that for eaves, i.e.

220 mm, and the depth 720 mm. Again making the centres of holes (A) equal to 110 mm, then the end-plate thickness is calculated from:

$$t_p = \sqrt{[(4 \times 150.1 \times 44.8)/(0.265 \times 246.8)]} = 20.3 \text{ mm}$$

Use 220 mm × 20 mm plate × 720 mm long.

As the basis of the calculations for the flange and web welds is identical to those determined for the eaves connection, then make:

Flange weld **10 mm FW.**
Web weld **6 mm FW.**

12.9.2.4 CHECK FOR REVERSED MOMENT CONDITION

Assuming that the size of end-plate and bolt diameter are unchanged, then the only check required is for the 'tension' region. The moment capacity of the two bolts in this zone (Fig. 12.19) is:

$$M = 2 \times 166.5 \times 0.600 = 200 \text{ kNm} > 57 \text{ kNm}$$

Therefore the apex connection as detailed in Fig. 12.19 is satisfactory.

12.10 GABLE FRAMING

When it is specified that there are to be no extensions to a building in the future, and bearing in mind that the gable framing has to support only half the load carried by an intermediate main frame, then the gable arrangement shown in Fig. 12.20 is commonly used. Basically, the gable framing consists of inclined beam members (supporting the purlins and some gable sheeting), spanning between the vertical gable posts.

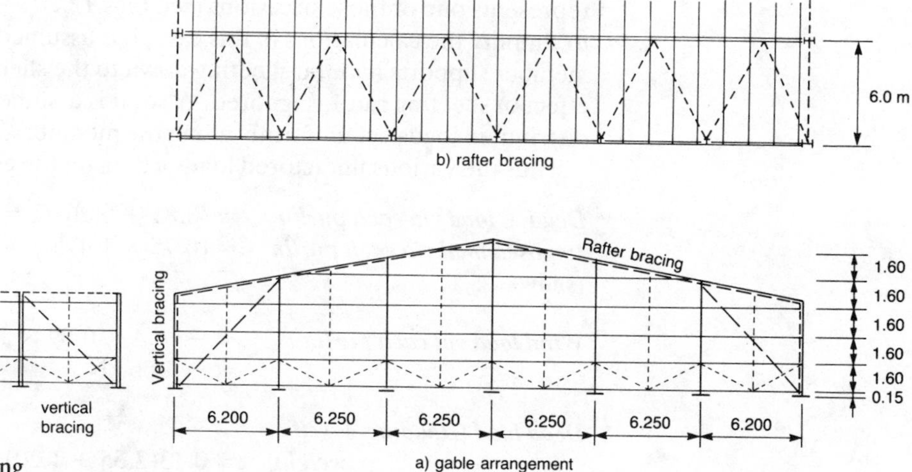

Fig. 12.20 Gable framing

In deciding the spacing of the gable posts, any dominant openings in the gables would have to be taken into account. However, in this example, it has been assumed there are no openings and it is proposed that the gable posts be positioned at approximately 6 m apart, i.e. the span for which the sheeting rail was originally chosen. The gables are subject to a maximum wind pressure of 1.0×0.59 kN/m^2 (see Section 12.10.1.2). A quick check on the load capacity of the sheeting rails (Multibeam B170/155) indicates that the proposed gable post spacing of 6.250 m for the four internal spans and $6.000 + 0.200(\text{est}) = 6.200$ m for the two outer spans is acceptable.

The in-plane stability for this relatively flexible gable framing is achieved by incorporating vertical gable bracing into the end bays of the gable (see Fig. 12.20a). The bracing members are designed as struts, resisting the side wind load acting on the corner gable posts. By triangulating the bracing as shown, additional wind load is induced into the outer edge members in the gable end bays.

12.10.1 Gable edge beams

For design purposes, consider the gable beams adjacent to the ridge of the building as being typical of the edge members. Such members are usually assumed to be simply supported, being designed to carry the gable cladding and wind loads, as well as the purlin loads, back to the gable posts. The actual length of the member to be designed is $6.250/\cos 10.41° = 6.355$ m.

12.10.1.1 DESIGN LOADING

In most design situations the loads are usually relatively simple to evaluate, but on occasions the time spent in specifying a loading regime precisely (and consequently the design forces) is not worth the effort. In these circumstances, it is advantageous to make safe assumptions in order to effect a quick design solution. The assessment of the cladding weight acting on the edge beam and the wind load on that cladding represents one of those occasions (see Figs 12.20 and 12.21a). In order to simplify the calculations in this case, it is assumed that the edge member supports half the sheeting down to the sheeting rail 2, i.e. the effect of sheeting rail 1 is ignored. Also, it is assumed that the resulting distributed loads act uniformly along the member.

Thus the various unfactored loads acting on the edge member are:

Dead + load via each purlin $= (0.81 + 0.26)/2 = 0.54$ kN

Imposed load via each purlin (snow) $= (0.75 \times 1.475 \times 6.0)/2 = 3.32$ kN

Wind load via each purlin $= -(1.4 \times 0.59 \times 1.475 \times 6.0)/2$
$= -3.66$ kN

Dead load (cladding + self weight) $= 0.13(2.35 + 1.20) \times 6.250/4 + 2.3(\text{est})$
$= 0.72 + 2.3 = 3.02$ kN

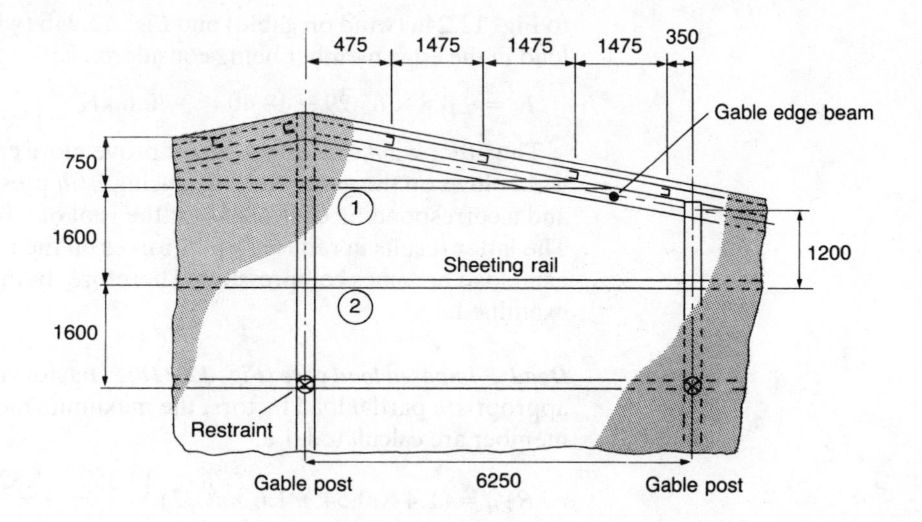

(a) Dimensions

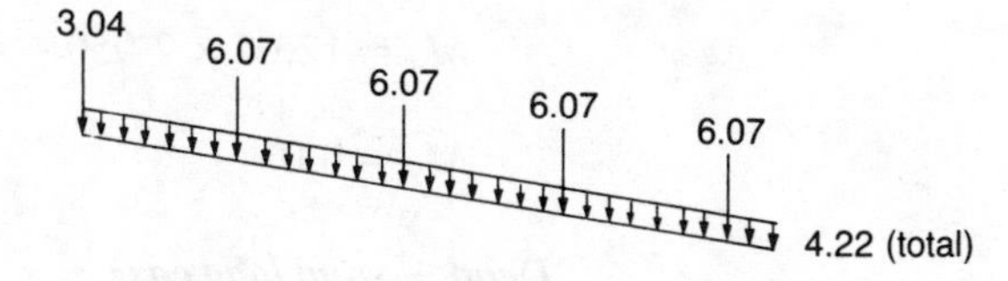

(b) Dead plus imposed loading (factored) (kN)

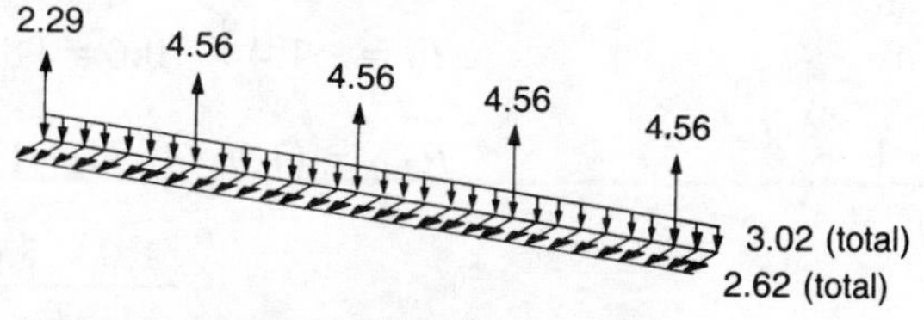

(c) Dead plus wind loading (factored) kN

Fig. 12.21 Loading for gable edge beams

Wind load acting on gable sheeting The coexistent wind loading on the gable, associated with the uplift coefficient of 1.4 on the roof (for wind on the side of the building) is 0.8 suction (see section 11.4.3 for explanation), hence the wind load on the relevant gable cladding is:

$$= -0.8 \times 0.59(2.35 + 1.20) \times 6.25/4 = -2.62 \text{ kN}$$

Axial load due to wind on gable Due to the wind drag and wind suction on the gable, loads are generated in the rafter bracing system (see Fig. 12.20b). As the edge members form part of this system, they have to carry axial load, the magnitude and nature of which depend on the particular wind condition occurring on the gable. For example, assuming that the wind load on the gable is $-0.8q$, then with reference to Fig. 12.24a (wind on gable) and Fig. 12.24b (wind drag), the axial load in the edge member being considered is:

$$F_t = -0.8 \times 65.29 - 18.40 = -70.6 \text{ kN}$$

The other wind case which might prove more critical occurs when the wind blows on the gable end, producing $1.0q$ pressure on the sheeting, and a corresponding coefficient for the roof of -0.5 (case C, Fig. 11.5). The latter results in reduced uplift forces on the roof members, but the axial load becomes compression. Therefore, both wind cases have to be examined.

Dead + imposed load case (Fig. 12.21b) Factoring the loads by the appropriate partial load factors, the maximum moments acting on the member are calculated, i.e:

$$R_{LH} = (1.4 \times 0.54 + 1.6 \times 3.32)\frac{[0.350 + 1.825 + 3.300 + 4.775]}{6.25}$$

$$+\frac{1.4 \times 3.02}{2}$$

$$= 6.07 \times 1.64 + 4.22/2 = 12.06 \text{ kN}$$

$$M_x = 12.06 \times 2.950 - 6.07 \times 1.475 - \frac{4.22 \times 2.950^2}{6.25 \times 2} = 23.7 \text{ kNm}$$

$$M_y = 0.0$$

Dead + wind load case

(i) Wind suction on gable ($-0.8q$): again factoring the loads with reference to Fig. 12.21c, the maximum design conditions associated with wind on the side of the building are calculated, i.e:

$$F_t = -1.4 \times 70.6 = 98.8 \text{ kN}$$

$$R_{LH} = (1.0 \times 0.54 - 1.4 \times 3.66)\frac{[0.350 + 1.825 + 3.300 + 4.775]}{6.25}$$

$$+\frac{1.0 \times 3.02}{2}$$

$$= -4.58 \times 1.64 + 3.02/2 = -6.00 \text{ kN}$$

$$M_x = -6.00 \times 2.950 + 4.58 \times 1.475 - \frac{3.02 \times 2.950^2}{6.25 \times 2} = -13.0 \text{ kNm}$$

$$M_y = -1.4 \times 2.62 \times 6.25/8 = -2.87 \text{ kNm}$$

(ii) Wind pressure on gable ($1.0q$): using the appropriate wind coefficients for this condition, then by similar calculations as in (i):

$$F_c = 117.2 \text{ kN}$$
$$M_x = -2.0 \text{ kNm}$$
$$M_y = -3.6 \text{ kNm}$$

12.10.1.2 MEMBER SIZE

As the member is loaded between the positional restraint provided by the 'simple' connections, then $m = 1.0$ (BS table 13) and therefore the member need only satisfy the following criterion for the two design cases:

$$\frac{M_x}{M_b} + \frac{M_y}{p_y Z_y} \leqslant 1.0$$

Unlike the purlin loading on intermediate portal frames where the load supported by the purlins is balanced about the vertical plane of the portal frame, the loading supported by the purlins attached to the edge member is not balanced. This represents a destabilizing condition and therefore $n = 1$ (BS table 13).

As the member has been assumed to be simply supported, then the effective length, $L_{Ex} = 1.0 \times 6.355 = 6.355$ m. For the dead plus imposed load case, the top flange of the edge member is in compression, and it is restrained at intervals by the purlins, therefore the effective length, $L_{Ey} = 1.0 \times 1.500 = 1.500$ m. However, the reverse is true for the dead plus wind load case, i.e. the bottom flange, being in compression, is not restrained between the connections, hence:

$$L_{Ey} = 1.0 \times 6.355 = 6.355 \text{ m}$$

For lightly loaded gable edge beams, a channel section is commonly used, as it can be bolted directly (when suitably notched) onto the outside of the gable posts. This enables the cleats supporting the sheeting rails to be positioned in the same vertical plane without the use of special cleats. However, it has been decided to use a universal column section which has better properties than a channel (weight for weight). The beams are to be positioned on the centrelines of the posts, which might result in special cleats for the sheeting rails. An alternative to using special cleats is to arrange the centres of the sheeting rails on the gables so that the rails are supported by the posts and not the edge members.

The design process can be shortened by making use of the tabulated values for the bending and compression resistances, given in the SCI guide[(6)], for the different hot-rolled sections for both grade 43 and 50 steels. Therefore, with reference to pp. 32 and 176 of the guide[6], a **152 × 152 × 30 UC** is chosen and now has to be checked.

Dead + imposed load case

$$F = 0.0 \text{ kN}$$
$$M_x = 23.7 \text{ kNm}$$
$$M_y = 0.0$$
$$L_{Ey} = 1.500 \text{ m}$$
$$\lambda = 39$$

From p. 176 of the SCI guide for the chosen section, $M_b = 67.0$ kNm, hence:

$$\frac{23.7}{67.0} + 0.0 = 0.354 < 1.0$$

Dead + wind load case (i)

$$F_t = 98.8 \text{ kN}$$
$$M_x = 13.0 \text{ kNm}$$
$$M_y = 2.87 \text{ kNm}$$
$$L_{Ey} = 6.355 \text{ m}$$
$$\lambda = 166$$

From p. 176 of the guide, $M_b = 35.6$ kNm and $p_y Z_y = 20.0$ kNm, hence a safe estimate of the member buckling resistance is:

$$\frac{13.0}{35.6} + \frac{2.87}{20.0} = 0.365 + 0.144 = 0.509 < 1.0$$

Dead + wind load case (ii)

$$F_c = 117.2 \text{ kN}$$
$$M_x = 2.0 \text{ kNm}$$
$$M_y = 3.6 \text{ kNm}$$
$$L_{Ey} = 6.355 \text{ m}$$
$$\lambda = 166$$

In addition the values of M_b and $p_y Z_y$ already obtained, the compression resistance is required; therefore, from p. 176, $P_c = 220$ kN, hence:

$$\frac{117.2}{220} + \frac{2.0}{35.6} + \frac{3.6}{20} = 0.533 + 0.056 + 0.180 = 0.769 < 1.0$$

Use 152 × 152 × 30 UC.

Use the same section for all gable edge beams, checking that a more severe design condition does not exist. A typical beam–post intersection is detailed in Fig. 12.22.

12.10.2 Gable posts

The central gable post is to be designed as it has to sustain the worst design condition of all the posts. The posts are assumed to be simply supported between the base and the positional restraint provided by the rafter bracing (see Fig. 12.20).

For the wind suction condition (-0.8×0.59 kN/m^2), it would appear that the inner compression flange is unrestrained between the base and the positional restraint of the rafter bracing. The benefit of the sheeting rail restraint on the outer tension flange could be taken into account by using the stability clauses of appendix G, BS 5950 as was done in checking the wind condition for the main rafter apex region (Section 12.8.2.3). However, in the latter case, the likelihood of the purlins being removed permanently is very remote. However, there is a greater possibility that the owner (who may be different from the original developer) may require other arrangements with respect to openings in the future.

Therefore it is decided (for simplicity) to ignore this potential benefit from the rails. Nevertheless, it is felt that any openings would probably not extend above the eaves level, with the result that it is proposed to restrain laterally the inner flange of the five internal gable posts at eaves level, by bracing back to the sheeting rail, i.e. 4.95 m from the ground. Therefore, the design assumes that the gable posts are unrestrained up to the sheeting rail at 'eaves level' in which case the worse wind condition is the wind pressure of $1.0 \times 0.59\ \mathrm{kN/m^2}$.

12.10.2.1 DESIGN LOADING

The axial load includes the gable sheeting, plus self weight of the rails and post, together with the end reactions from the appropriate edge members, plus the apex purlins loading (not included in end reactions of edge members). However, the axial load is usually relatively small, the main loading being the bending action induced into the post by the wind loading acting in the gable.

Dead + imposed load case

$$F_c = 1.4[(\text{cladding} + \text{insulation}) + (\text{post} + \text{rails}) + (\text{end reactions})]$$

$$= 1.4[0.13 \times 8.62 \times 6.25 + 0.6 \times 8.90$$

$$+ \left(0.54 + \frac{1.6}{1.4} \times 3.32\right)(1 + 2 \times 1.64) + 3.02]$$

$$= 1.4(7.00 + 5.34 + 18.55 + 3.02) = 47.5\ \mathrm{kN}$$

$$M_x = 0.0$$

$$M_y = 0.0$$

Dead + wind load case The wind loads used in these calculations are based on the wind pressure condition, i.e. $1.0 \times 0.59\ \mathrm{kN/mm^2}$. Note that in this combination of dead + wind, both the partial load factors are equal to 1.4, as uplift is not the condition being examined.

$$F_c = 1.4[0.13 \times 8.62 \times 6.25 + 0.6 \times 8.90$$

$$+ (0.54 - 3.66)(1 + 2 \times 1.64) + 3.02]$$

$$= 1.4(7.00 + 5.34 - 13.35 + 3.02) = 2.8\ \mathrm{kN}$$

Wind load on a typical sheeting rail is:

$$= 1.4(1.0 \times 0.59 \times 1.60 \times 6.25) = 8.26\ \mathrm{kN}$$

Again, ignoring the small local effect from rail 1 (Fig. 12.21a), then by proportion the wind load on rail 2 is:

$$= 8.26[(2.35 + 1.20)/4 + 0.80]/1.60 = 8.71\ \mathrm{kN}$$

and similarly for the bottom rail, the wind load is:

$$= 8.26(1.60 + 0.15)/(2 \times 1.60) = 4.52\ \mathrm{kN}$$

Therefore, the end reaction at the top of the post is:

$$R_{TP} = \frac{4.52 \times 0.15 + 8.26(1.75 + 3.35 + 4.95) + 8.71 \times 6.55}{8.90}$$

$$= 15.81 \text{ kN}$$

$$M_x = 15.81 \times 3.95 - 8.71 \times 3.20 - 8.26 \times 1.60 = 21.4 \text{ kNm}$$

$$M_y = 0.0$$

12.10.2.2 MEMBER SIZE

By ignoring any restraint from the rails below eaves level, then:

$$L_{Ey} = 1.0 \times 4.95 = 4.95 \text{ m}$$

With the experience gained in Section 12.10.1.2, it is advantageous intially to design the gable post for the dead plus wind load cases, and then check its adequacy for the dead plus imposed load case.

Referring to p. 134, SCI guide[6], it can be seen that a **254 × 102 × 28 UB** could be suitable. Therefore, check the adequacy of this section. The buckling resistance M_b of the universal section, for an effective length of 4.95 m, is 27.3 kNm and from p. 85 the compression resistance is 917 kN, hence checking the **dead + wind load case** gives:

$$\frac{2.8}{917} + \frac{21.4}{27.3} = 0.003 + 0.784 = 0.787 < 1.0 \quad \text{Section adequate}$$

Use 254 × 102 × 28 UB

If the restraint afforded by the rails is taken into account, then the wind suction load case (0.8×0.59 kN/m^2) would be the worst design condition, i.e. causing the inner flange to go into compression, in which case a 254 × 102 × 25 UB would probably prove satisfactory.

As there is only 47.5 kN axial load acting on the gable posts for the **dead + imposed load case**, then clearly the section is more than adequate, i.e. $47.5/917 = 0.052 < 1.0$. Figure 12.22 shows a typical detail at the top of the gable post.

The same section size can be used for all other gable posts. However, a check should be made on the corner gable posts, which, though supporting only half the load, is subject to wind loads acting simultaneously about the major and minor axes.

12.10.3 Corner gable posts

It is common practice to arrange the outer flange of the corner post to be in the same vertical plane as the outer flange of the portal leg, i.e. corner post is rotated through 90° compared with other gable posts. This means the worst design condition occurs when the wind direction is normal to the gable, resulting in a wind pressure of 1.0×0.59 kN/m^2 on the gable sheeting and a wind suction of -0.2×0.59 kN/m^2 acting on the side walls.

The typical wind load on a gable sheeting rail acting on the corner gable post is:

$$= 1.4(1.0 \times 0.59 \times 3.1 \times 1.60) = 4.10 \text{ kN}$$

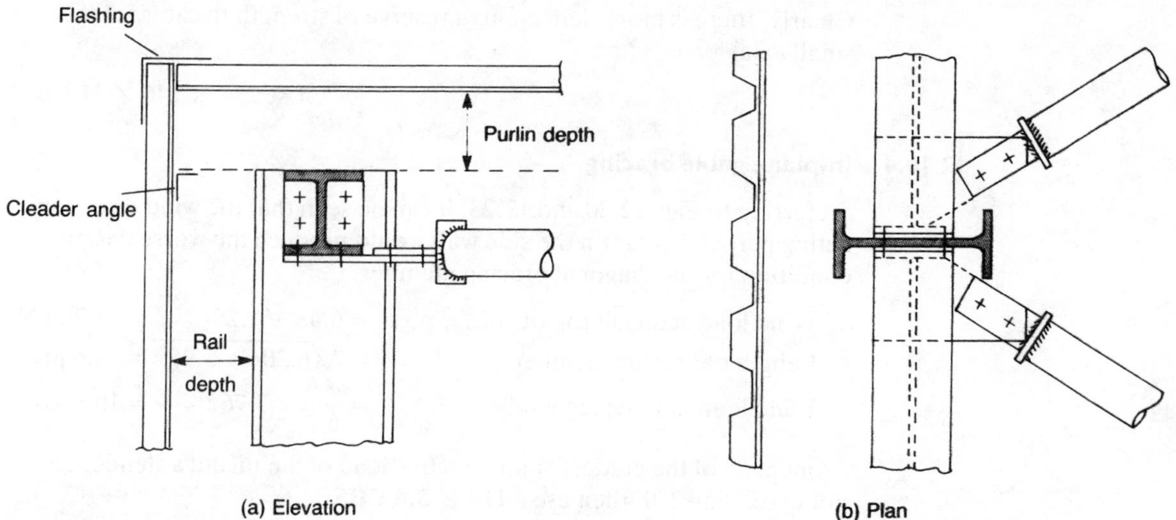

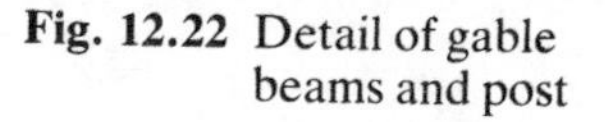

Fig. 12.22 Detail of gable beams and post

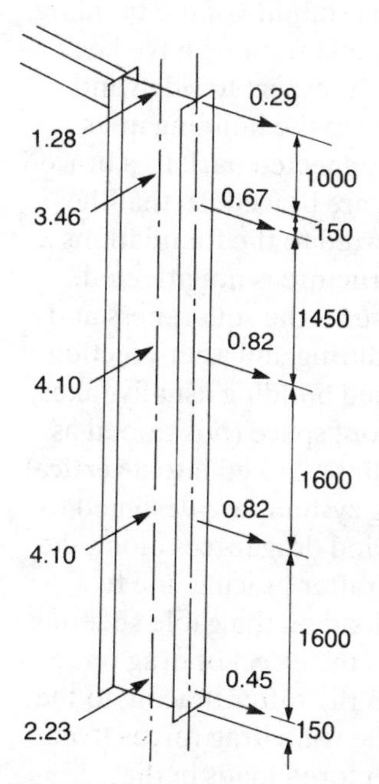

Fig. 12.23 Loading in corner gable post

The other rail loads are obtained by proportion, to give the loading pattern shown in Fig. 12.23. The end reaction at the top of the post about its minor axis for the loads acting normal to the web of the post is:

$$R_{TP_y} = \frac{2.23 \times 0.15 + 4.10(1.75 + 3.35) + 3.46 \times 4.95}{5.95} = 6.48 \text{ kN}$$

$$M_y \;= 6.48 \times 2.6 - 3.46 \times 1.60 = 11.3 \text{ kN}$$

Also, the reaction about the post's major axis is:

$$R_{TP_x} = \frac{0.45 \times 0.15 + 0.82(1.75 + 3.35) + 0.67 \times 4.80}{5.95} = 1.25 \text{ kN}$$

$$M_x \;= 1.25 \times 2.6 - 0.67 \times 1.45 = 2.28 \text{ kN}$$

As was the case with the internal posts, the effect of the axial load (which includes the load induced by bracing) is extremely small and can be ignored. Check the section size used for the internal posts. Therefore, as the member is loaded between end restraints, $m = 1.0$ and the design criterion again becomes:

$$\frac{M_x}{M_b} + \frac{M_y}{p_y M_y} \leqslant 1.0$$

$$\frac{2.28}{27.3} + \frac{11.3}{0.275 \times 34.9} = 0.084 + 1.177 \not< 1.0 \quad \text{Inadequate}$$

Try a 254 × 146 × 31 UB section. From pp 29–134, of the SCI guide[(6)], $Z_p = 61.5 \text{ cm}^3$ and $M_b = 37.0$ respectively, hence:

$$\frac{2.28}{45.5} + \frac{11.3}{0.275 \times 61.5} = 0.050 + 0.668 = 0.718 < 1.0$$

Clearly, there is more than enough reserve of strength to cater for the small axial load.

Use 254 × 146 × 31 UB.

12.10.4 In-plane gable bracing

Referring to Figs 12.20 and 12.23, it can be seen that the wind direction acting perpendicular on the side wall would produce the worse design condition for the diagonal bracing member, i.e:

$$\text{Wind load acting at top of corner post} = 6.48 + 1.28 = 7.76\text{ kN}$$

$$\text{Length of bracing member} = \sqrt{6.20^2 + 5.95^2} = 8.6\text{ m}$$

$$\text{Wind load in bracing member} = \frac{8.6}{6.2} \times 7.76 = 10.8\text{ kN}$$

From p. 85 of the guide, for an effective load of 8.6 m and a slenderness not exceeding 250, then use a **114 × 3.6 CHS**.

12.11 OVERALL STABILITY OF BUILDING

The designer must always ensure the structural stability of the building. At this stage, both the portal frames and the gable framing have been designed for in-plane stability, particularly with respect to side wind loading. However, in order to provide stability to the building in its longitudinal direction, all frames need to be connected back to a braced bay. Generally, the **end bay(s)** of the building are braced, so that the wind loads acting on the gables can be transferred to the foundations as soon as possible and thereby the rest of the structure is not affected. Another function of a braced bay is that it ensures the squareness and verticality of the structural framework, both during and after erection.

The typical bracing system for a portal framed building usually takes the form of **rafter bracing** in the plane of the roof space (positioned as close to the top flange without fouling the purlins), linked into a **vertical bracing** system (see Fig. 12.20). These bracing systems are designed to cater for wind loading on the gable, plus the wind drag forces along the building. Figure 12.24a gives the forces in the rafter bracing due to a wind pressure of $1.0q$, assuming that half the load on the gable sheeting is taken by the bracing, while Fig. 12.24b gives the effect of drag on the rafter bracing. These forces are transferred via the rafter bracing to the vertical bracing system, which also transfers the wind drag forces from the side cladding. Figure 12.25a gives the unfactored loads in the vertical bracing transferred from the rafter bracing due to a unit wind pressure ($1.0q$) acting on the gable. Figure 12.25b indicates the unfactored loads in the vertical bracing due to wind drag forces acting on the roof and sides. From these two sets of loads, the factored loads for the different design cases can be deduced. The design calculations for both bracing systems for this example are not given, as they are similar to the appropriate detailed calculations in Section 11.8 given in

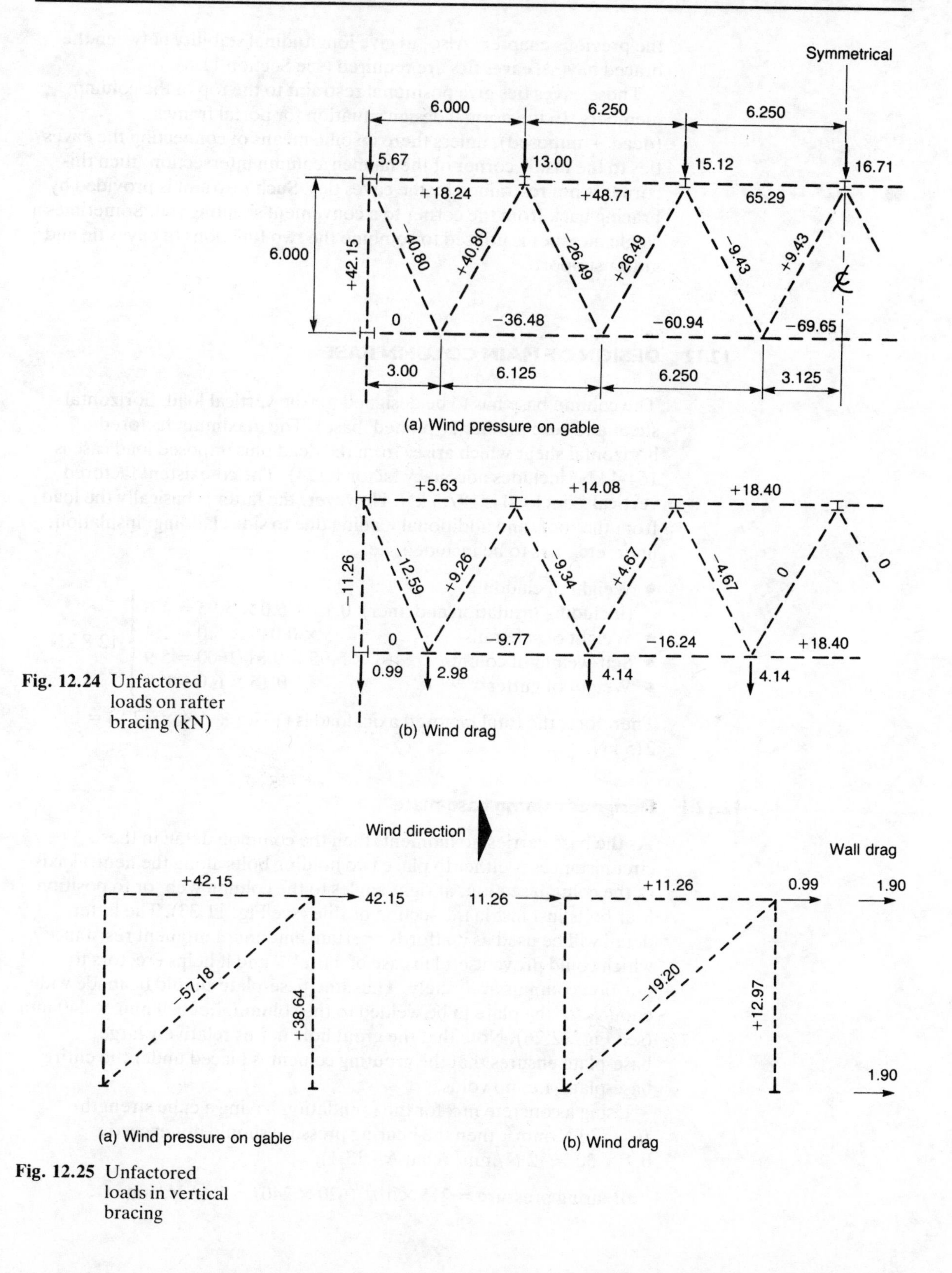

Fig. 12.24 Unfactored loads on rafter bracing (kN)

Fig. 12.25 Unfactored loads in vertical bracing

the previous chapter. Also, to give longitudinal stability between the braced bay(s), **eaves ties** are required (see Section 11.8).

Those eaves ties give positional restraint to the top of the column members. In the normal design situation for portal frames (dead + imposed), unless there is some means of connecting the eaves ties to the inside corner of the haunch/column intersection, then this corner is not restrained by the eaves ties. Such restraint is provided by bracing back from the corner to a convenient sheeting rail. Sometimes a single member is utilised to combine the two functions of eaves tie and gutter support.

12.12 DESIGN OF MAIN COLUMN BASE

The column base has to be designed for the vertical load, horizontal shear and zero moment ('pinned' base). The maximum factored horizontal shear which arises from the dead plus imposed load case is 164.4 kN (includes adequacy factor 1.124). The co-existent factored vertical axial load is 197.5 kN. However, the latter is basically the load from the roof, and additional loading due to side cladding, insulation, liner, etc., has to be included, i.e:

- Weight of cladding (including insulation and liner) $0.13 \times 6.0 \times 5.95 = 4.6$
- Weight of side rails $5 \times 0.045 \times 6.0 = 1.4$
- Self-weight of column $101 \times 5.95 \times 9.81/1000 = 5.9$
- Weight of gutter[10,11] $0.15 \times 6.0 = 0.9$

} 12.8 kN

Therefore, the total factored axial load is $(197.5 + 1.4 \times 12.8) =$ 215 kN.

12.12.1 Design of column base-plate

As the base carries no moment, then the common detail in these circumstances is either to place two holding bolts along the neutral axis of the column section, at right angles to the column web, or to position four bolts just inside the section profile (see Fig. 11.33). The latter detail will be used as it affords a certain amount of moment resistance which could prove useful in case of a fire[15] and it helps erectors to position columns accurately. Thus, the base-plate should be made wide enough for the plate to be welded to the column, i.e. 620 mm × 240 mm (see Fig. 12.26). Note that the grout hole in this relatively large base-plate ensures that the grouting cement is placed under the entire base-plate, i.e. no voids.

Using a concrete mix for the foundation having a cube strength $f_{cu} = 30\ \text{N/mm}^2$, then the bearing pressure should not exceed $0.4 \times 30 = 12\ \text{N/mm}^2$ (clause 4.13.1):

$$\text{Bearing pressure} = 215 \times 10^3/(620 \times 240) = 1.44\ \text{N/mm}^2$$

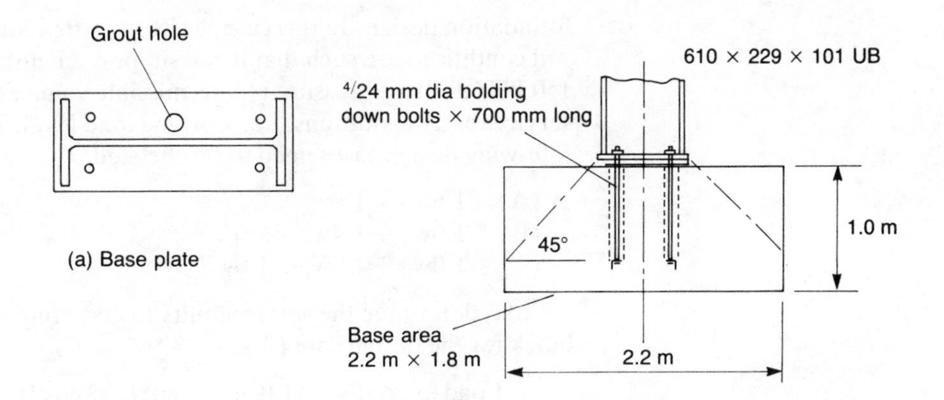

Fig. 12.26 Base details

As the projections of the base-plate beyond the profile of the column section are minimal, then the formula given in clause 4.13.2.2 cannot be used as it would result in a small value for the plate thickness. Therefore it is recommended that the base-plate thickness > flange thickness. Use a 20 mm thick base-plate (grade 43 steel), being the minimum practical thickness used in the construction industry for this size of base.

The welds connecting the column member to the base-plate need to transfer a horizontal shear of 164.4 kN. If a fillet weld was placed continuously around the profile of the column section then its length would be approximately 2 m, hence required design strength of weld is 164.4/2000 = 0.08 kN/mm, i.e. nominal size required, use 6 mm FW.

Use 240 mm × 20 mm plate × 620 mm long, 6 mm FW.

12.12.2 Sizing of holding down bolts

Where the axial load is transmitted by the base-plate (without moment) then nominal holding down bolts are required for location purposes; see Section 8.2 and reference (14). Assume **four** 24 mm diameter bolts, grade 4.6 steel, as smaller diameter bolts are more prone to damage. Nevertheless, these bolts may need to transfer the horizontal shear of 164.4 kN into the concrete foundation block if the bond between steel base and grout fails:

Shear/bolt = 164.4/4 = 41.1 kN
Shear capacity of bolt = 0.160 × 353 = 56.5 kN

Use four 24 mm diameter bolts (grade 4.6 steel).

12.13 DESIGN OF FOUNDATION BLOCK

The design of the foundations for any structure is very dependent on the ground conditions that exist on site. It is important that the engineer has this data available or some reasonable basis for formulating the foundation design. In this case, a site investigation has indicated that the soil conditions are such that it can support a bearing pressure of 150 kN/m^2. This pressure is a **permissible** value and is applicable to serviceability conditions, i.e. working load level. Therefore, the following design cases need to be checked:

(A) $1.0w_d + 1.0w_i$
(B) $1.0w_d + 1.4w_w$
(C) $1.0w_d + 1.0w_i + 1.0w_w$

First, determine the serviceability loads acting on the foundation block for the design case (A):

$$\begin{aligned}\text{Load ex roof} &= \alpha[1.0(w_d + w_i)L/(2\cos 10.41°)]\\ &= 1.124[1.0(1.60 + 4.44)37.0]/(2 \times 0.984)\\ &= 1.124(30.1 + 83.5) = 127.6\ \text{kN}\\ \text{Column loading} &= 12.8\ \text{kN}\\ F_v &= 127.6 + 12.8 = 140.4\ \text{kN}\end{aligned}$$

This vertical load should be used in conjunction with the appropriate horizontal shear, which is evaluated by multiplying the factored horizontal load by the ratio of the unfactored vertical load to the factored vertical load, i.e:

$$F_h = 164.4 \times 127.6/(351/2) = 120\ \text{kN}$$

To be strictly correct, the horizontal shear obtained from an elastic analysis using unfactored loads should have been used, i.e. 117.5 kN. However, for portal frames with shallow pitched roofs, the two values are almost identical(1), e.g. 120 kN compared with 117.5 kN. Therefore, the adjusted value from the plastic analysis is acceptable. The design case (A) loading is shown in Fig. 12.27.

As shown in Section 11.11, the design case (B) tends to govern the design of the foundation block. In order to maximise the wind uplift condition (dead plus wind), α is taken as unity, hence:

$$\begin{aligned}F_v &= 30.1 - 1.4[0.59 \times 6.0 \times 18.5(1.4 \times 0.75 + 0.6 \times 0.25)] + 12.8\\ &= +42.9 - 110.0 = -67.1\ \text{kN}\end{aligned}$$

Again using the horizontal shear from the plastic analysis, adjusted for severiceability conditions, i.e:

$$F_h = 162.3(30.1 - 110.0)/[2.375(1.4 \times 30.1 - 110.0)] = -80.5\ \text{kN}$$

Figure 12.27b shows the loading for design case (B)

In the design of the foundation block for the column members in the previous chapter (Section 11.11) the block was made square in shape. However, with the column member size being larger in this example, it might prove more economic to use a rectangular shaped base. Again, use mass concrete of sufficient thickness to spread the vertical load at 45° through the concrete block to the sub-strata. Therefore, try initially a foundation block of 2.2 m × 1.8 m × 1.0 m proportions, which weighs 2.2 × 1.8 × 1.0 × 23.7 = 78.2 kN (see Fig. 12.26b). Hence, the bearing

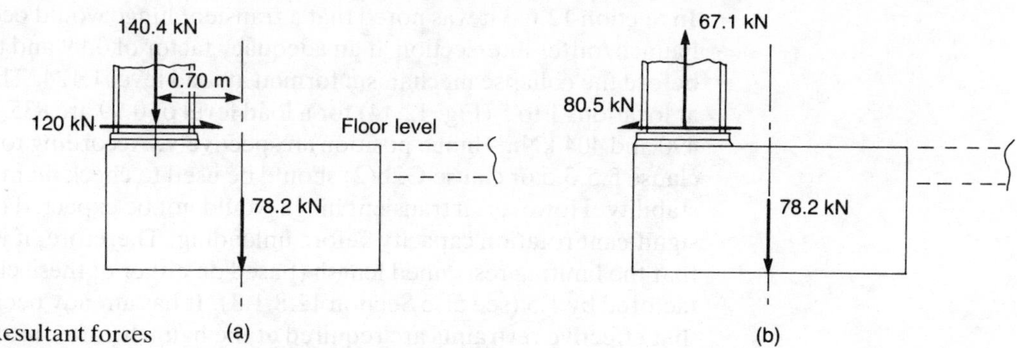

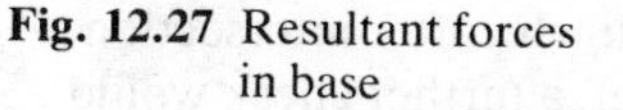
Fig. 12.27 Resultant forces in base

pressure at the concrete/soil interface for the maximum vertical load condition is (140.4 + 78.2)(2.2 × 1.8) = 55 kN/m² which is satisfactory.

The combination of vertical and horizontal loads on the base should be considered in design. The proposed foundation block now has to be checked to show the resultant soil bearing force lies within the middle third of the base length, i.e. not more than $L/6$ (= 0.367 m) from the base centreline. Both load combinations (A) and (B) are now examined. Anticipating the magnitude of the overturning moments, the column is positioned at 0.7 m from the centreline of the proposed block (see Fig. 12.27a). This causes the resultant forces to act at the concrete/soil interface for the two cases of:

$$\text{case (A)} \quad \frac{140.4 \times 0.7 - 120 \times 1.0}{140.4 + 78.2} = 0.101 \text{ m (satisfactory)}$$

$$\text{case (B)} \quad \frac{80.5 \times 1.0 - 67.1 \times 0.7}{78.2 - 67.1} = 3.02 \text{ m} \gg 0.367 \text{ m}$$

Case (B) is not satisfactory: had the margin been small, then one could either increase the size of the block or allow some tension at the concrete/soil interface, i.e. causing the resultant to act outside the $L/6$ dimension. However, in this particular case it would be economic to eliminate the moment due to the horizontal shear by tying the foundation block to the floor slab (assuming the slab is at the same level). That is, the floor slab acts as a tie and absorbs the horizontal shear. Reinforcement would need to be incorporated at floor slab level in both the slab and the block. Detailed design can be found in good concrete design textbooks.

A special block will be required for the penultimate frame to counterbalance the additional wind uplift from the vertical bracing system (see Fig. 12.25 and also Section 11.11).

12.14 OTHER CONSIDERATIONS

In designing a single-storey building the designer should be aware of other considerations which might affect the final design and the reader is referred to Section 11.12 of the previous chapter.

In Section 12.6.3 it was noted that a transient hinge would occur at the haunch/rafter intersection at an adequacy factor of 0.99 and then unload before the collapse mechanism formed at load level 1.124. The moments at locations 1 to 5 (Fig. 12.14) for a load level of 0.99 are 835, 690, 600, 490 and 404 kNm (hinge position) respectively. According to BS 5950, clause 5.5.3.5 or clause G2b(2) should be used to check member stability. However, a transient hinge would not be expected to develop significant rotation capacity before unloading. Therefore, it is suggested that the limiting restrained length (based on either of these clauses) be factored by 1.5 (see also Section 12.8.1.3). It has already been shown that effective restraints are required at the haunch/rafter intersection (transient hinge position) at collapse load level; a further check would indicate that no further restraints are necessary in the haunched region for conditions at a load level of 0.99.

STUDY REFERENCES

Topic	*Reference*
1. Comparative costs	**Horridge, J.F. & Morris, L.J.** (1986) Comparative costs of single-storey steel framed structures, *Structural Engineer*, vol. 64A (no. 7), pp. 177–81
2. Plastic design	**Morris, L.J. & Randall, A.L.** (1979) Plastic design. Steel Construction Institute
3. Plastic design	**Horne, M.R. & Morris, L.J.** (1981) *Plastic Design of Low-Rise Frames*. Collins
4. Plastic design	**Baker, J.F., Horne, M.R. & Heyman, J.** (1965) *The Steel Skeleton* vol. 2, *Plastic Behaviour and Design*. Cambridge University Press
5. Plastic design	**Horne, M.R. & Chin, M.W.** (1966) *Plastic Design of Portal Frames in Steel to BS 968.* BCSA Ltd
6. Section properties	(1985) *Steelwork Design* vol. 1, Section properties, member properties. Steel Construction Institute
7. Wind loading	BS 6399 *Loading for Buildings* Part 2: *Wind loads* (to be published; presently CP3 Ch. V Part 2
8. Stiffness method	**Coates, R.C., Coutie, M.G. & Kong, K.C.** (1988) *Structural Analysis.* Van Nostrand Reinhold
9. Plastic design	**Horne, M.R.** (1979) *The Plastic Design of Columns*, issued as supplement to reference (2). Steel Construction Institute
10. Rafter stability	**Horne, M.R., Shakir Khalil, H. & Akhtar, S.** (1979) *The Stability of Tapered and Haunched Beams*, Proc. ICE, vol. 67 (no. 9), pp. 677–94

11. Lateral restraint	**Morris, L.J. & Nakane, K.** (1983) Experimental behaviour of haunched members. In Morris (ed.) *Instability and Plastic Collapse of Steel Structures*, Proc. of Int. Conf., pp. 547–59. Granada Publishing
12. Lateral restraint	**Morris, L.J.** (1981) A commentary on portal frame design, *Structural Engineer,* vol. 59A (no. 12), pp. 394–404
13. Bolt strength	**Godley, M.H.R. & Needham, F.H.,** (1982) Comparative tests on 8.8 and HSFG bolts in tension and shear, *Structural Engineer,* vol. 60A (no. 3), pp. 94–9
14. Column web panel	**Morris, L.J. & Newsome, C.P.** (1981) Bolted corner connections subjected to an out-of-balance moment – the behaviour of the column web panel. In Howlett, Jenkins, and Stainsby (eds) *Joints in Structural Steelwork,* Proc. Int. Conf. Pentech Press
15. Fire boundary	(1980) *The Behaviour of Steel Portal Frames in Boundary Conditions.* Steel Construction Institute
16. Holding down bolts	(1980) *Holding Down Systems for Steel Stanchions.* BCSA publication 8/80

13

DESIGN OF AN OFFICE BLOCK – COMPOSITE CONSTRUCTION

In earlier chapters the design of individual elements such as beams, columns and composite floors, has been described. Complete multi-storey structures consist of a number of these elements fitted together to form a framework. In addition to the design of individual elements, the engineer must ensure that the complete structure is stable under all loading conditions. For example, the structure must be capable of withstanding some horizontal loading either actual, e.g. wind (see Section 2.3), or notional (see Section 10.2). As emphasized in Section 1.5, when bringing together structural elements into a framework the designer must ensure proper load paths, i.e. reactions from one element form loads on the supporting elements, and so on until the loads are transferred to the foundations.

13.1 LAYOUT AND BASIC CHOICES

An eight-storey block, for general office occupancy, is to be designed in structural steelwork for a site on the outskirts of Newcastle upon Tyne. The principal dimensions are shown in Fig. 13.1. The arrangement of each floor is similar, allowing the steelwork layout to be the same on each floor, and on the roof as well (with minor modifications).

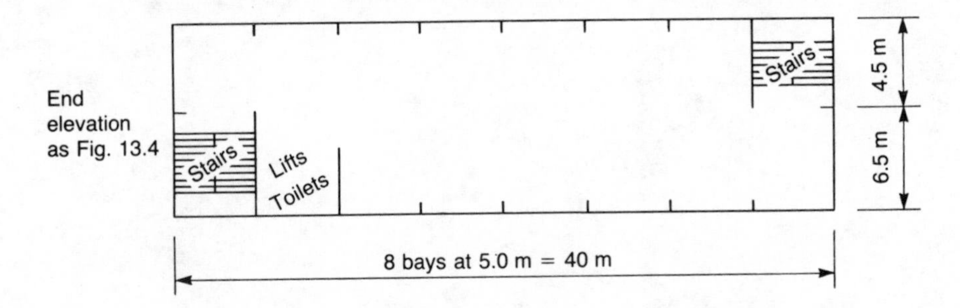

Fig. 13.1 Office block floor plan

The designer has to make basic choices with regard to:

- floor construction
- frame construction
- stair construction
- resistance to wind loading
- architectural details
- integration of structure with building services

These choices will be made taking into account:

- the economy of construction, which may require specialist advice from, e.g., quantity surveyors;
- the speed of construction, which may require liaising with contractors;
- details of possible finishes, which will generally be decided in conjunction with an architect.

All these factors affect the final cost and quality of the building, and the design team must produce a combination of these which is satisfactory to the client. A review of the factors affecting multi-storey steel frame construction is given by Mathys[1].

13.1.1 Floor construction

The floor construction could be *in situ* reinforced concrete, precast concrete, or composite construction. For speed of construction a composite flooring using a profiled steel formwork is chosen. This form of construction has been discussed in Section 9.6, and type CF60 by PMF[2] has been chosen for the present design, as shown in Table 13.1.

Table 13.1 CF60 – 1.2 mm with LWAC
(Courtesy of PMF Ltd)

Maximum spans in metres

Concrete thickness in mm	Imposed loading in kN/m²							
	3.0	4.0	5.0	6.0	7.0	8.0	9.0	10.0
120	3.80	3.80	3.80	3.60	3.35	3.10	2.90	2.70
140	3.60	3.60	3.60	3.60	3.55	3.40	3.20	3.05
160	3.45	3.45	3.45	3.45	3.45	3.45	3.40	3.35
180	3.30	3.30	3.30	3.30	3.30	3.30	3.30	3.30
200	3.20	3.20	3.20	3.20	3.20	3.20	3.20	3.20
220	3.00	3.00	3.00	3.00	3.00	3.00	3.00	3.00

This floor is capable of spanning up to 3.3 m with a lightweight aggregate concrete, and design and construction details are given by Lawson[3]. For a fire rating of 1 hour, mesh reinforcement type A193 is recommended[4], giving a cross-section for the floor construction as shown in Fig. 13.2.

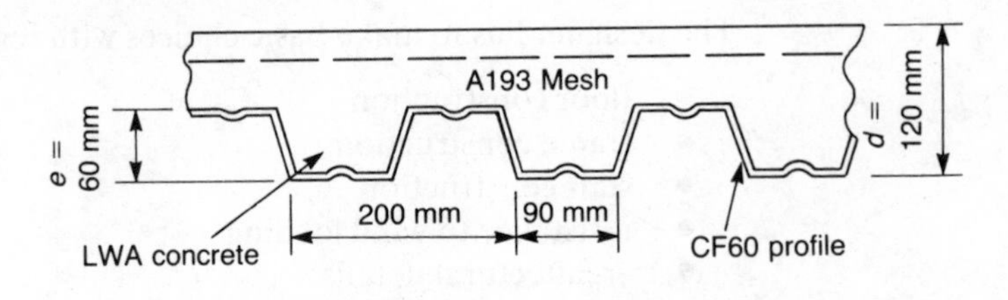

Fig. 13.2 Floor construction

13.1.2 Frame construction

The design of a multi-storey steel frame may use the method known as rigid design (clause 2.1.2.3) or simple design (clause 2.1.2.2). In rigid design connections are assumed capable of developing the required strength and stiffness for full continuity. In simple design the connections are assumed not to develop significant moments, i.e. beams are designed assuming they are simply supported. The choice between the two forms of construction is generally economic, and is outside the scope of this chapter. The present design assumes simple construction.

As discussed in Section 9.1, it is advantageous to make the slab and beams act compositely, and such an arrangement is possible with profiled steel sheeting. Fire protection is required for the steel beams and a lightweight system such as Pyrotherm is chosen (see Section 14.4).

While it is possible to design the columns to act compositely with a concrete casing, it may be preferred not to involve the process of shuttering and *in situ* casing. In the present design lightweight casing for fire protection is used, of the same type as for the beams.

13.1.3 Stair construction

A number of methods of stair construction are possible, some of which influence the speed of construction in general, and the access of operatives during construction.

Generally, concrete construction is chosen rather than an all steel arrangement, due to the complexity of the steelwork fabrication. The concrete may be *in situ* or precast, or a combination of both. The choice of method may affect the supporting steelwork arrangement, and possible alternatives are shown in Fig. 13.3. For the present design, flights and half landings are supported separately.

13.1.4 Resistance to wind loading

The horizontal loading due to wind may be resisted either by frame action, in which all the beams and columns act together, or by designing specific parts of the structure to resist these forces. In rigid frame design, wind loading would be included as one of the load systems, and the frame analysed accordingly. This is discussed further in Section 13.8.

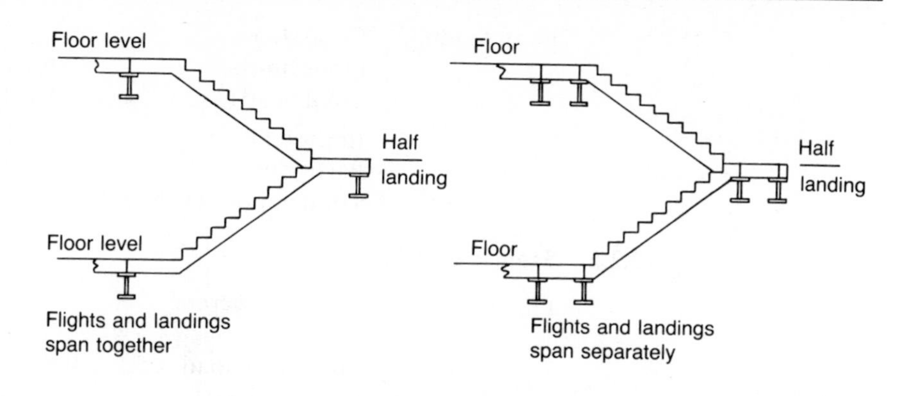

Fig. 13.3 Stair construction

The alternative to frame action is to transfer the wind forces to wind towers, shear walls or bracing located at specific points in the structure. These wind resisting parts of the structure may be constructed in either steel (as a framework described in Section 10.3) or in concrete (as shear walls or shafts). In the present design a wind bracing framework has been chosen (Section 13.7), with a brief comparison as to the effects of frame action (Section 13.8). The final choice is based on both economic and architectural considerations, as above six to eight storeys the use of a wind frame becomes cost-effective, but its presence in the structure may affect both the façade and the building layout.

13.1.5 Architectural details

All details of the steelwork frame affect the appearance and layout of the building and the design team must be aware of the results of each others' actions. Some further choices relate to external façade construction, internal partitions, floor and ceiling finishes.

In the present design a precast wall unit (below sill level) with glazing above, is chosen, giving loadings as in Section 13.2. The same unit is used at roof level as a parapet. Internal partitions are not defined in position, and an allowance for movable lightweight partitions is made in the imposed floor loading. A screeded floor finish is allowed for, together with a lightweight suspended ceiling.

Imposed loads and wind loads are obtained from the appropriate British Standard[5].

13.2 LOADING

13.2.1 Roof and floor loading

Roof loading:	CF60 slab	3.0 kN/m^2
	Roof finishes	1.8 kN/m^2
	Total dead load = 3.0 + 1.8 =	4.8 kN/m^2
	Imposed load	1.5 kN/m^2

Floor loading:	CF60 slab	3.0 kN/m^2
	Floor finishes	1.2 kN/m^2
	Total dead load = 3.0 + 1.2 = 4.2 kN/m^2	
	Imposed load	5.0 kN/m^2
	Partitions	1.0 kN/m^2
	Total imposed load = 5.0 + 1.0 = 6.0 kN/m^2	

13.2.2 Stair loading

Flights:	Precast concrete	5.5 kN/m^2
	Finishes	0.8 kN/m^2
	Total dead load = 5.5 + 0.8 = 6.3 kN/m^2	
	Imposed load	4.0 kN/m^2
Landings:	Precast concrete	3.5 kN/m^2
	Finishes	1.2 kN/m^2
	Total dead load = 3.5 + 1.2 = 4.7 kN/m^2	
	Imposed load	4.0 kN/m^2

13.2.3 Wall unit and glazing

Roof parapet:	Precast unit	2.0 kN/m
Floor wall unit:	Precast unit	2.0 kN/m
	Glazing	0.3 kN/m
	Total dead load = 2.0 + 0.3 = 2.3 kN/m	

13.2.4 Wind loading

The following notation and method may be found in reference (5).

Basic wind speed V (Newcastle upon Tyne)	46 m/s
Topography factor S_1	1.0
Ground roughness (outskirts of city)	Type (3)
Building size (max. dimension 40 m)	Class B
Factor S_2 increases with height	
Statistical factor S_3 (50-year exposure)	1.0

Design wind speed $V_s = S_1 S_2 S_3 V$ m/s
Dynamic pressure $q = 0.613 V_s^2$ N/m^2
Force coefficient $C_f = 1.3$ (for $l/w = 3.6$, $h/b = 0.8$)

The wind speed and pressure vary with height, and the appropriate values are shown in Fig. 13.4. The wind pressures may be resolved in forces at each floor level, which are also shown in Fig. 13.4, giving values for one bay width of 5 m only.

13.3 ROOF BEAM DESIGN

A suitable arrangement of beams for all floors is shown in Fig. 13.5. Roof beams are denoted R1, R2, etc., and typical floor beams (floors 1 to 7) are denoted T1, T2, etc. Using the composite slab (type CF60) spanning 3.3 m maximum, secondary beams type 1 to 4 must be

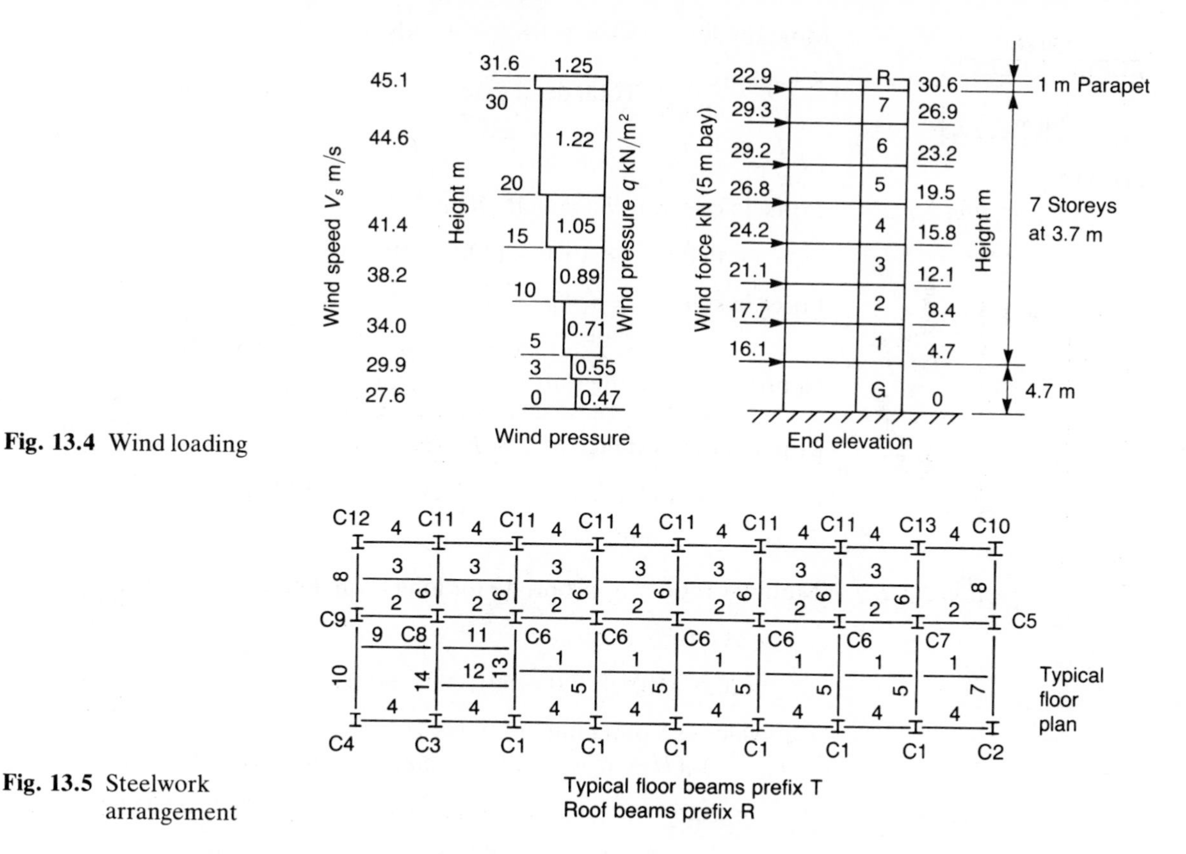

Fig. 13.4 Wind loading

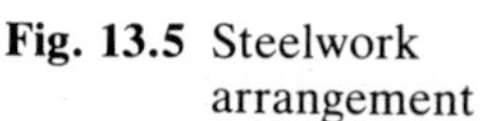
Fig. 13.5 Steelwork arrangement

provided to support the slab, at a spacing not greater than 3.3 m. These beams are supported on main beams 5 and 6, which are in turn supported by the columns. In the region of the stair wells special beams may be required such as 9 and 10, and in the vicinity of lift shafts there will be additional requirements and loadings affecting beams 11 to 14.

(a) Roof beam R1 – 152 × 89 × 16 UB

The design generally follows the recommendations of the CIRIA report[6] and section references are given for guidance where appropriate. (See Figs 13.6, 13.7.)

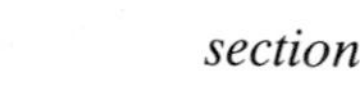
section 2.1[6]

Effective breadth b_s = span/5 = 1000 mm
Concrete cube strength = 30 N/mm²
Steel design strength = 275 N/mm²
Beam spacing = 3.25 m

Loading:	dead,	$4.8 \times 5.0 \times 3.25 =$	78 kN
	own weight,	$0.16 \times 5.0 =$	1 kN
	fire casing,	$0.2 \times 5.0 =$	1 kN
			80 kN
	imposed	$1.5 \times 5.0 \times 3.25 =$	24 kN

Design loading $1.4 \times 80 + 1.6 \times 24 = 150$ kN
Shear force F_v $150/2 = 75$ kN

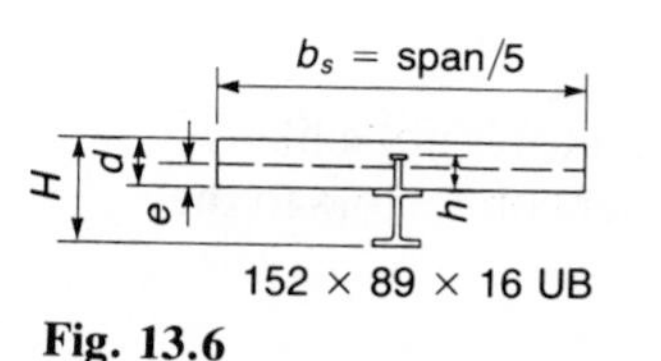

Fig. 13.6

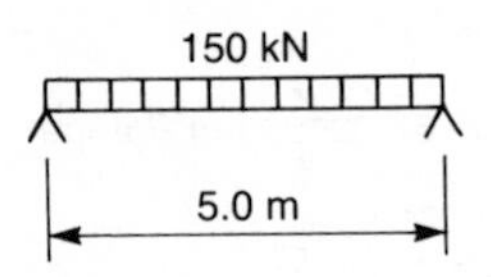

Fig. 13.7

Moment $M_x = 150 \times 5.0/8 = 94$ kNm

Reactions (unfactored):
$R_d = 40$ kN
$R_i = 12$ kN
(See Fig. 13.8)

section 2.1.2(6) Force in concrete $F_c = 0.4 f_{cu} b_s (d - e)$

$$= 0.4 \times 30 \times 1000 \times (120 - 60) \times 10^{-3} = 720 \text{ kN}$$

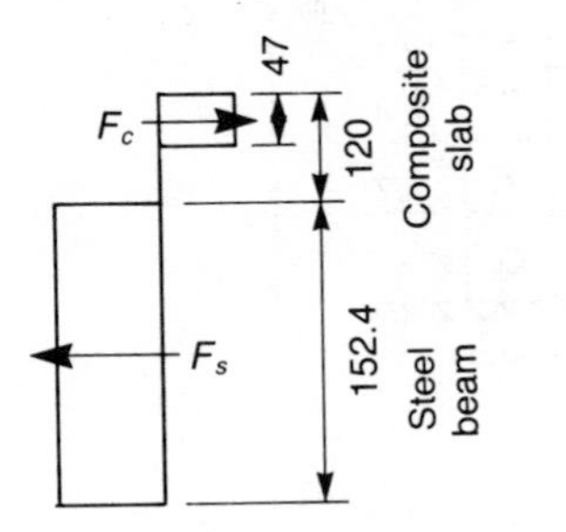

Force in steel $F_s = p_y A_b$

$$= 275 \times 20.5 \times 10^{-1} = 564 \text{ kN}$$

Neutral axis is in the slab as $F_c > F_s$ and
$x_p = 564 \times 10^3/(0.4 \times 30 \times 1000) = 47$ mm
Moment of resistance $M_{pc} = F_s(H/2 + d - x_p/2)$

$$= 564\,(152.4/2 + 120 - 47/2)\,10^{-3} = 97 \text{ kNm}$$

$$M_x/M_{pc} = \mathbf{0.97}$$

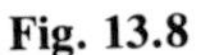

Fig. 13.8

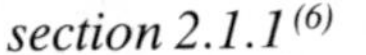

section 2.1.1(6) Modular ratio(6) appropriate for long term deflections = 15

$$r = A_b/(d - e)b_s$$

$$= 20.5 \times 10^2/[(120 - 60)1000] = 0.0342$$

Equivalent second moment of area
$I_{bc} = A_b(H + d + e)^2/4(1 + mr) + b_s(d - e)^3/12m + I_b$

$$= 20.5 \times 10^2(152.4/2 + 120 + 60)^2/[4(1 + 15 \times 0.0342)]$$

$$+ 1000(120 - 60)^3/(12 \times 15) + 838 \times 10^4$$

$$= 4700 \times 10^4 \text{ mm}^4 = 4700 \text{ cm}^4$$

Unfactored imposed load = 24 kN

section 2.3.3(6) Deflection $= 5 \times 24 \times 5^3/(384 \times 205 \times 4700 \times 10^{-5}) =$ **4.1 mm**
(based on unfactored imposed load)

Limit of deflection $= 5000/360 = 13.9$ mm

section 2.3.5(6) Use 19 mm diameter studs 100 mm high as shear connectors with design strength (P_d) of 65 kN

section 2.2.3(6) Reduction due to profile introduces a factor
$0.85/\sqrt{n} \times w/e \times (h - e)/e$

$$= 0.85/1 \times 113/60 \times (100 - 60)/60 = 1.07$$

which should not exceed 1

section 2.1.2(6) $N_{sc} = F_u/P_d$

$$= 564/65 = 9$$

Total of **18 studs** along beam

section 2.2.4(6) Longitudinal shear transfer (through the concrete) is not a likely problem where the sheeting is attached by shear connectors to the beam.

Vertical shear capacity P_v

$$= 0.6 \times 275 \times 4.6 \times 152.4 \times 10^{-3} = 116 \text{ kN } F_v/P_v = \mathbf{0.65}$$

(b) Roof beam R2 – 152 × 89 × 16 UB

The design is as beam R1
Beam spacing = 2.75 m

Loading:	dead,	$4.8 \times 5.0 \times 2.75$	=	66 kN
	own weight			1 kN
	fire casing			1 kN
				68 kN
	imposed,	$1.5 \times 5.0 \times 2.75$	=	21 kN
Design loading		$1.4 \times 68 + 1.6 \times 21$	=	129 kN
Moment	M_x	$129 \times 5.0/8$	=	81 kNm

Reactions (unfactored):
$R_d = 34$ kN
$R_i = 11$ kN

(c) Roof beam R3 – 152 × 89 × 16 UB

The design is as beam R1
Beam spacing = 2.25 m

Loading	dead $4.8 \times 5.0 \times 2.25$	= 54 kN
	own weight + casing	2 kN
		56 kN
	imposed $1.5 \times 5.0 \times 2.25$	= 17 kN

Reactions (unfactored):
$R_d = 28$ kN
$R_i = 8$ kN

(d) Roof beam R4 – 152 × 89 × 16 UB

For an edge beam the effective breadth $b_s = 1000/2 = 500$ mm

section 2.1[6]

Loading	dead $4.8 \times 5.0 \times 3.25/2$	= 39 kN
	own weight + casing	= 2 kN
	parapet 2.0×5.0	= 10 kN
		51 kN
	imposed $1.5 \times 5.0 \times 3.25/2$	= 12 kN
Design loading	$1.4 \times 51 + 1.6 \times 12$	= 91 kN
Moment M_x	$91 \times 5.0/8$	= 57 kNm

Reactions unfactored:
$R_d = 25$ kN
$R_i = 6$ kN

section 2.1.2[6]

$F_c = 0.4 \times 30 \times 500 \times (120 - 60) \times 10^{-3} = 360$ kN

$F_s = 275 \times 20.5 \times 10^{-1} = 564$ kN

Neutral axis is in the steel as $F_s > F_c$

$$M_{pc} = [360\,(120 + 60)/2 + 564 \times 152.4/2]10^{-3} = 75 \text{ kNm}$$
$$M_x/M_{pc} = \mathbf{0.76}$$
$$N_{sc} = 360/65 = 6$$

Total of **12 studs** along beam

(e) Roof beam R5 – 254 × 146 × 37 UB

This beam carries reactions from two type R1 beams as well as some distributed load (See Fig. 13.9.)

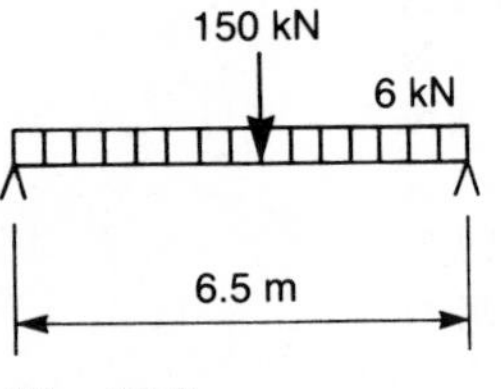

Fig. 13.9

Point load:	dead, 2 × 40	= 80 kN
	imposed, 2 × 12	= 24 kN
Distributed load:	own weight, 0.37 × 6.5	= 2 kN
	casing, 0.3 × 6.5	= 2 kN
Design loading (point)	= 1.4 × 80 + 1.6 × 24	= 150 kN
Design loading (distributed)	= 1.4 × 4	= 6 kN
Moment	$M_x = 150 \times 6.5/4 + 6 \times 6.5/8$	= 248 kNm
Shear force	$F_v = (150 + 6)/2$	= 78 kN

Reactions (unfactored):

$R_d = 42$ kN

$R_i = 12$ kN

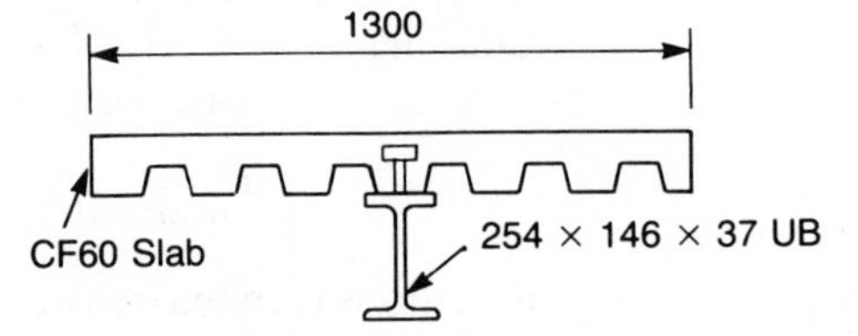

Fig. 13.10

(See Fig. 13.10.)

section 2.1(6) Effective breadth $b_s = 6500/5 = 1300$ mm

section 2.1.2(6) $F_c = 0.4 \times 30 \times 1300\,(120 - 60) = 936$ kN

$F_s = 275 \times 47.5 \times 10^{-1} = 1306$ kN

$M_{pc} = [936(120 + 60)/2 + 1306 \times 156/2]10^{-3} = 251$ kNm

$M_x/M_{pc} = \mathbf{0.99}$

section 2.1.1(6) $r = 47.5 \times 10^2/1300\,(120 - 60) = 0.0609$

Following the calculation for beam R1:

$I_{bc} = 17\,500 \text{ cm}^4$

Deflection $= 24 \times 6.5^3/(48 \times 205 \times 17\,500 \times 10^{-5}) = 3.8$ mm

section 2.1.2(6) $N_{sc} = 936/65 = 15$

Total of **30 studs** along beam

$P_v = 0.6 \times 275 \times 6.4 \times 256 = 270$ kN

$F_v/P_v = \mathbf{0.29}$

(f) Roof beam R6 – 254 × 146 × 37 UB

This beam carries reactions from two type R3 beams as well as some distributed load

Point load: dead, $2 \times 28 = 56$ kN
imposed, $2 \times 8 = 16$ kN
Distributed load: $= 4$ kN

Reactions (unfactored):
$R_d = 30$ kN
$R_i = 8$ kN

Design is the same as beam R5

(g) Roof beam R7 – 254 × 146 × 37 UB

This beam carries reactions from one type R1 beam as well as distributed load

Point load: dead 40 kN
imposed 12 kN
Distributed load:
own weight + casing 4 kN
parapet 2.0×6.5 13 kN
Design loading (point): $1.4 \times 40 + 1.6 \times 12 = 75$ kN
Design loading (distributed): $1.4 \times 17 = 24$ kN
Moment $M_x = 75 \times 6.5/4 + 24 \times 6.5/8 = 141$ kNm
Reactions (unfactored):

$R_d = 28$ kN

$R_i = 6$ kN

section 2.1[6] Effective breadth $b_s = 1300/2 = 650$ mm

section 2.1.2[6]

$$F_c = 0.4 \times 30 \times 650\,(120 - 60) = 468 \text{ kN}$$

$$F_s = 275 \times 47.5 \times 10^{-1} = 1306 \text{ kN}$$

$$M_{pc} = [468(120 + 60)/2 + 1306 \times 256/2]10^{-3} = 209 \text{ kNm}$$

$$M_x/M_{pc} = \mathbf{0.67}$$

$$N_{sc} = 468/65 = 8$$

Total of **16 studs** along beam

(h) Roof beam R8 – 254 × 146 × 37 UB

This beam carries reaction from type R3 beam as well as distributed load

Point load: dead $= 28$ kN
imposed $= 8$ kN
Distributed load
own weight + casing $= 4$ kN
parapet $= 13$ kN

Reactions (unfactored):
$R_d = 22$ kN
$R_i = 4$ kN

Design is the same as beam R7

Roof beams over the stair well will support additional loads where stairs exit on to the roof. The typical stair beams are included as beams

T9 and T10 in Section 13.4. Beams in the lift shaft area will be designed to suit detailed layouts and loadings for the lift motor room, etc., and are not included in this exercise.

13.4 TYPICAL FLOOR BEAM DESIGN

The same arrangement is used for the steelwork on the typical floor (floors 1 to 7 inclusive) as that used for the roof. Some variation may be needed in the vicinity of the stair wells and the lift shaft.

In general the number of beam sizes used is kept to a minimum to ease ordering and fabrication. Two sizes only were used for the roof steelwork. Four sizes will be used for the typical floor, which is, of course, repeated seven times.

The design calculations follow the layout in Section 13.3.

(a) Typical floor beam T1 – 254 × 102 × 25 UB

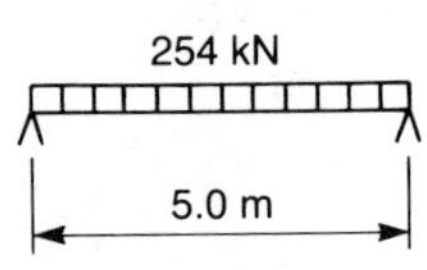

Fig. 13.11

Beam spacing 3.25 m

Loading: dead, $4.2 \times 5.0 \times 3.25$	= 68 kN
own weight, 0.25×5.0	= 1 kN
casing, 2.0×5.0	= 1 kN
	70 kN
imposed, $6.0 \times 5.0 \times 3.25$	= 98 kN

(See Fig. 13.11.)

Design loading: $1.4 \times 70 + 1.6 \times 98 = 254$ kN

Shear force $F_v = 254/2 = 127$ kN

Moment $M_x = 254 \times 5.0/8 = 159$ kNm

Reactions unfactored:

$R_d = 35$ kN

$R_i = 49$ kN

(See Fig. 13.12.)

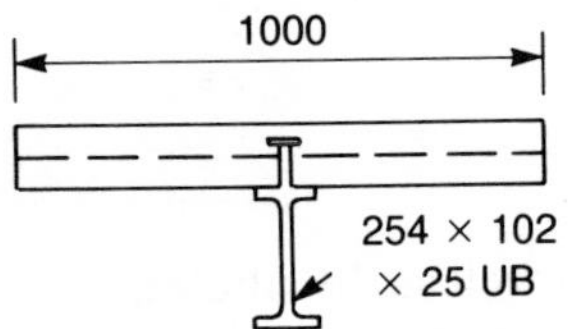

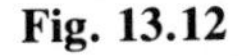

Fig. 13.12

Effective breadth $b_s = 1000$ mm

$F_c = 0.4 \times 30 \times 1000\,(120 - 60)10^{-3} = 720$ kN

$F_s = 275 \times 32.2 \times 10^{-1} = 886$ kN

$M_{pc} = [720(120 + 60)/2 + 886 \times 257/2] \times 10^{-3} = 178$ kNm

$M_x/M_{pc} = \mathbf{0.89}$

$r = 0.0537$

$I_{bc} = 12\,040$ cm^4

Deflection $= 5 \times 98 \times 5.0^3/(384 \times 205 \times 12\,040 \times 10^{-5}) = 6.5$ mm

$N_{sc} = 720/65 = 11$

Total of **22 studs** along beam

$P_v = 0.6 \times 275 \times 6.1 \times 257 = 259$ kN

$F_v/P_v = \mathbf{0.49}$

(b) Typical floor beam T2 – 254 × 102 × 25 UB

Beam spacing 2.75 m

Loading: dead, $4.2 \times 50 \times 2.75 = 58$ kN
own weight + casing $= 2$ kN
60 kN
imposed, $6.0 \times 5.0 \times 2.75 = 83$ kN

Reactions (unfactored):
$R_d = 30$ kN
$R_i = 41$ kN

Design is the same as beam T1

(c) Typical floor beam T3 – 254 × 102 × 25 UB

Beam spacing 2.25 m
Loading: dead, $4.2 \times 5.0 \times 2.25 = 47$ kN
own weight + casing $= 2$ kN
49 kN
imposed, $6.0 \times 5.0 \times 2.25 = 68$ kN

Reactions (unfactored):
$R_d = 25$ kN
$R_i = 34$ kN

Design is the same as beam T1

(d) Typical floor beam T4 – 254 × 102 × 25 UB

Loading: dead, $4.2 \times 5.0 \times 3.25/2 = 34$ kN
own weight + casing $= 2$ kN
wall + glazing, $2.3 \times 5.0 = 12$ kN
48 kN
imposed, $6.0 \times 5.0 \times 3.25/2 = 49$ kN
Design loading: $1.4 \times 48 + 1.6 \times 49 = 146$ kN
Moment $M_x = 146 \times 5.0/8 = 91$ kNm

Reactions (unfactored):
$R_d = 24$ kN
$R_i = 25$ kN

$b_s = 500$ mm
$F_c = [0.4 \times 30 \times 500\,(120 - 60)] \times 10^{-3} = 360$ kN
$F_s = 275 \times 32.2 \times 10^{-1} = 886$ kN
$M_{pc} = [360\,(120 + 60)/2 + 886 \times 257/2] \times 10^{-3} = 146$ kNm
$M_x/M_{pc} = \mathbf{0.62}$
$N_{sc} = 360/65 = 6$
Total of **12 studs** along beam

(e) Typical floor beam T5 – 356 × 171 × 57 UB

This beam carries two reactions from two type T1 beams as well as some distributed load

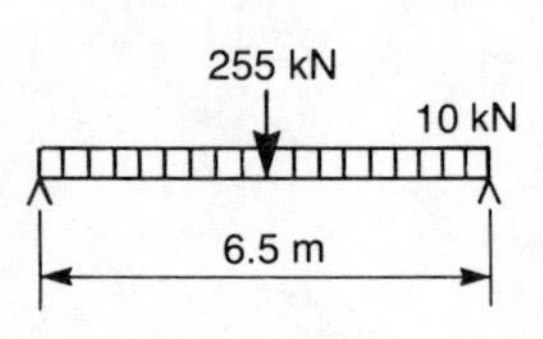

Fig. 13.13

Point load: dead, $2 \times 35 = 70$ kN
imposed, $2 \times 49 = 98$ kN
Distributed load: own weight, $0.57 \times 6.5 = 4$ kN
casing, $0.4 \times 6.5 = 3$ kN

(See Fig. 13.13.)
Design loading (point): $1.4 \times 70 + 1.6 \times 98 = 255$ kN
Design loading (distributed): $1.4 \times 7 = 10$ kN
Shear force $F_v = (255 + 10)/2 = 133$ kN
Moment $M_x = 255 \times 6.5/4 + 10 \times 6.5/8 = 422$ kNm

Reactions (unfactored):
$R_d = 38$ kN
$R_i = 49$ kN

$$b_s = 1300 \text{ mm}$$
$$F_c = 936 \text{ kN}$$
$$F_s = 275 \times 27.2 \times 10^{-1} = 1986 \text{ kN}$$
$$M_{pc} = [936\,(120 + 60)/2 + 1986 \times 358.6/2]10^{-3} = 440 \text{ kNm}$$
$$M_x/M_{pc} = \mathbf{0.96}$$
$$r = 0.0926$$
$$I_{bc} = 38\,200 \text{ cm}^4$$

Deflection $= 98 \times 6.5^3/(48 \times 205 \times 38\,200 \times 10^{-5}) = 7.2$ mm
$N_{sc} = 936/65 = 15$
Total of **30 studs** along beam
$P_v = 0.6 \times 275 \times 8.0 \times 358.6 = 473$ kN
$F_v/P_v = \mathbf{0.28}$

(f) Typical floor beam T6 – 356 × 127 × 33 UB

This beam carries two reactions from two type T3 beams as well as distributed load

Point Load: dead, $2 \times 25 = 50$ kN
imposed, $2 \times 34 = 68$ kN
own weight + casing = 4 kN
Design loading (point): $1.4 \times 50 + 1.6 \times 68 = 179$ kN
Design loading (distributed): $1.4 \times 4 = 6$ kN
Moment $M_x = 179 \times 4.5/4 + 6 \times 4.5/8 = 205$ kNm

Reactions (unfactored):
$R_d = 27$ kN
$R_i = 34$ kN

$$b_s = 4500/5 = 900 \text{ mm}$$
$$F_c = 0.4 \times 30 \times 900\,(120 - 60) = 648 \text{ kN}$$
$$F_s = 275 \times 41.8 \times 10^{-1} = 1150 \text{ kN}$$
$$M_{pc} = [645\,(120 + 60)/2 + 1150 \times 348.5/2]10^{-3} = 259 \text{ kNm}$$
$$M_x/M_{pc} = \mathbf{0.79}$$
$$N_{sc} = 648/65 = 10$$

Total of **20 studs** along beam

(g) Typical floor beam T7 – 356 × 127 × 33 UB

Span 6.5 m

Point load:	dead	= 35 kN
	imposed	= 49 kN
Distributed load:	own weight + casing	= 4 kN
	wall + glazing, 2.3 × 6.5	= 15 kN
Design loading (point):	1.4 × 35 + 1.6 × 49	= 127 kN
Design loading (distributed):	1.4 × 19	= 26 kN
Moment	$M_x = 127 \times 6.5/4 + 26 \times 6.5/8 = 228$ kNm	

Reactions (unfactored):

$R_d = 27$ kN

$R_i = 25$ kN

$$b_s = 650 \text{ mm}$$
$$F_c = 468 \text{ kN}$$
$$F_s = 275 \times 41.8 \times 10^{-1} = 1150 \text{ kN}$$
$$M_{pc} = [468\,(120 + 60)/2 + 1150 \times 348.5/2]10^{-3} = 243 \text{ kNm}$$
$$M_x/M_{pc} = \mathbf{0.94}$$
$$N_{sc} = 468/65 = 8$$

Total of **16 studs** along beam

(h) Typical floor beam T8 – 356 × 127 × 33 UB

Span 4.5 m

Point load (T3):	dead	= 25 kN
	imposed	= 34 kN
Distributed load:	own weight + casing	= 4 kN
	wall + glazing, 2.3 × 4.5	= 10 kN
Design loading (point):	1.4 × 25 + 1.6 × 34	= 89 kN
Design loading (distributed):	1.4 × 14	= 20 kN
Moment	$M_x = 89 \times 4.5/4 + 20 \times 4.5/8 = 111$ kNm	

Reactions (unfactored):

$R_d = 19$ kN

$R_i = 17$ kN

$$b_s = 450 \text{ mm}$$
$$F_c = 324 \text{ kN}$$
$$F_s = 1150 \text{ kN}$$
$$M_{pc} = 230 \text{ kNm}$$
$$M_x/M_{pc} = \mathbf{0.48}$$
$$N_{sc} = 324/65 = 6$$

Total of **12 studs** along beam

(i) Typical floor beam T9 – 254 × 146 × 37 UB

This is a non-composite beam in the stair well supporting precast flight and landing units

Span 5.0 m

Loading: dead: $6.3 \times 5.0 \times 2.5/2 = 39$ kN

$4.7 \times 5.0 \times 2.0/2 = 24$ kN

own weight + casing $= 4$ kN

67 kN

imposed, $4.0 \times 5.0 \times 4.5/2 = 45$ kN

Design loading: $1.4 \times 67 + 1.6 \times 45 = 166$ kN

$F_v = 166/2 = 83$ kN

Moment $M_x = 166 \times 5.0/8 = 104$ kNm

Reactions (unfactored):

$R_d = 34$ kN

$R_i = 23$ kN

clause 4.2.3 Shear resistance $P_v = 0.6 \times 275 \times 6.4 \times 256 = 270$ kN

$F_v/P_v = \mathbf{0.31}$

clause 4.2.5 Moment capacity $M_c = 275 \times 485 \times 10^{-3} = 133$ kNm

$M_x/M_c = \mathbf{0.78}$

Deflection $= 5 \times 45 \times 50^3/(384 \times 205 \times 5560 \times 10^{-5}) = \mathbf{6.4\ mm}$

Limit of deflection $= 5000/360 = 13.9$ mm

(j) Typical floor beam T10 – 356 × 171 × 57 UB

This is a non-composite beam in the stair well supporting a reaction from beam T9 as well as some distributed load. Lateral restraint along the beam is provided only by beam T9. The design of such unrestrained beams is discussed in greater detail in Section 3.2.

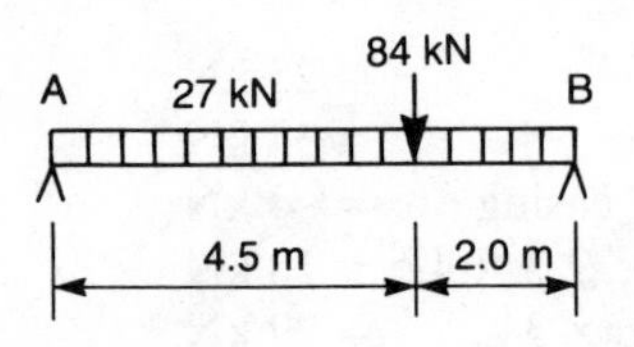

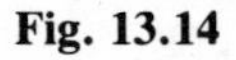

Fig. 13.14

Span 6.5 m

Point load (T9):	dead	= 34 kN
	imposed	= 23 kN
Distributed load:	own weight + cladding	= 4 kN
	wall + glazing, 2.3×6.5	= 15 kN

(See Fig. 13.14.)

Design loading (point): $1.4 \times 34 + 1.6 \times 23 = 84$ kN

Design loading (distributed): $1.4 \times 19 = 27$ kN

Maximum shear $F_v = 84 \times 4.5/6.5 + 27/2 = 72$ kN

Moment $M_x = 84 \times 2.0 \times 4.5/6.5 + 27 \times 6.5/8 = 138$ kNm

Reactions (unfactored):

At A: $R_d = 20$ kN

$R_i = 7$ kN

At B: $R_d = 28$ kN

$R_i = 16$ kN

BS table 9 Effective length $L_E = 4.5$ m

BS table 13 $m = 1.0$
BS table 16 $n = 0.94$

$\lambda = 4500/39.2 = 115$

$\lambda/x = 115/28.9 = 4.0$

BS table 14 $v = 0.86$

$\lambda_{LT} = 0.94 \times 0.884 \times 0.86 \times 115 = 82$

BS table 11 $p_b = 161 \text{ N/mm}^2$

$M_b = 161 \times 1010 \times 10^{-3} = 163 \text{ kNm}$

$M_x/M_b = \mathbf{0.85}$

$P_v = 0.6 \times 275 \times 8.0 \times 358.6 = 473 \text{ kN}$

$F_v/P_v = \mathbf{0.15}$

13.5 COLUMN DESIGN

Loads for each column must be calculated, and in the present design three columns are selected as typical: an external (side) column C1; a corner column C2; and an internal column C6. Further columns could be designed individually if desired.

Column loads are best assembled from the unfactored beam reactions, with dead and imposed loads totalled separately. The imposed loads may be reduced where a column supports more than one floor. The reduction is 10% per floor until a maximum of 40% is reached, and 40% thereafter. This applies to buildings up to 10 storeys, and is detailed in BS 6399[7]. Column loads must include an allowance for self weight and fire casing.

The design load condition is $1.4W_d + 1.6W_i$.

(a) Column loads

COLUMN C1

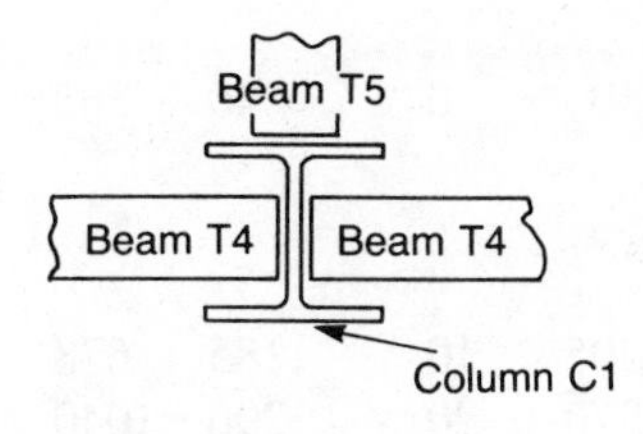

Fig. 13.15

Reactions R4, R5, T4, T5 are taken from previous Sections 13.3(d), etc., and tabulated below. (See Fig. 13.15.)

Column length: 7th floor–roof:

Total $W_d = 25 + 25 + 42 + 4 = 96$ kN

Total $W_i = 6 + 6 + 12 = 24$ kN

Design load $F = 1.4 \times 96 + 1.6 \times 24 = 173$ kN

Column length: 6th–7th floors:

Total $W_d = 96 + 24 + 24 + 38 + 4 = 186$ kN

Total $W_i = 24 + 25 + 25 + 49 = 123$ kN

Reduced $W_i = 0.9 \times 123 = 111$ kN

Design load $= 1.4 \times 186 + 1.6 \times 111 = 438$ kN

In the same way each table may be completed.

Column length	Beams	Reactions			Totals			Reduced	
		R_d (kN)	R_i (kN)	Own Weight (kN)	W_d (kN)	W_i (kN)	Reduction (%)	W_i (kN)	Design load F (kN)
COLUMN C1									
7–R	R4	25	6	4					
	R4	25	6						
	R5	42	12		96	24	0	24	173
6–7	T4	24	25	4					
	T4	24	25						
	T5	38	49		186	123	10	111	438
5–6	ditto	86	99	4	276	222	20	178	671
4–5	ditto	86	99	4	366	321	30	225	872
3–4	ditto	86	99	4	456	420	40	252	1040
2–3	ditto	86	99	4	546	519	40	311	1260
1–2	ditto	86	99	4	636	618	40	371	1480
G–1	ditto	86	99	6	728	717	40	430	1710
COLUMN C2									
7–R	R4	25	6	4					
	R7	28	6		57	12	0	12	99
6–7	T4	24	25	4					
	T7	27	25		112	62	10	56	246
5–6	ditto	51	50	4	167	112	20	90	378
4–5	ditto	51	50	4	222	162	30	113	492
3–4	ditto	51	50	4	277	212	40	127	591
2–3	ditto	51	50	4	332	262	40	157	716
1–2	ditto	51	50	4	387	312	40	187	841
G–1	ditto	51	50	6	444	362	40	217	969
COLUMN C6									
7–R	R2	34	10	4					
	R2	34	10						
	R5	42	12						
	R6	30	8		144	40	0	40	265
6–7	T2	30	41	4					
	T2	30	41						
	T5	38	49						
	T6	27	34		273	205	10	185	678
5–6	ditto	125	165	4	402	370	20	296	1040
4–5	ditto	125	165	4	531	535	30	375	1340
3–4	ditto	125	165	4	660	700	40	420	1600
2–3	ditto	125	165	4	789	865	40	519	1930
1–2	ditto	125	165	4	918	1030	40	618	2270
G–1	ditto	125	165	6	1049	1195	40	717	2610

(b) Column C1 design: Ground–1st floor – 203 × 203 × 86 UC

Over this length the column carries an axial load of 1710 kN. The reaction from beam T5 at first floor level is eccentric to the column:

clause 4.7.6 $e = 100 + 222.3/2 = 211$ mm

$R_{T5} = 133$ kN

(See Fig. 13.16.)

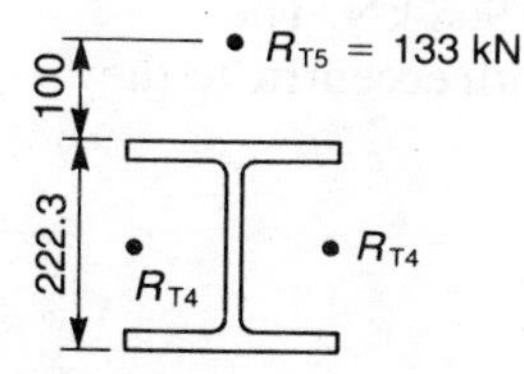

Fig. 13.16

The nominal moment is divided between column lengths above and below first floor equally, assuming approximately equal column stiffnesses:

clause 4.7.7 $M_x = 133 \times 0.211/2 = 14$ kNm

BS table 6 $p_y = 265$ N/mm^2

BS table 24 $L_E = 0.85 \times 4.7 = 4.00$ m

$\lambda = L_E/r_y = 4000/53.2 = 75$

BS tables 25, 27c $p_c = 167$ N/mm^2

$P_c = p_c A_g = 167 \times 110 \times 10^{-1} = 1840$ kN

clause 4.7.7 $m = 1.0$

clause 4.7.7 $\lambda_{LT} = 0.5L/r_y = 0.5 \times 4700/53.2 = 44$

BS table 11 $p_b = 244$ N/mm^2

$M_b = p_b S_x = 244 \times 979 \times 10^{-3} = 239$ kNm

clause 4.8.3.3 Overall buckling check (simplified):

$$F/P_c + mM_x/M_b \ngtr 1$$

$$1700/1840 + 1.0 \times 14/239 = \mathbf{0.98}$$

Using the same method other lengths of the column may be designed and the results tabulated. Where floor beams providing directional restraint are substantial, and are not required to carry more than 90% of their moment capacity, the effective length may be taken as 0.7 of actual length for column lengths above first floor. The design strength, p_y is 265 N/mm^2 for column sections greater than 16 mm thick and 275 N/mm^2 for sections less than 16 mm thick.

Column Length	Size	F (kN)	M_x (kNm)	λ	p_c (N/mm^2)	P_c (kN)	λ_{LT}	p_b (N/mm^2)	M_b (kNm)	Check
G–1	203 × 203 × 86 UC	1710	14	75	167	1840	44	244	239	0.97
1–2	203 × 203 × 71 UC	1480	14	60	201	1830	35	265	213	0.87
2–3	203 × 203 × 60 UC	1260	14	61	199	1510	36	271	177	0.91
3–4	203 × 203 × 46 UC	1040	13	62	197	1160	36	271	135	0.99
4–5	ditto	872	13	satisfactory						
5–6	ditto	671	13	satisfactory						
6–7	152 × 152 × 30 UC	438	12	82	157	600	48	243	60	0.93
7–R	ditto	173	14	satisfactory						

It is the practice, in the interests of economy, for columns to be fabricated in two-storey lengths, and assembled on site using splices. It is not therefore good practice to change column size for every storey, and in this case the same size (203 × 203 × 86 UC) would probably be used between ground and second floors, and one size (203 × 203 × 60 UC) between second and fourth floors.

(c) Column C2 design: Ground–1st floor – 203 × 203 × 60 UC

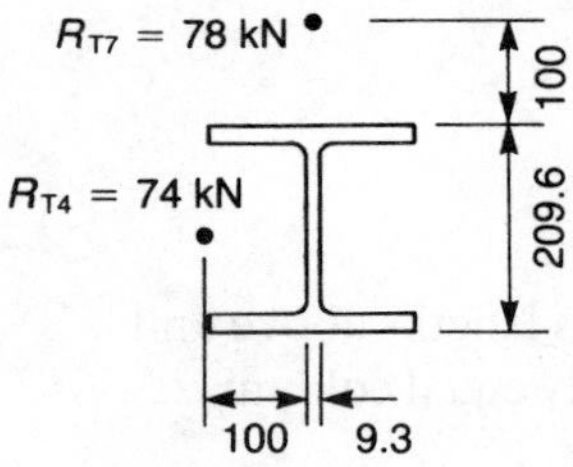

R_{T7} = 78 kN
R_{T4} = 74 kN
100
209.6
100
9.3

Fig. 13.17

Over this length the column carries an axial load of 968 kN. The reactions from beams T4 and T7 at first floor are both eccentric to the column. (See Fig. 13.17.)

For T7: $e = 100 + 209.6/2 = 205$ mm

For T4: $e = 100 + 9.3/2 = 105$ mm

$R_{T7} = 1.4 \times 27 + 1.6 \times 25 = 78$ kN

$R_{T4} = 1.4 \times 24 + 1.6 \times 25 = 74$ kN

$M_x = 78 \times 0.205/2 = 8.0$ kNm

$M_y = 74 \times 0.105/2 = 3.9$ kNm

BS table 6 $p_y = 275$ N/mm^2

BS table 24 $L_E = 0.85 \times 4.7 = 4.00$ m

$\lambda = 4000/51.9 = 77$

BS tables 25, 27c $p_c = 167$ N/mm^2

$P_c = 167 \times 75.8 \times 10^{-1} = 1260$ kN

clause 4.7.7 $m = 1.0$

$\lambda_{LT} = 0.5 \times 4700/51.9 = 46$

BS table 11 $p_b = 248$ N/mm^2

$M_b = 248 \times 652 \times 10^{-3} = 161$ kNm

clause 4.8.3.3 Overall buckling check (simplified):

$$F/P_c + mM_x/M_b + mM_y/p_yZ_y \ngtr 1$$

$968/1260 + 1.0 \times 8.0/161 + 1.0 \times 3.9/(275 \times 199 \times 10^{-3}) = \mathbf{0.89}$

Using the same method other lengths of the column may be designed and the results tabulated.

Column Length	Size	F (kN)	M_x (kNm)	M_y (kNm)	λ	p_c (N/mm²)	P_c (kN)	λ_{LT}	p_b (N/mm²)	M_b (kNm)	Check
G–1	203 × 203 × 60 UC	968	8.0	3.9	77	167	1260	46	248	161	0.89
1–2	203 × 203 × 46 UC	841	7.9	3.9	62	197	1160	36	271	135	0.86
2–3	ditto	716	7.9	3.8	satisfactory						
3–4	ditto	591	7.9	3.8	satisfactory						
4–5	152 × 152 × 37 UC	491	7.1	3.8	81	159	754	48	243	75	0.90
5–6	152 × 152 × 30 UC	378	7.0	3.8	82	157	600	48	243	60	0.94
6–7	ditto	246	6.2	3.4	satisfactory						
7–R	ditto	99	8.7	4.6	satisfactory						

As for column C1, two-storey lengths at the same size are preferred.

(d) Column C6 design: Ground–1st floor – 254 × 254 × 132 UC

Over this length the column carries an axial load of 2610 kN. The reactions from the beams at first floor level are eccentric, but will tend to balance each other. The difference between the reactions from T5 and T6 will, however, give a net moment about the major axis. Note

clause 4.7.7 that the effect of the absence of imposed load on any beam (pattern loading) is not taken into account, and all beams are considered fully loaded (See Fig. 13.18.)

Column Length	Size	F (kN)	M_x (kNm)	λ	p_c (N/mm²)	P_c (kN)	λ_{LT}	p_b (N/mm²)	M_b (kNm)	Check
G–1	254 × 254 × 132 UC	2610	3.7	60	195	3300	35	265	496	0.80
1–2	254 × 254 × 89 UC	2270	3.6	48	217	2470	28	265	326	0.93
2–3	254 × 254 × 73 UC	1930	3.5	49	222	2060	29	275	272	0.95
3–4	ditto	1600	3.5	satisfactory						
4–5	203 × 203 × 60 UC	1340	3.5	61	199	1510	36	271	177	0.91
5–6	203 × 203 × 46 UC	1040	3.6	62	197	1160	36	271	135	0.92
6–7	152 × 152 × 37 UC	678	3.5	81	159	754	48	243	75	0.95
7–R	152 × 152 × 30 UC	265	4.1	82	157	600	48	243	60	0.51

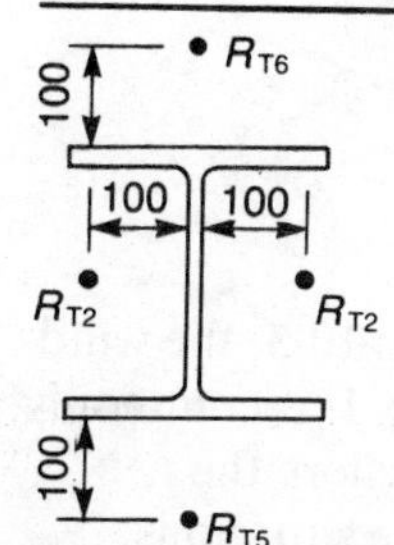

Fig. 13.18

As previously, two-storey lengths at the same size are preferred.

13.6 CONNECTIONS

The design of typical beam to column connections is given in Section 3.7(g). A typical connection for the present design is detailed for beam T5 to column C6. (See Fig. 13.19)

Beam reaction (Section 13.4(e)) = 133 kN

BM = 133 × 0.05 = 6.7 kNm

Use 9 no. 20 mm bolts grade 4.6

Use **2 no 90 × 90 × 10 angle** cleats.

Fig. 13.19

(a) Column bolts

Shear/bolt = 133/6 = **22.2 kN**

clause 6.3.2 Shear capacity P_s = 160 × 245 × 10^{-3} = **39.2 kN**

clause 6.3.3 Bearing capacity of bolts P_{bb} = 20 × 9.4 × 435 × 10^{-3} = **81.8 kN**

where 9.4 is the column flange thickness

(b) Beam bolts

Vertical shear/bolt = 133/3 = 44.3 kN

Horizontal shear due to eccentric bending moment

= $Md_{max}/\Sigma d^2$ = 6.7 × 0.10/(2 × 0.10^2) = 33.5 kN

Resultant shear/bolt = $\sqrt{(44.3^2 + 33.5^2)}$ = **55.5 kN**

Shear capacity (double shear) P_s = 160 × 2 × 245 × 10^{-3} = 78.4 kN

Bearing capacity of bolt P_{bb} = 20 × 8.0 × 435 = **73.1 kN**

(c) Angle cleat

Shear area of cleats $= 2 \times 0.9(300 \times 10 - 3 \times 22 \times 10) = 4210\ \text{mm}^2$
Shear capacity $P_v = 0.6 \times 275 \times 4210 \times 10^{-3} = \mathbf{695\ kN}$
Shear force $F_v = 133\ \text{kN}$

Similar connections may be designed for all other beams. Where a beam is designed for composite action, such as T1, T2, T3 and T4, no load is considered to be transferred to the column by the slab, and the cleat and bolts should carry all the beam reaction.

Splices connect the ends of each section of column together so that loads are transmitted between them satisfactorily. Such connections are proportioned in accordance with empirical rules as shown in *Steel Designers' Manual*[8]. Typical splice details are given by Needham[9] and Pask[10].

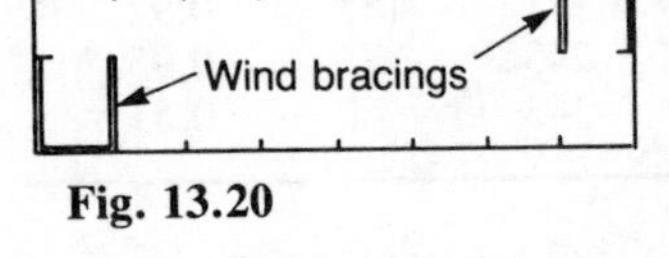

Fig. 13.20

13.7 WIND BRACING

As discussed in Section 13.1.4, and previously in Section 10.3, the wind loading may be designed to be carried by a wind bracing. It is commonly convenient to locate the wind bracing at stair/lift wells where the diagonal members may be hidden by brickwork. In some situations, such as industrial frameworks, it may be satisfactory to leave the wind bracing exposed.

An arrangement for the bracing is shown in Fig. 13.20. The stair wells provide four frames in the lateral direction (two frames in the longitudinal direction) as shown.

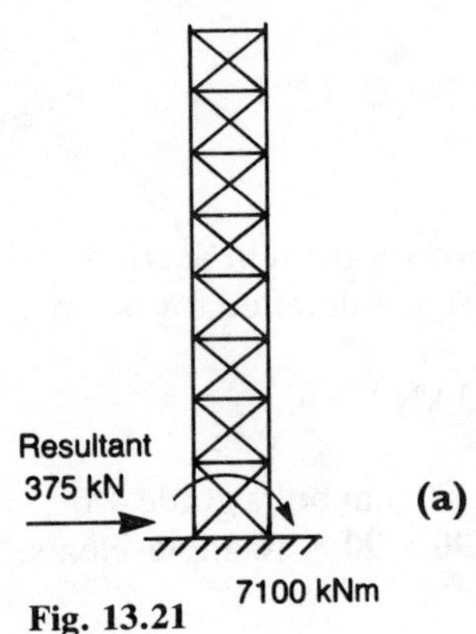

Fig. 13.21

(a) Loading and forces

Force above ground floor level is the sum of the forces shown in Fig. 13.4 = 187.3 kN (for one 5 m bay)
Force on each lateral wind bracing $W_w = 187.3 \times 8/4 = 375\ \text{kN}$
Moment of wind forces about ground level is the sum of forces × heights shown in Fig. 3.4 = 3550 kNm
Moment on each lateral wind bracing $M_w = 3550 \times 8/4 = 7100\ \text{kNm}$
(See Fig. 13.21.)
Considering the part of the frame between ground and first floors and analysing as a pin-jointed frame: (See Fig. 13.22.)

$$R_v = 7100/4.5 = 1580\ \text{kN}$$
$$R_h = 375\ \text{kN}$$
$$F_{C13} = 1580\ \text{kN compression}$$
$$F_{diagonal} = 375/\cos 46° = 540\ \text{kN tension}$$
$$F_{C7} = 1580 - 540 \sin 46° = 1190\ \text{kN tension}$$

As discussed in Section 10.2 cross-bracing allows a tension only design for the diagonals. For this arrangement wind from either direction produces tension in the appropriate diagonal, but tension or compression in the columns.

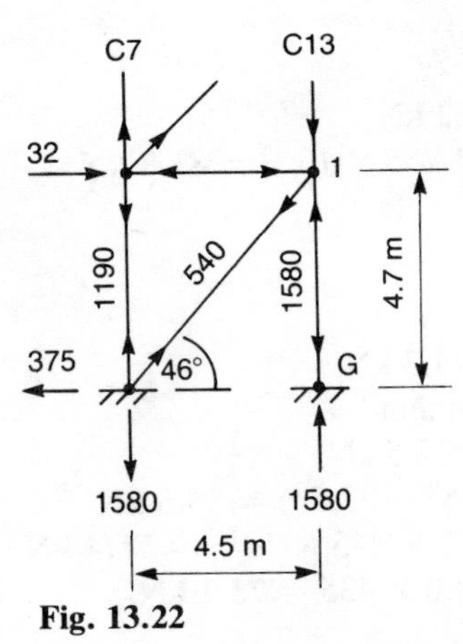

Fig. 13.22

(b) Column C7 (G–1) – 254 × 254 × 167 UC

The forces in column C7 will include dead and imposed loads similar to C6 (Section 13.5):

$W_d = 1049$ kN
$W_i = 717$ kN
$W_w = 1580$ kN compression or 1190 kN tension

Load combinations for maximum compression:

BS table 2 Either $1.4W_d + 1.4W_w = 1.4 \times 1049 + 1.4 \times 1580 = 3680$ kN

Or $1.2W_d + 1.2W_i + 1.2W_w = 1.2\,(1049 + 717 + 1580) = 4020$ kN

Load combination for maximum tension:

$$1.0W_d + 1.4W_w = 1.0 \times 1049 - 1.4 \times 1190 = -617 \text{ kN}$$

$F_t = -\,617$ kN (tension)
$F_c = 4020$ kN
$M_x = 3.7$ kNm (see note below)
$\lambda = 4000/67.9 = 59$

BS table 27c $p_c = 197$ N/mm^2
$P_c = 197 \times 212 \times 10^{-1} = 4180$ kN
$\lambda_{LT} = 0.5 \times 4700/67.9 = 35$

BS table 11 $p_b = 265$ N/mm^2
$M_b = 265 \times 2420 \times 10^{-3} = 641$ kNm

Overall bucking check:

clause 4.8.3.3 $4020/4180 + 3.7/641 = \mathbf{0.97}$

Note that $M_x = 3.7$ kNm is used as in Section 13.5. This value could be reduced to take account of the lower values of γ_f used here ($1.2W_d + 1.2W_i$ in place of $1.4W_d + 1.6W_i$ in Section 13.5), but this would have little effect.

(c) Column C13 (G–1) – 254 × 254 × 167 UC

Taking the dead and imposed loads as similar to those in column C1:

$W_d = 728$ kN
$W_i = 430$ kN
$W_w = 1580$ kN compression or 1190 kN tension

Maximum compression:

$1.4W_d + 1.4W_w = 1.4\,(728 + 1580) = 3230$ kN
$1.2W_d + 1.2W_i + 1.2W_w = 1.2\,(728 + 430 + 1580) = 3290$ kN

Maximum tension:

$$1.0W_d + 1.4W_w = 1.0 \times 728 - 1.4 \times 1190 = -938 \text{ kN}$$

$F_t = -\,938$ kN (tension)
$F_c = 3290$ kN
$M_x = 10.5$ kNm (see note after Section 13.7(b))
$P_c = 4180$ kN
$M_b = 641$ kNm

Overall buckling check:

$$3290/4180 + 10.5/641 = \mathbf{0.80}$$

(d) Diagonal (G–I) – 203 × 89 channel

The force due to wind only W_w = 540 kN tension
$1.4W_w$ = 756 kN tension
Net area of web (allowing 2 no. 24 diameter holes across section) $= 203.2 \times 8.1 - 2 \times 24 \times 8.1 = 1260 \text{ mm}^2$

clause 4.6.3 Area of flanges $= 3790 - 1260 = 2530 \text{ mm}^2$
Multiplier $= 3 \times 1260/(3 \times 1260 + 2530) = 0.60$
Effective area A_e $= 1260 + 2530 \times 0.60 = 2780 \text{ mm}^2$

clause 4.6.1 Tension capacity P_t $= A_e p_y$
$= 2780 \times 275 \times 10^{-3} = 765 \text{ kN}$

$$F/P_t = 756/765 = \mathbf{0.99}$$

The design of all members in the bracing system follows the method outlined. The bracing system in the direction at right angles is designed in a similar manner.

13.8 WIND RESISTANCE BY FRAME ACTION

Previous design codes (e.g. BS 499) permitted a simplified frame action for wind resistance, and design methods for this appear in, e.g., *Steel Designers Manual*[(11)]. The method makes a number of assumptions regarding shear distribution and points of contraflexure. Although these methods once enjoyed wide application they are no longer sanctioned under BS 5950: Part 1.

Frame action to resist wind loading requires the frame elements to be connected by rigid joints, and the design is thus controlled by section five of BS 5950: Part 1. Plastic or elastic design is permitted, but the horizontal loads must be applied to the whole frame and forces analysed accordingly (clause 5.6.4.2).

An alternative design method using simple connections for vertical loading, but recognizing their stiffness in the design for wind loading has been examined by Nethercot[(12)]. This is shown to give some advantages, but may lead to some overstressing.

Frame action for wind resistance has the disadvantage economically of more complex connections, as well as increased column sizes generally. These costs must be offset against the saving of the wind bracing in any economic comparison.

STUDY REFERENCES

Topic	*Reference*
1. Steel buildings	**Mathys, J.H.** (1987) Multistorey steel buildings - a new generation, *Structural Engineer*, vol. 65A no. 2 pp. 47–51

2. Profiled Sheeting — (1985) Profiles for composite flooring, *Profiles for Concrete.* Precision Metal Forming Ltd

3. Composite slabs — **Lawson, R.M.** (1983) Composite slabs using profiled steel sheeting, *Composite Beams and Slabs with Profiled Steel Sheeting,* pp. 8–20. CIRIA Report no. 99

4. Fire resistance — (1986) *Fire Resistance of Composite Slabs with Steel Decking.* CIRIA Special Publication 42

5. Loading — BS 6399 *Loading for buildings*
Part 1: *Dead and imposed loads* (1984)
Part 2: *Wind loads* (to be published; presently CP3 Ch. V Part 2)

6. Composite beams — **Lawson, R.M.** (1983) Composite beam design, *Composite Beams and Slabs with Profiled Steel Sheeting,* pp. 21–39. CIRIA Report No. 99

7. Imposed load reduction — Reduction in total imposed floor loads. BS 6399 *Loading for buildings*
Part 1: *Dead and imposed loads* (1984), clause 5

8. Column splices — (1972) Design of multi-storey stanchions, *Steel Designers' Manual,* pp. 809–48. Crosby Lockwood Staples

9. Column splices — **Needham, F.H.** (1980) Connections in structural steelwork for buildings, *Structural Engineer,* vol. 58A no. 9 pp. 267–77.

10. Column splices — **Pask, J.W.** (1982) Bolted column splices, *Manual on Connections,* pp. 80–3. BCSA Ltd

11. Frame action — (1972) Wind on multistorey buildings, *Steel Designers' Manual,* pp. 847–67. Crosby Lockwood Staples

12. Frame action — **Nethercot, D.A.** (1985) Joint action and the design of steel frames, *Structural Engineer,* vol. 63A no. 12 pp. 371–9

14

DETAILING PRACTICE AND OTHER REQUIREMENTS

14.1 FABRICATION PROCESSES

The designer needs to have an understanding of the processes involved in the fabrication and erection of structural steelwork. This understanding is necessary to ensure that:

(a) all the details shown by the designer are capable of fabrication;
(b) the effects of the fabrication processes on the design are allowed for, e.g. corrosion traps, plate distortion in cropping and bending;
(c) the details shown do not involve unnecessarily complex, time-consuming and hence costly processes;
(d) the responsibilities of the fabricator are clear e.g. what assembly of cleats is required prior to delivery to site.
(e) the details chosen should allow a safe means of erection.

The processes involved in structural steelwork fabrication, and the requirements of good design, are described by Taggart(1). Other publications are available giving fuller descriptions of steelwork fabrication(2).

The processes may be summarized as follows.

14.1.1 Surface preparation and priming

Surface preparation is usually carried out either by blast cleaning or by use of mechanical tools. In blast cleaning an abrasive material is projected at high speed at the surface to be cleaned. The abrasive material can be metallic ('shot' blasting) or non-metallic such as slag or other minerals ('sand' blasting).

Alternatively preparation may be carried out by a variety of mechanical tools such as wire brushes and sanders, or by mechanical chisels and needle guns. These are usually less effective than blast cleaning but may be used in smaller fabrication works and on site prior to the final painting.

Priming of the steel surface is carried out immediately after cleaning with the surface clean and free from moisture. A number of different primers are available, and their use should take account of the processes which are to follow. In particular some primers may give rise to hazardous fumes during subsequent flame cutting and welding. In addition some primers may interfere with the welding processes which it is intended to use.

14.1.2 Cutting and drilling

The steel sections or plates are cut to length and size by guillotining, sawing or flame cutting. Guillotining is a process of shearing steel plates to the required length and width, and cropping is a similar process but which may be applied to steel sections. The method may be limited in its use for a particular fabrication by minor distortion and burring requiring subsequent correction.

Sawing may be carried out by circular saws, hacksaws or bandsaws. Clean, accurate, straight cutting may be achieved.

The thermal cutting processes ('flame' cutting) involve a number of different systems which may be process controlled, and used to produce steel plates cut to a predetermined profile.

Drilling of the required holes in steelwork may be carried out using single and multi-spindle machines which may be set to produce a pattern of holes determined by a template. Punching is also used for making holes, but these have a limited use due to embrittlement of the edge of the hole and possible edge cracking. They should not, for example, be permitted in connections which are required to develop plastic hinges.

14.1.3 Bending and forming

In more complex steelwork assembly bending of sections and plates to specified shapes may be needed. Sections can be bent to circular and other profiles as required, but with the local radius of curvature limited by the proportions of the sections. Presses are used in bending plates to form sections of specified shape. Examples are shown in Fig. 14.1.

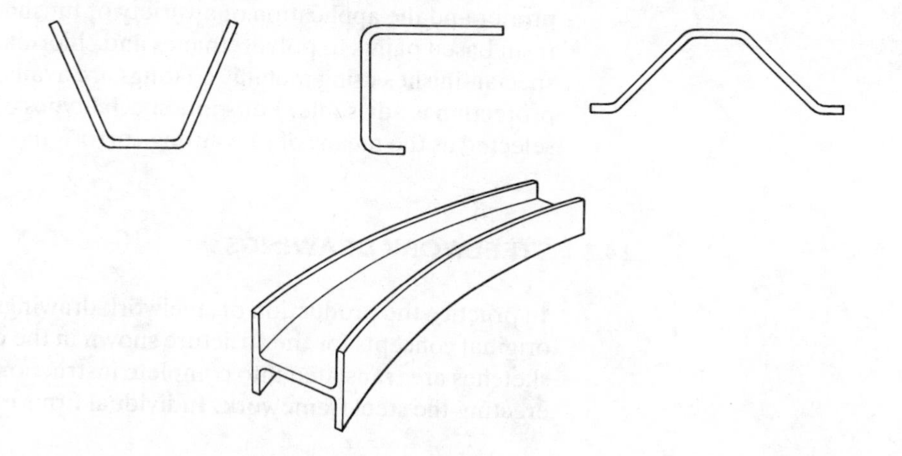

Fig. 14.1 Bending and forming

14.1.4 Welding

Many welding processes are available, but metal arc welding is the one which is normally permitted for steelwork fabrication. Manual metal arc welding is used for attaching end-plates, cleats etc., to steel members, while automatic gas shielded processes are used for the fabrication of plate girders. Different types of weld (Fig. 14.2) are used in different situations and further details may be found in BS 5315 and specialist literature[2,4].

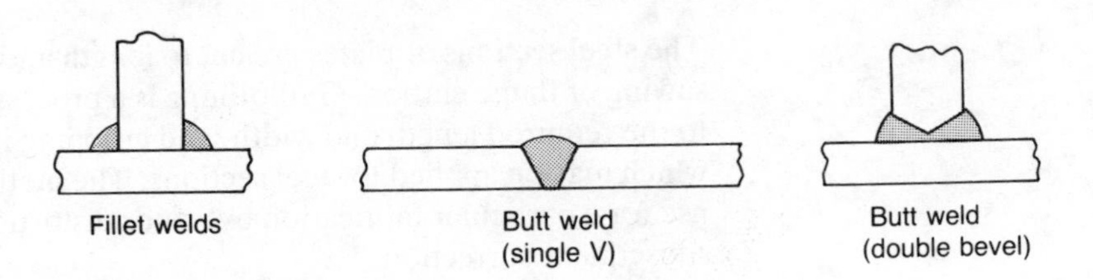

Fig. 14.2 Types of weld

14.1.5 Inspection and protection

Inspection of the steelwork is carried out at several points in the fabrication processes. The most important of these are at entry and after welding. Before commencement of fabrication the steel sections or plates are checked for straightness, and where steel is subject to stresses applied perpendicular to the direction of rolling, the material should be examined by ultrasonic techniques to detect hidden defects. Defects discovered at a later stage can prove costly to rectify, and may involve rejection of a finished item. Welding is tested on either a sample basis (say 5%) or fully. The methods used may be ultrasonic or radiographic, or may involve use of dyes or magnetic particles. The choice of method will depend on the quality control required, accessibility, and the relative importance of the weld to the overall structure. Tension welds are normally tested to a greater frequency, and sometimes all welds in tension are required to be tested.

Final surface protection of the steelwork is carried out both in the fabrication works and on site. It will involve the retreatment of damaged primer and the application of a variety of finishing paints from oil and resin based paints to polyurethanes and chlorinated rubbers. In addition special finishes using metallic coatings are available where additional protection is advisable. For some special types of structure galvanizing is selected as the means of preventing corrosion (see also Section 14.5).

14.2 STEELWORK DRAWINGS

In practice the production of steelwork drawings is to ensure that the original concepts for the structure shown in the calculations and sketches are translated into complete instructions for fabricating and erecting the steel framework. Individual firms maintain varying practices for detailing structural steelwork, but will include some or all of the items described in this section. Some guidance regarding detailing practice is given in a BCSA publication[3].

For the student the preparation of scale drawings will assist in:

(a) visualizing the structure being designed;
(b) bringing a recognition of relative size, e.g. the slenderness of a particular member, or the relationship of a bolt hole to the member size and edge distance;
(c) adopting a discipline in providing complete information both in drawing and in calculation.

14.2.1 Construction drawings

The first drawing produced by the designer is one showing the overall relationship of the steel framework to the other building components. This general arrangement drawing will have plans and sections showing clearly the relative positions of floor slabs, cladding, walls, windows, foundations, etc., to the steel framework. Multi-level plans as shown in Fig. 14.3 are useful in condensing much information into one plan view. For the designer the general arrangement is essential to ensure that he has included all the client's requirements, that the steel framework is fully compatible with the other building components, and that the structure is shown to be inherently stable with discernable load paths to the foundations.

14.2.2 Steelwork general arrangement

The layout of all the steelwork members and their relationship to each other must be shown on one drawing. This will be used by the steelwork erectors to assemble the framework with its connections on site. For the simplest of structures it may be possible to show this information on the construction drawings, but more usually a special drawing of the steelwork layout is necessary.

This layout (Fig. 14.4) will show only the steelwork with principal dimensions and grid lines. It will incorporate a numbering system for each steelwork member and may give member sizes and other details. This latter information is, however, usually placed in a steelwork schedule on the layout drawing, or on a separate sheet (Fig. 14.5). In addition this general arrangement drawing will need to show the information required to enable the connections to be designed and detailed. This takes the form of end reactions (and moments where appropriate), beam levels and eccentricities. Forces and moments must be clearly indicated as factored or unfactored.

14.2.3 Fabrication drawings

The fabrication process requires drawings of the steelwork members in full detail showing precise sizes, lengths, positions of holes, etc. While in practice these drawings are often produced by the fabricators

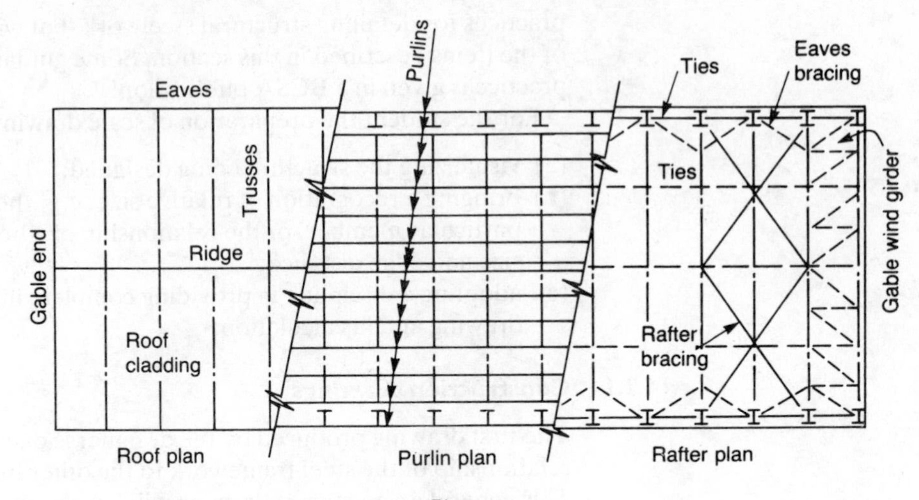

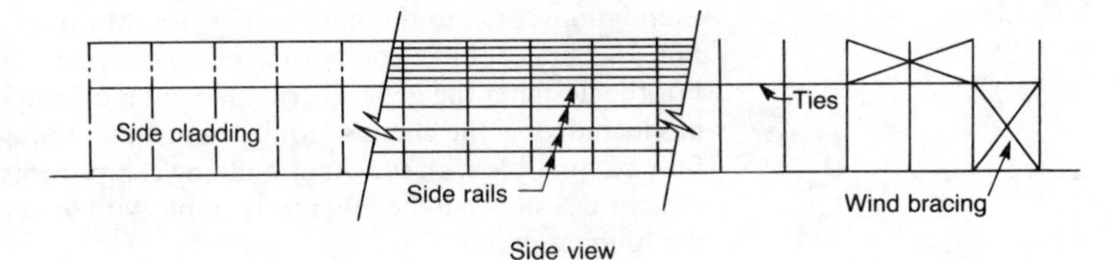

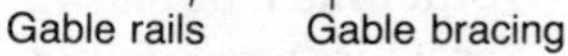

Fig. 14.3 Construction drawing

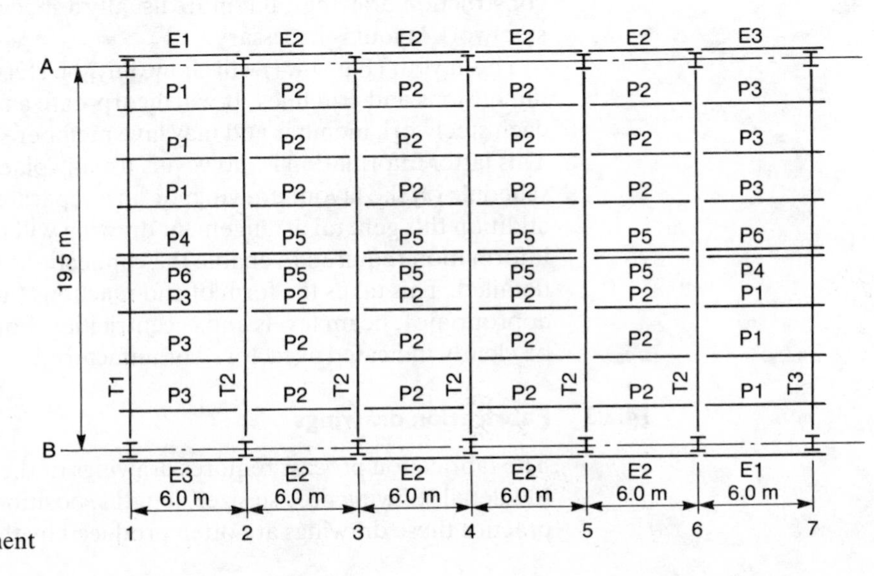

Fig. 14.4 Steelwork general arrangement

Drawing	Ref.	Section	Size	No. off	Length (mm)	Remarks
471/2	A1	UB	533 x 210 x 92	4	6750	
"	A2	UB	ditto	4	6750	Handed A1
"	A3	UB	305 x 127 x 37	6	3520	
"	B1	UC	203 x 203 x 60	6	3520	
"	C1	UC	203 x 203 x 86	2	3710	
"	C2	ASC	152 x 76	18	2890	
471/3	E1	UB	533 x 210 x 92	2	5210	
"	E2	UB	ditto	2	5210	Handed E1
"	E3	UB	305 x 127 x 37	2	4790	
"	E4	UC	203 x 203 x 60	4	3520	Bracket attached

Fig. 14.5 Steelwork schedule

themselves, it is useful for the student to draw fabrication details. This ensures that the design concepts are practicable and develops the discipline of conveying these concepts with precision.

An example of a beam fabrication drawing is shown in Fig. 14.6. The information required will include:

- size of member and steel grade;
- precise length, allowing for clearance at each end;
- size of notches and any other special shaping;
- size and position of bolt holes, but not bolt sizes;
- welds types, size and length where appropriate;
- parts (such as cleats) to be connected during fabrication, and in this case only the bolt sizes if appropriate;
- notes giving number of members required, any handed (mirror image) members required, reference to painting preparation and specification.

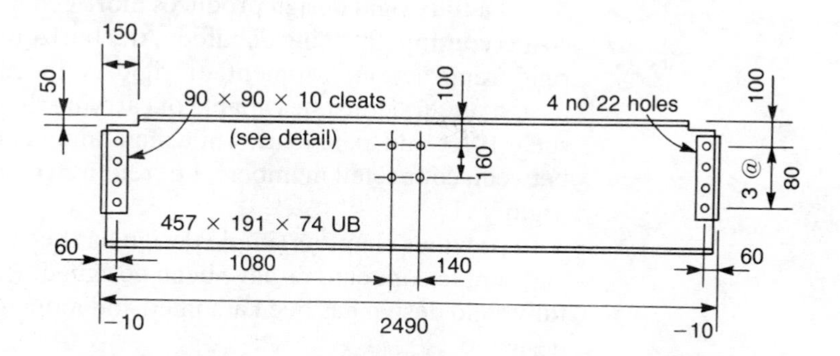

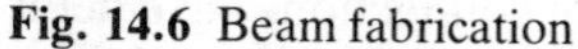
Fig. 14.6 Beam fabrication

14.2.4 Connection details

Reference has been made to connection details previously (Sections 3.6, 6.6 and 8.3) and the overall need for neat and balanced solutions to design problems has already been emphasized (Section 1.1). Guidance regarding the arrangement of connections and their design is given by

Pask[5] and Needham[6] for beam and column construction with I and H sections. The Cidect publication[7] gives guidance for the use of hollow sections including connection arrangements. Morris[8] deals specifically with moment connections for single-storey structures.

Drawings of connections may originate with the designer, particularly when a special arrangement has been assumed in the calculations, and a sketch detail has been given. It is important that sketches of such details are conveyed to the fabricators when they are responsible for connection design. Typical connection details showing bolt sizes, packs and clearances similar to those referenced above[5,6] may be needed where site erection is not the responsibility of the fabricators.

It is useful for the student to detail some of the assembled connections required by the design, so that he/she becomes aware of the difficulties a particular arrangement may cause, and its effect on design capacity. Where the fabricator is required to design the connections, details of forces and moments (factored or unfactored) must be provided (see Section 14.2.2).

The precise behaviour of particular connections is complex and the subject of ongoing research. The performance of connections is generally defined by the three permitted design methods. These design methods are defined as simple design, semi-rigid design and fully rigid design. The 'simple' method is based on the assumption that beams are simply supported and therefore implies that beam-to-column connections must be sufficiently flexible so as to restrict the development of end fixity. Any horizontal forces have to be resisted by bracing or other means.

The design code permits the use of both elastic design and plastic design within the context of fully rigid design. This method, based on full continuity at the connections, gives the greatest rigidity and economy (in terms of weight of steel) for a given framework. Whether or not a fully rigid design produces more economic structures in terms of cost is continually being debated. Any horizontal loads are resisted by rigid frame action. 'Moment' or 'rigid' connections required by this design method must be capable of carrying the design bending moment, shear force and axial load, while maintaining more or less the angle between connected members, i.e. required connection to behave 'rigidly'.

In previous chapters simple design has been assumed in most cases, and simple connections have been designed. In Chapter 12, however, fully rigid design has been assumed and moment connections have been designed.

14.2.5 Movement joints

Movements occur in buildings due to changes of temperature, moisture content, foundation arrangements, etc., and in buildings of unusual shape or size it may be necessary to accommodate these movements by provision of a joint. For single-storey construction the provision of an expansion joint should be considered when the length of the structure exceeds about 150 m. For multi-storey construction, expansion joints may be required at lesser lengths (say 70 m), but in addition settlement joints may be necessary at major height changes, and for unusual plan shapes.

The simplest joint is provided by dividing the structure at the joint and placing columns on either side of the joint. Joint details involving members sliding on a bearing, or the use of slotted holes are less desirable due to their greater complexity and uncertain lifespan.

14.3 COST CONSIDERATIONS

Previous discussion (Section 1.1) showed briefly that a minimum weight of steelwork would not necessarily produce the minimum cost structure. Clearly simplicity in fabrication and repetition of member types affects total cost significantly.

In multi-storey construction comparative costs will be influenced by choice of floor system, especially the use of composite steel deck floors (Chapter 9). In addition simplicity of layout and connections has its effect on speed of construction, and hence of the considerable cost of servicing the capital involved in such a building project. The arrangements for wind bracing and staircases also have cost effects, and these and other factors are discussed in reference (9).

Comparative costs of single-storey construction will depend on both spacing and span of frames as well as the form of structure chosen. The most common forms of structure are included in a cost comparison by Horridge and Morris[10]. The charts produced may be used for settling the basic layout and choosing the best spacing and span for a single-storey structure. They also provide guidance on the cost effects of the choice of a particular form for the roof structure.

14.4 FIRE PROTECTION

The protection of a building structure from the effects of fire is required by regulations to provide adequate time prior to collapse in order to:

- allow for any occupants to leave
- allow for fire fighting personnel to enter if necessary
- delay the spread of fire to adjoining property

To achieve these requirements it is often necessary to cover the bare steelwork with a protective coating. This may be simply concrete cast around the steel with a light steel mesh to prevent spalling, or any one of a number of proprietory systems. These may be sprayed on to the steel surface, or may take the form of prefabricated casings clipped round the steel section. Examples of each of these are shown in Fig. 14.7. Details of a wide range of these systems are set out by Elliott[11,12].

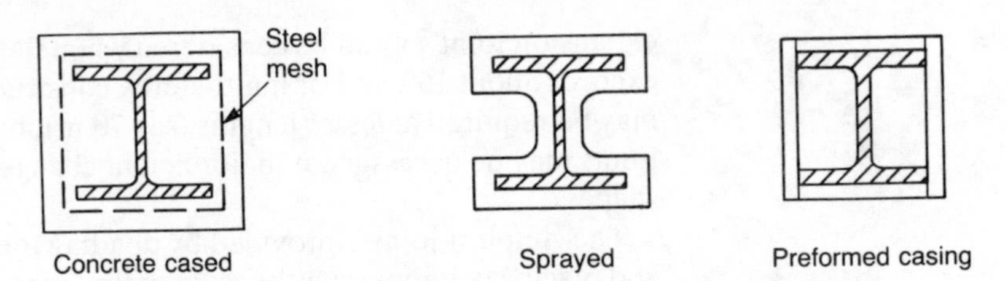

Fig. 14.7 Fire protection

Systems of fire protection are designed and tested by their manufacturers to achieve the fire resistance periods specified in the Building Regulations. These periods are somewhat inflexible and a more fundamental design approach is possible using structural fire engineering(12,13). In this design approach an assessment of the maximum atmosphere temperature is made from the fire load, ventilation and other conditions. The heating curve of the steel member is then estimated, allowing for the location of the steel and its protection. And finally the effects of temperature on the structural capacity of the steelwork are determined.

Special requirements apply to steel portal frames where they form part of the fire barrier to adjacent buildings. A method of designing portal frames to ensure the integrity of the boundary wall in severe fires is given in reference (14).

14.5 CORROSION PROTECTION

The detailing of steelwork can affect the manner and speed of corrosion. Care should therefore be taken to minimize the exposed surface, and to avoid ledges and crevices between abutting plates or sections which may retain moisture. Protective coatings are dependent for their effectiveness on their type, quality and thickness, but most of all on the degree of care taken in the preparation of the steel and in the application of the coating.

The mechanisms of corrosion form a special study area, and this is basic to the proper protection of steelwork. This area together with information on coatings, surface preparation, inspection and maintenance is discussed in a CIRIA report(15). The choice of a protective system usually involves consultation with experts in this area. The cost of protection varies with the importance of the structure, its accessibility for maintenance and the frequency at which this can be permitted without inconvenience to the users.

STUDY REFERENCES

Topic	*Reference*
1. Steelwork fabrication	**Taggart, R.** (1986) Structural steelwork fabrication, *Structural Engineer* vol. 64A no. 8 pp. 207–11

2. Steelwork fabrication	**Davies B.J. & Crawley, E.J.** (1980) *Structural Steelwork Fabrication* vol. 1. BCSA Ltd
3. Detailing practice	(1979) *Metric Practice for Structural Steelwork.* BCSA Ltd
4. Welding	**Pratt J.L.** (1979) *Introduction to the Welding of Structural Steelwork.* Steel Construction Institute
5. Connections	**Pask, J.W.** (1982) *Manual on Connections.* BCSA Ltd
6. Connections	**Needham F.H.** (1980) Connections in structural steelwork for buildings, *Structural Engineer* vol. 58A no. 9 pp. 267–77.
7. Connections	Design of joints under static loading, *Construction with Hollow Steel Sections,* pp. 129–52. British Steel Corporation (Tubes Division)
8. Connections	**Morris L.J.** (1981) A commentary on portal frame design, *Structural Engineer* vol. 59A no. 12 pp. 384–404
9. Steelwork costs	**Gray B.A. & Walker H.B.** (1985) *Steel Framed Multi-storey Buildings. The economics of construction in the UK.* Steel Construction Institute
10. Steelwork costs	**Horridge J.F. & Morris L.J.** (1986) Comparative costs of single-storey steel framed structures, *Structural Engineer* vol. 64A no. 7 pp. 177–81
11. Fire protection	**Elliott, D.A.** (1974) *Fire and Steel Construction.* Steel Construction Institute
12. Fire protection	**Elliott, D.A.** (1983) *An Introduction to the Fire Protection of Steelwork.* Steel Construction Institute
13. Fire Engineering	**Kirby, B.R.** (1985) *Fire Resistance of Steel Structures.* British Steel Corporation
14. Fire Engineering	(1980) *The Behaviour of Steel Portal Frames in Boundary Conditions.* Steel Construction Institute
15. Corrosion protection	**Haigh, I.P.** (1982) *Painting Steelwork.* CIRIA Report no. 93

APPENDIX A

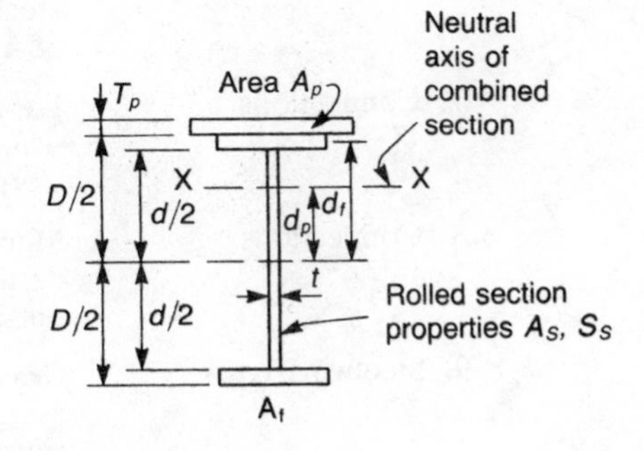

Fig. A1

PLASTIC SECTION PROPERTIES

(See Fig. A1.)
Area $A = A_s + A_p$
Neutral axis at position of equal area above and below it. Hence:

$$A_s/2 + d_p t = A/2 - d_p t + A_p$$

$$d_p = A_p/2t$$

$$S_s = 2A_f d_f + td^2/4$$

$$S_x = A_f(d_f - d_p) + A_f(d_f + d_p) + t(d/2 - d_p)^2/2 + t(d/2 + d_p)^2/2 + A_p(D/2 + T_p/2 - d_p)$$

$$= 2A_f d_f + td^2/4 + td_p{}^2 + A_p(D/2 + T_p/2 - d_p)$$

$$= S_s + td_p{}^2 + A_p(D/2 + T_p/2 - d_p)$$

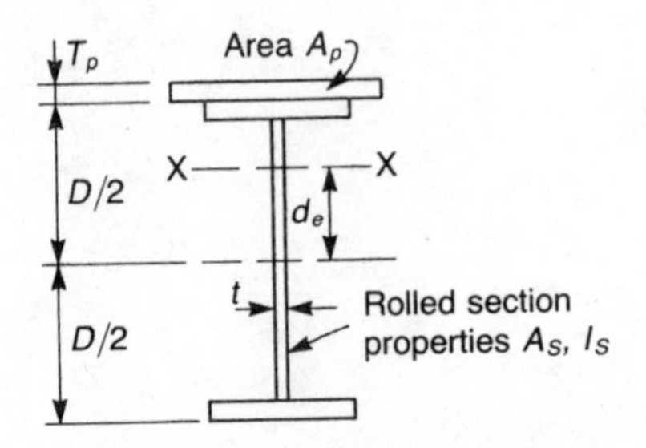

Fig. A2

ELASTIC SECTION PROPERTIES

(See Fig. A2.)
Area $A = A_s + A_p$
Neutral axis at centroid, hence:

$$Ad_e = A_p(D/2 + T_p/2)$$

$$d_e = A_p(D + T_p)/2A$$

$$I_x = I_s + A_s d_e^{\,2} + A_p(D/2 + T_p/2 - d_e)^2$$

$$Z_x = I_x/(D/2 + d_e)$$

Note: these fomulae only apply for when the neutral axis of the combined section lies in the web depth. If the neutral axis lies with the flange of the I-section, the various section properties would need to be determined from first principles.

APPENDIX B

A RAPID METHOD FOR ASSESSMENT OF n_t FACTOR

The following method gives assessment of n_t (the stress distribution factor) for a typical British haunch detail where a UB cutting of same section is used as the haunch, producing a haunch depth of approximately twice that of the basic section. The resulting value of n_t can be used in the various formulae dealing with member buckling, BS 5950.

$$n_t \sqrt{\left[\frac{1}{12M_y}\left\{\frac{N_1}{R_1} + \frac{3N_2}{R_2} + \frac{4N_3}{R_3} + \frac{3N_4}{R_4} + \frac{N_5}{R_5} + 2\left(\frac{N_S}{R_S} - \frac{N_E}{R_E}\right)\right\}\right]}$$

where

N_i = factored moments (kNm)
$M_y = p_y Z_x$ (basic section)
R_i = coefficient applied to uniform rafter moment capacity to produce an estimate of moment capacity at point i
$= [1.60 + 0.65(p^* - 0.20) + 0.60(p^* - 0.45)]$

$$p^* = \frac{\text{length from intersection to point } i \text{ within haunch}}{\text{total length of haunch}}$$

NOTES

(a) All compound terms have to be positive, otherwise make zero.
(b) $R_i = 1.0$ when N_i is located in the uniform part of the rafter.
(c) If quarter point coincides with intersection, $R_i = R_{inter} = 1.0$.
(d) If intersection lies between two quarter points then N_i nearest to intersection in the uniform section becomes N_{inter} with $R_{inter} = 1.0$.
(e) Particular values of $(N_i/R_i)Z_x$ (basic) **cannot be used** to evaluate individual stresses.
(f) If $L^*(= L_k/cn_t)$ is within 5% of actual length then revert to normal calculation of n_t; see Section 12.8.1.2.

Examples

1. Special case, when restrained length equals haunch length, with reference to Fig. B1:

$R_5 = R_{inter} = 1.00$
$R_4 = [1.60 + 0.65(0.25 - 0.20) + 0.60(0.25 - 0.45)] = 1.63$
$R_3 = [1.60 + 0.65(0.50 - 0.20) + 0.60(0.50 - 0.45)] = 1.83$
$R_2 = [1.60 + 0.65(0.75 - 0.20) + 0.60(0.75 - 0.45)] = 2.14$
$R_1 = [1.60 + 0.65(1.00 - 0.20) + 0.60(1.00 - 0.45)] = 2.45$

2. With reference to Fig. B2:

$R_5 = 1.00$
$R_4 = 1.00$

$$R_3 = [1.60 + 0.65(0.33 - 0.20) + 0.60(0.33 - 0.45)] = 1.68$$
$$R_2 = [1.60 + 0.65(0.67 - 0.20) + 0.60(0.67 - 0.45)] = 2.04$$
$$R_1 = [1.60 + 0.65(1.00 - 0.20) + 0.60(1.00 - 0.45)] = 2.45$$

3. With reference to Fig. B3:

$$R_5 = 1.00$$
$$R_4 = R_{inter} = 1.00$$
$$R_3 = [1.60 + 0.65(0.25 - 0.20) + 0.60(0.25 - 0.45)] = 1.63$$
$$R_2 = [1.60 + 0.65(0.63 - 0.20) + 0.60(0.63 - 0.45)] = 1.99$$
$$R_1 = [1.60 + 0.65(1.00 - 0.20) + 0.60(1.00 - 0.45)] = 2.45$$

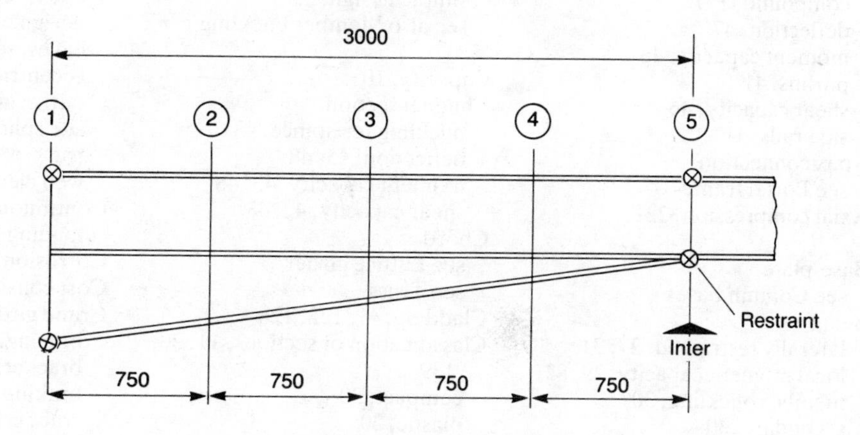

Fig. B1

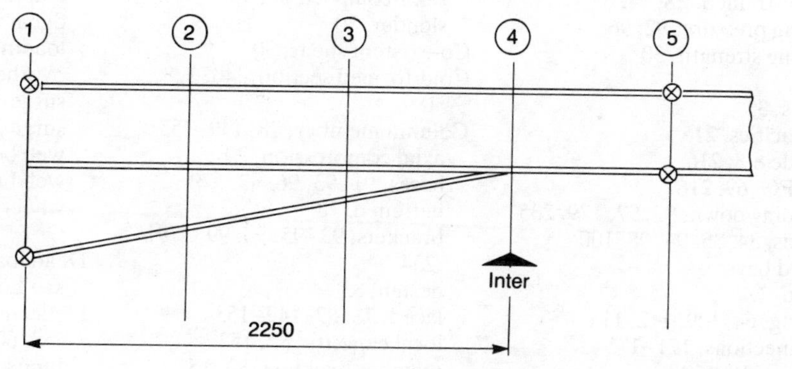

Fig. B2

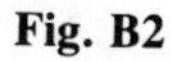

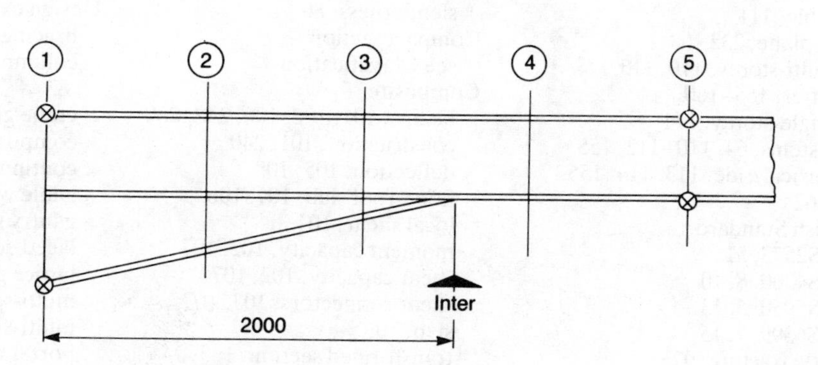

Fig. B3

INDEX